Function Spaces and Applications

Function Spaces and Applications

EDITORS
David E. Edmunds
Pawan K. Jain
Pankaj Jain
Alois Kufner
Lars-Erik Persson
Saburou Saitoh

CRC Press
Boca Raton London New York Washington, D.C.

Narosa Publishing House
New Delhi Chennai Mumbai Calcutta

EDITORS

David E. Edmunds
Centre for Mathematical Analysis
and Its Applications
School of Mathematical Sciences
University of Sussex
Falmer Brighton BN1 9QH, UK
e-mail: d.e.edmunds@sussex.ac.uk

Pawan K. Jain
Department of Mathematics
University of Delhi
Delhi-110 007, India
e-mail: pkjain@nda.vsnl.net.in

Pankaj Jain
Department of Mathematics
Desh Bandhu College
(University of Delhi)
New Delhi-110 019, India
e-mail: pankajkrjain@hotmail.com

Alois Kufner
Mathematical Institute
Academy of Sciences
of the Czech Republic
115 67 Praha 1, Zitna 25, Czech Republic
e-mail: kufner@matsrv.math.cas.cz

Lars-Erik Persson
Department of Mathematics
Luleå University of Technology
S-97187, Luleå, Sweden
e-mail: larserik@sm.luth.se

Saburou Saitoh
Department of Mathematics
Faculty of Engineering
Gunma University
Kiryu 376, Japan
e-mail: ssaitoh@eg.gunma-u.ac.jp

Library of Congress Cataloging-in-Publication Data:

A catalog record for this book is available from the Library of Congress.

Exclusive distribution in North America only by CRC Press LLC

Direct all inquiries to CRC Press LLC, 2000 N.W. Corporate Blvd., Boca Raton, Florida 33431. E-mail: orders@crcpress.com

International Standard Book Number 0-8493-0938-7
Printed in India

Preface

It is now widely accepted that Functional Analysis is a powerful tool in mathematical problems arising from day-to-day life. In particular, the concepts of generalized functions (distributions) have been used to obtain weak solutions of Partial Differential Equations or Boundary Value Problems where there might not be any solution in the classical sense. It gave rise to a study of various function spaces. The most important spaces are Sobolev Spaces with or without weights and their generalizations such as Orlicz-Sobolev Spaces. These spaces allow us to study solutions of Boundary Value Problems for Partial Differential Equations and help us to estimate errors arising in numerical analysis. The aspects of function spaces are density of smooth functions, imbedding properties, inequalities, trace theory, extension theorems, dual theory, interpolation theory, entropy and approximation numbers, measures of non-compactness and many others. Tremendous amount of work has been done and those topics are growing on in these aspects.

An international conference on "Function Spaces and Applications to Partial Differential Equations" (December 15–19, 1997) was held under the directorship of Pawan K. Jain and with Adimurthi (India), H.P. Dikshit (India), David E. Edmunds (UK), S. Kesavan (India), Alois Kufner (Czech Republic), Bohumir Opic (Czech Republic), Giorgio Talenti (Italy) and William P. Ziemer (USA) as Scientific Advisors, at the University of Delhi. The conference, attended by 62 participants (46 from India and 16 from abroad), was a great success and it was decided that the papers presented, after getting refereed by experts, should be published in the form of a book. This volume is mainly the proceedings of this conference. Furthermore, it also contains articles by the leading mathematicians who could not attend the conference.

We express our sincere gratitude to the contributors for submitting their papers and to those who extended their help and cooperation in refereeing the articles of this volume. It would never have been possible to complete this book without their kind cooperation.

The conference was partly supported by Council of Scientific and Industrial Research, Department of Science & Technology (DST), National Board for Higher Mathematics and the University of Delhi (as it was organized in the Platinum Jubilee Year of the University which was also the Golden Jubilee Year of the Department of Mathematics). We record our gratitude to all those concerned.

EDITORS

Contents

Function Spaces and Applications
D.E. Edmunds et al (Eds)

1. Limiting Behavior of Solutions of a Sequence of Non-Homogeneous Boundary Value Problems

G.S. Balashova

MEI (TU), 111250 Moscow, Krasnokazarmennaja Str. 14

1. Introduction

Let Ω be a domain in $\mathbf{R}^n$, $n \geq 1$, $\alpha = (\alpha_1, \dots, \alpha_n)$ is a multi-index consisting of integers, $|\alpha| = \sum_{i=1}^{n} \alpha_i$ is the length of the multi-index. If $u(x)$ is a function defined in the domain Ω, then $D^\alpha u = \dfrac{\partial^{|\alpha|} u}{\partial x_1^{\alpha_1} \partial x_2^{\alpha_2} \dots \partial x_n^{\alpha_n}}$ denotes its partial derivative. The number of all n-dimensional multi-indexes of length not greater than k is equal to $s = \dfrac{(n+k)!}{n!k!}$. At the same time s is the number of all (partial) derivatives of the function $u(x)$ of n variables from order 0 to k inclusively. Let the symbol $\delta_k u$ denote the vector-function composed of all derivatives of the function $u(x)$ of orders $0, 1, 2, \dots, k$:

$$\delta_k u = \{D^\alpha u\}_{|\alpha| \leq k} = \left\{ u, \frac{\partial u}{\partial x_1}, \dots, \frac{\partial u}{\partial x_n}, \frac{\partial^2 u}{\partial x_1^2}, \dots, \frac{\partial^k u}{\partial x_n^k} \right\}$$

(the vector $\delta_k u$ has s components which are ordered lexicographically). The symbol $\hat{\delta}_k u$ denotes the vector of all derivatives of order k of the function $u(x)$:

$$\hat{\delta}_k u = \{D^\alpha u\}_{|\alpha| = k}$$

This vector is called the main part of the vector $\delta_k u$. Let $A_\alpha = A_\alpha(x, \xi)$ be functions of $(n + s)$ variables defined for $x \in \Omega$, $\xi \in \mathbf{R}^s$. We will consider equations of the following type

$$\sum_{|\alpha|=0}^{\infty} (-1)^{|\alpha|} D^\alpha A_\alpha(x, \delta_k u(x)) = h(x). \tag{1}$$

This is an equation of infinite order written in divegence form. Here is a model example:

$$\sum_{|\alpha|=0}^{\infty} (-1)^{|\alpha|} D^{\alpha}(a_{\alpha} \mid D^{\alpha} u \mid^{p-2} D^{\alpha} u) = h(x), \tag{2}$$

with constants $a_\alpha \geq 0$ and $1 < p < \infty$.

The "energy" space for the equation (1) is the Sobolev space of infinite order

$$W^{\infty}\{a_\alpha, p\}_{(\Omega)} \equiv \{u(x) \in C^{\infty}(\Omega) : \rho(u) = \sum_{|\alpha|\geq 0} a_\alpha \| D^\alpha u \|_{L_p}^p < \infty\}. \tag{3}$$

The number p, $1 < p < \infty$ defines the growth of the coefficients A_α. Generally speaking, p may take on different values, but we always consider (or choose) the least possible, for to the lesser p corresponds the larger space. And, in addition, solving boundary value problems, we are interested in the most precise estimate of the coefficients $A_\alpha(x, \xi)$:

$$| A_\alpha(x, \xi) | \leq g_\alpha(x) + a_\alpha \sum_{|\beta|\leq|\alpha|} | \xi_\beta |^{p-1},$$

$g_\alpha(x) \in L_{p'}(\Omega)$, $a_\alpha \geq 0$ are constants, and the estimate not only from above, but also from below.

If we consider a linear differential operator, then the convenient space for finding a weak solution is the space $W^\infty\{a_\alpha, 2\}_{(\Omega)}$, because for it

$$A_\alpha(x, \xi) = \sum_{|\beta|\leq|\alpha|} a_{\alpha\beta}(x)\, \xi_\beta.$$

The space $W^{-\infty}\{a_\alpha, p'\}_{(\Omega)}$ of right hand side is conjugate to

$$\overset{\circ}{W}{}^{\infty}\{a_\alpha, p\}_{(\Omega)} = \{u(x) \in C_0^\infty(\Omega) : \rho(u) \equiv \sum_{|\alpha|\geq 0} a_\alpha \| D^\alpha u \|_{L_p}^p < \infty\}$$

and consists of generalized functions of the following form:

$$W^{-\infty}\{a_\alpha, p'\}_{(\Omega)}$$

$$= \Big\{ h(x) = \sum_{|\alpha|\geq 0} (-1)^{|\alpha|} a_\alpha D^\alpha h_\alpha(x),\ h_\alpha(x) \in L_{p'}(\Omega)$$

$$p' = \frac{p}{p-1},\ \rho'(h) = \sum_{|\alpha|\geq 0} a_\alpha \| h_\alpha \|_{p'}^{p'} < \infty \Big\}. \tag{4}$$

Therefore the right hand side of the equation (1) may be a generalized function of singularity of infinite order.

The duality of elements $h(x) \in W^{-\infty}\{a_\alpha, p')_{(\Omega)}$ and $v(x) \in \overset{\circ}{W}{}^{\infty}\ \{\alpha, p\}_{(\Omega)}$ is defined by the formula

$$\langle h, v \rangle = \sum_{|\alpha|=0}^{\infty} a_\alpha \int_\Omega h_\alpha(x)\, D^\alpha v(x)\, dx,$$

which is correct according to (4).

Before solving a problem for the equation (1) we have to find out whether the space (3) contains functions not equal to zero; in other words, whether it is nontrivial. Conditions of notriviality of the space (3) depend on the type of the domain Ω and are examined, to a sufficient extent, in [1].

Considering boundary value problems for the equation (1) with the following conditions on the boundary $\partial\Omega$ of the domain Ω

$$D^{\omega}u\,|_{\delta\Omega} = f_{\omega}(x'),\ x' \in \delta\Omega,\ |\,\omega\,| = 0, 1, \ldots . \tag{5}$$

First of all it is necessary to learn whether a function satisfying the boundary conditions (5) might exist in the corresponding space of functions. Differing from the case with classical Sobolev spaces, this question proved to be quite complicated one, even for the one-dimensional space. Conditions of existence of functions in the space (3) such that satisfy the conditions (5) are studied, in G.S. Balahova's works [2], [3], [4] sufficiently deeply.

Definition 1 Let $\Omega \subset R^n$ be a domain with the boundary $\partial\Omega$. If a function $F(x) \in C^{\infty}(\Omega)$ such that

$$D^{\omega}F(x)\,|_{\delta\Omega} = f_{\omega}(x'),\ x' \in \partial\Omega;\ |\,\omega\,| = 0, 1, \ldots,$$

exists, than we call the function $f_{\omega}(x')$ a trace of the function $F(x)$ on $\delta\Omega$, and the function $F(x)$ itself an extension of the trace $f_{\omega}(x')$ in this space.

If for the problem (1), (5) the corresponding space (3) is nontrivial, the conditions can be extended inside the domain Ω, the right hand side $h \in W^{-\infty}\{a_{\alpha}, p'\}_{(\Omega)}$, then the process of finding a solution includes two stages:

1^0. Solving truncated problems of order $2l$ (or other more general equations of order $2l$).

2^0. Passing to the limit as $l \to \infty$.

2. Limiting Behavior of Solutions of a Sequence of Nonhomogeneous Boundary Value Problems

This section establishes conditions under which one can go to the limit in a sequence of nonhomogeneous boundary value problems

$$L_m(u) = \sum_{|\alpha|=0}^{t_m} (-1)^{|\alpha|} D^{\alpha} A_{\alpha m}(x; \delta_k u(x)) = h^m(x), x \in \Omega, \tag{6}$$

$$D^{\omega}u\,|_{\partial\Omega} = \Psi_{\omega m}(x'),\ x' \in \partial\Omega,\ |\,\omega\,| = 0, 1, \ldots, t_m - 1, \tag{7}$$

where t_m, $m = 1, 2, \ldots$, are certain natural numbers or infinity. We shall in fact assume that all $t_m = \infty$, simply defining in the case of finite t_m that $A_{\alpha m}(x, \xi) \equiv 0$ for all $\alpha : |\,\alpha\,| > t_m$. Suppose, that as $m \to \infty$ the sequences of operators L_m and functions $h^m(x)$ converge to an operator L and a function $h(x)$. We wish to determine conditions under which the set of solutions $u_m(x)$, $m = 1, 2, \ldots$, of the problems (6), (7) has an accumulation of the limiting problem

$$L(u) = \sum_{|\alpha|=0}^{t} (-1)^{|\alpha|} D^{\alpha} A_{\alpha}(x; \xi) = h(x),, \tag{8}$$

$$D^{\omega} u \,|_{\partial\Omega} = \Psi_{\omega}(x'),\ x' \in \partial\Omega,\ |\,\omega\,| = 0, 1, \ldots, t-1. \tag{9}$$

The conditions for this to be true differ according to whether the limiting operator is of infinite ($t = \infty$) or finite ($t < \infty$) order.

1^0. Limiting Differential Operator of Infinite order ($t = \infty$).

We formulate the following conditions:

(a) $A_{\alpha m}(x, \xi^m)$ are continuous functions of the arguments $x \in \Omega$, $\xi^m \in R^s$ and satisfy the inequalities

$$|\sum_{|\alpha|=0}^{\infty} A_{\alpha m}(x, \xi^m)\eta_{\alpha}^m\,| \le K \sum_{|\alpha|=0}^{\infty} a_{\alpha m}\,|\,\xi_{\alpha}^m\,|^{p-1}\,|\,\eta_{\alpha}^m\,|,\ m = 1, 2, \ldots.$$

Here $\{\alpha_{\alpha m} \ge 0\}$ is some sequence of numbers with $a_{\alpha m} = 1$ for all m, $p > 1$. Henceforth K to denote absolute positive constants—not necessarily the same each time.

(b) for any $x \in \Omega$, $\xi^m \in R^s$,

$$\sum_{|\alpha|=0}^{\infty} A_{\alpha m}(x, \xi^m)\xi_{\alpha}^m \ge K \sum_{|\alpha|=0}^{\infty} a_{\alpha m}\,|\,\xi_{\alpha}^m\,|^p\,;$$

(c) $a_{\alpha m} \to a_{\alpha}$ as $m \to \infty$, and $a_{\alpha} > 0$ for infinitely many values of $|\,\alpha\,|$;

(d) if $\xi_{\gamma}^m \to \xi_{\gamma}$ as $m \to \infty$ for all $\gamma : |\,\gamma\,| \le |\,\alpha\,|$; then

$$A_{\alpha m}(x, \xi^m) \to A_{\alpha}(x, \xi)$$

uniformly on $x \in \Omega$;

(e) for any number $\varepsilon > 0$ exist $N_0(\varepsilon)$ such that for every m

$$\sum_{N=N_0}^{\infty} M_{Nm}^c (M_{N+1,m}^c)^{-1} < \varepsilon$$

where M_{Nm}^c is the convex regularization of the sequence

$$M_{Nm} = \begin{cases} (\sum_{|\alpha|=N} a_{\alpha m})^{-1/p}, & \text{if } \sum_{|\alpha|=N} a_{\alpha m} \ne 0; \\ \infty, & \text{if } \sum_{|\alpha|=N} a_{\alpha m} = 0; \end{cases}$$

by logarithms [6] (if $M_{Nm}^c = M_{N+1,m}^c = \infty$, then $M_{Nm}^c / M_{N+1,m}^c = 0$), for every ω, $\Psi_{\omega m}(x') \to \Psi_m(x')$ uniformly of $\delta\Omega$ as $m \to \infty$. In addition, for every m the trace $\{\Psi_{\omega m}(x')\}$ is extended in the space $W^{\infty}\{a_{\alpha m}, p\}_{(\Omega)}$, that its extension is such a function $\Psi_m(x) \in W^{\infty}\{a_{\alpha m}, p\}_{(\Omega)}$ that

$$D^{\omega}\Psi_m(x)\,|_{\delta\Omega} = \Psi_{\omega m}(x'),\ |\,\omega\,| = 0, 1, \ldots,$$

and $$\sum_{|\alpha|=0}^{\infty} a_{\alpha m} \| D^{\alpha} \Psi_m(x) \|^p_{L_p(\Omega)} \leq K;$$

(g) the right-hand sides in equations (6) $h^m(x) \in W^{-\infty}\{a_{\alpha m}, p'\}_{(\Omega)}$, and moreover $\rho'(h^m) \leq K$ and $h^m \to h \in W^{-\infty}\{a_\alpha, p'\}_{(\Omega)}$ as $m \to \infty$ in the sense that $\langle h^m, v_m \rangle \to \langle h, v \rangle$, where $\{v_m\}$ is any sequence from $\overset{\circ}{W}{}^{\infty}\{a_{\alpha m}, p\}_{(\Omega)}$ such that $v_m(x) \to v(x) \in \overset{\circ}{W}{}^{\infty}\{a_\alpha, p\}_{(\Omega)}$ in the sense of convergence in $C_0^\infty(\Omega)$ and for every $m = 1, 2, \ldots$

$$\sum_{|\alpha|>N(\varepsilon)} a_{\alpha m} \| D^\alpha v_m(x) \|^p_{L_p(\Omega)} < \varepsilon.$$

Theorem 2.1 If conditions (a)-(g) hold, then a sequence $\{u_m(x)\}$ of solutions of the problems (6), (7) has an accumulation point $u(x)$ in the sense of convergence in $C^\infty(\Omega)$ and moreover $u(x) \in W^\infty(a_\alpha, p\}_{(\Omega)}$ is a solution to the limiting problem (8), (9).

3. Limiting Differential Operator of Finite Order ($t < \infty$)

We introduce a few changes in the conditions formulated above:

(c*) $a_{\alpha m} \to a_\alpha$ as $m \to \infty$, with $a_\alpha = 0$ for all α such that $|\alpha| > t$, and all $a_\alpha > 0$ if $|\alpha| = t$;

(e*) for any number $\varepsilon > 0$ there exists m_0 such that for every $m > m_0$

$$\sum_{N>t} M^c_{Nm}(M^c_{N+1,m})^{(-1)} < \varepsilon;$$

(f*) for every m the trace $\{\Psi_{\omega m}(x')\}$ is extended in space $W^\infty\{a_{\alpha m}, p\}_{(\Omega)}$, that is its extension is a such function $\Psi_m(x) \in W^\infty\{a_{\alpha m}, p\}_{(\Omega)}$, that

$$D^\omega \Psi_m(x)\,|_{\partial\Omega} = \Psi_{\omega m}(x'),\ x' \in \partial\Omega, \quad \sum_{|\alpha|\geq t+1} a_{\alpha m} \| D^\alpha \Psi_m(x) \|^p_p < \varepsilon$$

as soon as $m > m_0$. In addition there exists such function $\Psi(x) \in C^t(\Omega)$ that $D^\omega \Psi_m(x) \to D^\omega \Psi(x)$ as $m \to \infty$ uniformly on $x \in \Omega$ for all ω with $|\omega| \leq t$;

(g*) the right-hand sides in equations (6) $h^m(x) \in W^{-\infty}\{a_{\alpha m}, p'\}_{(\Omega)}$ and there exist $h_\alpha(x) \in L_{p'}(\Omega)$ such that $\| h_\alpha - h^m_\alpha \|_{p'} \to 0$ as $m \to \infty$ for all α with $|\alpha| \leq t$ and $\sum_{|\alpha|\geq t+1} a_{\alpha m} \| h^m_\alpha \|^{p'}_{p'} < \varepsilon$ for all $m > M(\varepsilon)$;

(h*) the operators L_m, $m = 1, 2, \ldots$, are monotone, i.e. for any functions $u(x)$ and $v(x) \in W^\infty(a_{\alpha m}, p\}_{(\Omega)}$

$$\langle L_m(u) - L_m(v), u - v \rangle \geq 0.$$

Theorem 2.2 If conditions (a), (b), (c*), (d), (e*)-(h*) hold, then a sequence of solutions $\{u_m(x)\}$ of problems (6), (7) has a weak accumulation point $u(x) \in W^t_p(\Omega)$, which is a solution of the problems (8), (9).

4. Discussion of the Results

This paper [5] shows that in the onedimensional case conditions (e) and (e*) of Theorems 2.1 and 2.2 are essential for passing to the limit and cannot be weakened. In this section we will show that condition (e*) in Theorem 2.2 is in a certain sense best possible, i.e. for a certain class of problems it is also a necessary condition.

Theorem 3.1 A sequence of solution of problems

$$L_m(u) = (-1)^m a_{mm} D^{2m} u_m(x) - D^2 u_m(x) + u_m(x) = h^m(x),\ x \in (0, a),$$

$$D^k u_m(0) = D^k u_m(a) = 0,\ k = 0, 1, \ldots, m-1$$

where $h^m(x)$ are arbitrary functions in $L_2(\Omega)$ converging to $h(x) \in L_2(\Omega)$, has a weak accumulation point in $W_2^1(\Omega)$ which is a solution of the limiting problem

$$L(u) \equiv -D^2 u(x) + u(x) = h(x),\ x \in (0, a)$$

$$u(0) = u(a) = 0$$

if and only if

$$\lim_{m \to \infty} m a_{mm}^{1/2m} = 0$$

Remark Analogous example may be given in the case of a limiting equation of infinite order.

References

1. Dubinskij Ya.A. (1986), '*Sobolev spaces of infinite order and differential equations*', Leipzig.
2. Balashova G.S. (1982), Mat. Sb., **118(160)**, N 3, pp. 371–385, English transl. in Math. USSR., Sb. 46.
3. Balashova G.S. (1988), Izv. Acad. Nauk SSSR, Ser. Mat. **51**, N 6, pp. 129–1308, Eng. transl. in Math. USSR Izv. **31**.
4. Balashova G.S. (1991), Dokl. Acad. Nauk SSSR, **319**, N 2, pp. 267–270.
5. Balashova G.S. (1981), Differents. Uravn., **17**, N 2, pp. 256–269.
6. Mandelbrojt (1955), Adjoined series. Regularization of sequences. Applications. Moscow (Russian).

Function Spaces and Applications
D.E. Edmunds et al (Eds)

2. Some Separation Criteria and Inequalities Associated with Linear Second Order Differential Operators

R.C. Brown[1], D.B. Hinton[2] and M.F. Shaw[3]

[1]Department of Mathematics, University of Alabama Tuscaloosa, AL 35487–0350, USA

[2]Department of Mathematics, University of Tennessee Knoxville, TN 37996, USA

[3]Department of Electrical Engineering, University of Kentucky Lexington, KY 40506, USA

1. Introduction

Let $M[y] := -(py')' + qy$ be a symmetric second order differential expression defined on $I = (a, \infty)$, $a > -\infty$. We assume that M satisfies at least the so-called "minimal conditions" given in the book of Naimark [22], i.e., p and q are real and p^{-1}, q belong to $L_{\text{loc}}(I)$, the space of measurable locally integrable functions on I. Under these conditions if $x_0 \in I$ and $f \in L_{\text{loc}}(I)$ are an arbitrary point and function it can be shown that there is a unique solution y of the initial value problem

$$M[y] = f$$

$$y(x_0) = c_0$$

$$py'(x_0) = c_1$$

for any complex numbers c_0, c_1. Moreover, if $AC_{\text{loc}}(I)$ denotes the space of locally absolutely continuous complex-valued functions on I, both y and $py' \in AC_{\text{loc}}(I)$. In particular this result implies that the equation $M[y] = 0$ will have exactly two linearly independent solutions with the stated absolute continuity properties.

It also is known that M determines two densely defined closed operators which we denote by $T_0(M)$ and $T(M)$ in the Hilbert space $L^2(I)$ consisting of all complex square integrable measurable functions f with norm $\| f \|_I$ given by $\| f \|_I = \left(\int_I |f|^2 \right)^{1/2}$. These operators are constructed as follows. We begin by defining the "preminimal operator" $T_0'(M)$ to be the restriction of M to $\mathscr{D}_0' := \{y \in L^2(I) \cap AC_{\text{loc}}(I)$: y has compact support; $py' \in AC_{\text{loc}}(I)$; $M[y] \dot\in$

$L^2(I)\}$. If p is differentiable, p' is continuous, and $q \in L^2(I)_{\text{loc}}$ it is easy to see that $\mathcal{D}_0'$ contains the space $C_0^\infty(I)$ of infinitely differentiable functions on I. This is not necessarily true under minimal conditions on M. However $T_0'(M)$ is still densely defined (although not necessarily consisting of twice differentiable functions) and *closable* in the sense that the closure of its graph is an operator which we call the "minimal operator" T_0. $T(M)$, the "maximal operator" is given by M on

$$\mathcal{D} := \{y \in L^2(I) \cap AC_{\text{loc}}(I) : py' \in AC_{\text{loc}}(I);\ M[y] \in L^2(I)\}.$$

$T_0'(M)$, $T_0(M)$, and $T(M)$ enjoy the following properties

(i) $T_0(M) \subset T(M)$,
(ii) $T_0'(M)^* = T_0(M)^* = T(M)$,
(iii) $T(M)^* = T_0(M)$.

Thus both $T_0'(M)$ and $T_0(M)$ are symmetric.

If p^{-1}, $q \in L_{\text{loc}}([a, c])$ for $a < c < \infty$ we say that a is *regular;* otherwise it is *singular.* Unless we assert otherwise we assume that a is regular and signal this by writing I as a semi-closed interval $[a, \infty)$. We regard the end-point "∞" as singular. These assumptions will simplify the technical apparatus of the paper. However our definitions and results may be formulated for a singular end-point a or, indeed, on a general interval (a, b), $-\infty \le a < b \le \infty$ where one or both end-points are singular with slight (and obvious) changes in their statements.

M is said to be *limit-point* or *LP* at ∞ if there is at most one solution of $M[y] = 0$ which is in $L^2[a, \infty)$. M is *limit-circle* at ∞ if both solutions are L^2 integrable. The limit-point case at ∞ can further be shown equivalent to the property that $\mathcal{D}$ is exactly a two dimensional extension of $\mathcal{D}_0$; while if M is limit-circle at ∞, then $\mathcal{D}$ is a four dimensional extension of $\mathcal{D}_0$. That in our setting M must be either limit-point or limit-circle is called the Weyl alternative after the inventor of these concepts[1].

Finally, we remark that all the above theory can be developed in a *weighted* L^2 space on I, i.e., the space $L^2(w; I)$ with norm $\|(\cdot)\|_{w,I} := \|w^{1/2}(\cdot)\|_I$ where w is a positive measurable function. In this setting we work with the operator

$$M_w[y] := w^{-1}[-(py')' + qy]$$

and define $T_0'(M_w)$, $T_0(M_w)$, and $T(M_w)$ as in the unweighted case. A classic treatment of the theory of the (unweighted) maximal and minimal operators

---

[1]Likewise the nomenclature "limit-point" or "limit-circle" is due to Weyl and results from his technique which associates these cases with nested families of circles in the complex plane which converge respectively either to a point or a circle. See e.g. Coddington and Levinson [5, Chapter 9] for an account of Weyl's method.

$T_0(M)$ and $T(M)$ may be found in [22, § 17 to 18]. Further valuable information is contained in the books of Glazman [17], Kauffman, Read, and Zettl [20] and Weidmann [25]. A full treatment of the weighted setting seems not to be in the literature; however the arguments in the references we have cited may be routinely modified to allow $w \neq 1$.

Now suppose for the differential expression M_w that $y, z \in \mathscr{D}$ and $a < c < d < \infty$. One integration by parts gives the *Dirichlet* formula

$$\int_a^c py'\bar{z}' + qy\bar{z} = (py\bar{z}')(c) - (py\bar{z}')(a) - \int_a^c wM_w[y]\bar{z} \qquad (1.1)$$

A second integration by parts yields the *Lagrange identity*

$$\int_a^c wM_w[y]\bar{z} - \int_a^c wy\overline{M_w[z]} = \{y, z\}(c) - \{y, z\}(a) \qquad (1.2)$$

where the *Lagrange billinear form*

$$\{y, z\}(t) := (yp\bar{z}' - py'\bar{z})(t). \qquad (1.3)$$

Similar formulae remain true if either p or q are complex. In this case if p, q satisfy the minimal integrability conditions stated above the closed operators $T_0(M)$ and $T(M)$ may still be constructed; however, symmetry is lost. Some of the results of this paper (Theorems 6 and 7) will be formulated for complex p and q.

Since the left-hand integrals in (1.2) exist by the Cauchy-Schwartz inequality as $c \to \infty$, $\lim_{c\to\infty} \{y, z\}(c)$ exists Certainly

$$\{y, z\}(\infty) \equiv \lim_{c\to\infty} \{y, z\}(c) = 0 \qquad (1.4)$$

$$\{y, z\}(a) = 0 \qquad (1.4')$$

for all $y, z \in \mathscr{D}_0'$ and it is known that these zero end-point values of the billinear form are also true for all $y, z \in \mathscr{D}_0$. In fact $\mathscr{D}_0$ may be described as the set of functions $y \in \mathscr{D}$ such that $y(a) = py'(a) = 0$ and (1.4) is satisfied for all $z \in \mathscr{D}$. In this case (1.2) becomes the *Green's formula*

$$\int_I wM_w[y]\bar{z} - \int_I wy\overline{M_w[z]} = 0. \qquad (1.5)$$

A third and very important property that M_w be limit-point at ∞ (which is some-times taken as a definition) is that (1.4) hold for all $y, z \in \mathscr{D}$. Indeed, to see that (1.4) on $\mathscr{D}$ is equivalent to the statement that dim $(\mathscr{D}/\mathscr{D}_0) = 2$. Let ϕ_1, ϕ_2 be compact support functions in $\mathscr{D}$ such that

$$\phi_1(a) = 1 \qquad \phi_2(a) = 0$$

$$p\phi_1'(a) = 0 \qquad p\phi_2'(a) = 1. \tag{1.6}$$

(A proof of the existence of such functions under minimal conditions may be found in [22, §17.3, Lemma 2].) Then since ϕ_1, ϕ_2 give a two dimensional extension of $\mathscr{D}_0$, (1.4) is evidently true on $\mathscr{D}$ if dim $(\mathscr{D}/\mathscr{D}_0) = 2$. Conversely, given (1.4) on $\mathscr{D}$ if dim $(\mathscr{D}/\mathscr{D}_0) > 2$ there is a function $\psi \in \mathscr{D}$ such that ψ is not expressible in the form $\psi = c_1\phi_1 + c_2\phi_2 + y$ where $\psi \in \mathscr{D}_0$. On the other hand, $y, = \psi - c_1\phi_1 - c_2\phi_2$ has the property that $y(a) = py'(a) = 0$. By hypothesis $\{\psi, z\}(\infty) = 0$ for any $z \in \mathscr{D}$ so that by the remarks above $y \in \mathscr{D}_0$ which is a contradiction.

If $\lim_{c\to\infty} (py\bar{z})(c) = 0$ for all $y, z \in \mathscr{D}$ we say that M is *strong limit-point* or *SLP* at ∞. Obviously $SLP \Rightarrow LP$ but examples exist showing that the converse is not true (see Everitt [9]).

Looking now at the Dirichlet formula (1.1), we say following [9] that M_w satisfies *the Dirichlet (or D) condition* on I if in addition to satisfying minimal conditions, $p > 0$, and

(i) $p^{1/2}y'$ and $|q|^{1/2}y \in L^2(I)$,

(ii) $\displaystyle\int_I py'\bar{z}' + qy\bar{z} = -(y\bar{z})(a) + \int_I wy\overline{M_w[z]}$

for all y or $y, z \in \mathscr{D}$. It follows at once from this definition and that of *SLP* that $\mathscr{D} \Rightarrow SLP \Rightarrow LP$ (at ∞).

The properties of M_w we consider in this paper are the following.

Definition 1 $T(M_w)$ or $T_0(M_w)$ is said to be separated on $\mathscr{D}$ or $\mathscr{D}_0$ if whenever $y \in \mathscr{D}$ or $y \in \mathscr{D}_0$ then $w^{-1}|q|y \in L^2(w; I)$.

Definition 2 $T(M_w)$ or $T_0(M_w)$ satisfies a separation inequality on $\mathscr{D}$ or $\mathscr{D}_0$ if whenever $y \in \mathscr{D}$ or $y \in \mathscr{D}_0$ then there are constants $A, C, K > 0$, $B \geq 0$, and a constant L independent of y such that the inequality

$$A\,\| w^{-1}(py')'\|_{w,I}^2 + B\,\| w^{-1}\sqrt{pq}\,y'\|_{w,I}^2 + C\,\| w^{-1}qy\|_{w,I}^2 \leq K\,\| M_w[y]\|_{w,I}^2 + L\,\| y\|_{w,I}^2. \tag{1.7}$$

For $w = 1$ these ideas were pioneered by Everitt and Giertz in a series of fundamental papers [10–14], and also by Everitt, Giertz, and Weidmann [15], and by Atkinson [1]. We state here three significant results from this area which are representative of the types of theorems that have been obtained. Theorem *A* and Corollary *A* are from [1], Theorem *B* is from [13], and Theorem *C* is from [15].

Theorem A If $I = [a, \infty)$, $p = 1$, $q(x) > 0$ and locally L^1, and satisfies for constants A_1, A_2, B_1, and B_2

$$[\ln q(x)]_s^t \leq A_1 + B_1 \int_s^t q^{1/2}(x)\,dx, \quad \text{for } a \leq s \leq t \leq \infty,$$

$$[-\ln q(x)]_s^t \leq A_2 + B_2 \int_s^t q^{1/2}(x)\,dx, \quad \text{for } a \leq s \leq t \leq \infty$$

where

$$B_1 < \exp(-A_1/2)\,[\sqrt{16 + B_1^2} - B_1],$$

$$B_2 < \frac{1}{3}\exp(-A_1/2)\,[\sqrt{16 + B_1^2} - B_1],$$

then $M[y]$ is separated on $\mathscr{D}$.

Corollary A If $q(x) > 0$ is locally absolutely continuous on $I = [a, \infty)$, and satisfies

$$-B_2 \leq \frac{q'(x)}{q^{3/2}(x)} \leq B_1 < \frac{4}{\sqrt{3}}$$

where $3B_2 \leq \sqrt{16 + B_1^2} - B_1$, then $M[y]$ is separated on $\mathscr{D}$.
Note that the corollary holds for $B_1 = B_2 < 4\sqrt{15}$.

Theorem B If $I = [a, \infty)$, $p = 1$; $q(x) > 0$ is locally absolutely continuous and satisfies for some $\theta < 2$.

$$\left| \frac{q'(x)}{q^{3/2}(x)} \right| \leq \theta, \tag{1.8}$$

then $M(y]$ is separated on D.

Theorem C Let $p = 1$, $I = [0, \infty)$, and $q \in L^r(I)$ for some $1 \leq r < \infty$ and possibly complex q. Then

(i) $|q|^{1/2} y \in L^2(I)$, $y' \in L(I)$ for $y \in \mathscr{D}$;
(ii) $y, y' \in \mathscr{D}$ are essentially bounded on I and

$$\lim_{t\to\infty} y(t) = 0, \qquad \lim_{t\to\infty} y'(t) = 0;.$$

(iii) M satisfies the Dirichlet condition on $\mathscr{D}$ on I.

Further if q is real, then

(iv) $\int_I |q|^r |y|^2 < \infty$ for $y \in \mathscr{D}$,
(v) M is *SLP* at ∞,
(vi) M satisfies the Dirichlet condition on I,
(vii) M is sepatated if $r \geq 2$.

What is the connection between the idea of separation and the inequality (1.7)? Obviously if $T_0(M_w)$ or $T(M_w)$ satisfies (1.7) it is separated. That separation implies a special case of (1.7) was first shown by Kauffman [19]. The proof which is short depends on the Closed Graph Theorem. For completeness we include it here.

Proposition 1 If M_w is separated on $\mathscr{D}$ or $\mathscr{D}_0$ then (1.7) holds on these domains for $A = C = 1$, $B = 0$ and certain positive constants K, L.

Proof Since $T(M_w)$ is closed, the graph $G(T) = \{(y, T(M_w)(y)): y \in \mathscr{D}\}$ is a Hilbert space with respect to the norm $\| y \|_{w,I} + \| M[y] \|_{w,I}$. Let N_w be the expression $-w^{-1}(py')'$ and define $S: G(T) \to L^2(w; I)$ by $S(y) = N_w(y)$. Now the maximal operator $T(N_w)$ is closed. An easy computation using the hypothesis that $N_w(y) \in L^2(w; I)$ if $y \in \mathscr{D}$ shows that S is also closed. Thus S is a closed linear map defined on a Hilbert space. By the Closed Graph Theorem S is continuous and so

$$\| w^{-1}(py')' \|_{w;I} \le D\{ \| M_w [y] \|_{w,I} + \| y \|_{w,I}\}$$

for some positive constant D. However,

$$\| q(y) \|_{w,I} \le \| M_w[y] \|_{w,I} + \| w^{-1}(py')' \|_{w,I}$$

by the triangle inequality. Therefore upon squaring both these inequalities (1.7) will follow with $A = C = 1$, $B = 0$, $K = 6D^2 + 2$ and $L = 4D^2$.

Remark 1 It is obvious (as has been noted in [1] and [13]) that separation essentially depends only on the behavior of the coefficients p and q in a neighborhood of singular endpoints. For example if $q \in L^2_{\text{loc}}(I)$ and if (1.8) is replaced by

$$\limsup_{x\to\infty} \left| \frac{p^{1/2} q'}{q^{3/2}} \right| < 2$$

we can consider the interval $[c, \infty)$ for a sufficiently large c. Then separation will still occur on $\mathscr{D}_0$ since $\| qy \|_{[c,\infty)} < \infty \Rightarrow \| qy \|_I < \infty$. The triangle inequality then shows that $\| (py')' \|_I < \infty$. It follows additionally from of Proposition 1 that an inequality of form (1.7) will continue to hold but possibly with $B = 0$.

Another preliminary result concerns the connection between separation on $\mathscr{D}_0$ and $\mathscr{D}$.

Proposition 2 If M_w is regular at a and $T_0(M_w)$ is separated, then $T(M_w)$ is separated if and only if M_w is *SLP* at ∞.

Proof If M_w is *SLP* at ∞ it is *LP* at ∞, and therefore

$$\dim (\mathscr{D}/\mathscr{D}_0) = 2.$$

Hence if ϕ_1 and ϕ_2 are functions in $\mathscr{D}$ with compact support on I satisfying (1.6) then $\mathscr{D} = \mathscr{D}_0 \oplus S$ where $S = \text{span}\{\phi_1, \phi_2\}$. Suppose $y \in \mathscr{D}$. Since $T_0(M_w)$ is a closed operator, $\mathscr{D}_0$ is closed with respect to the graph norm topology on $\mathscr{D}$. Moreover S being finite dimensional is also closed. It follows that there exists a continuous (in the graph norm topology) projection $\mathcal{P} : \mathcal{D} \to \mathcal{D}_0$. Hence if the graph norm of y is denoted by $\| y \|_G$, we have that

$$\begin{aligned}
&\| w^{-1}(py')' \|_{w,I} + \| w^{-1} qy \|_{w,I} \\
&\le \| w^{-1}(p(\mathcal{P}y)')' \|_{w,I} + \| w^{-1}q(\mathcal{P}y) \|_{w,I} \\
&\quad + \| w^{-1}(p((I-\mathcal{P})y)')' \|_{w,I} + \| w^{-1}q\,(I-\mathcal{P})y \|_{w,I} \\
&\le K\{ \| w^{-1}M_w[\mathcal{P}y] \|_{w,I} + \| \mathcal{P}y \|_{w,I}\} + \| (I-\mathcal{P})y \|_G \\
&\le K \| \mathcal{P} \| \, \| y \|_G + K_1 \| (I-\mathcal{P}) \| \, \| y \|_G \\
&\le K_2\{ \| M_w[y] \|_{w,I} + \| y \|_{w,I}\}.
\end{aligned}$$

Here the first inequality follows from linearity of the operations and the triangle inequality, the second from the assumption that $T_0(M_w)$ is separated and Proposition 1, and the third from the continuity of the projections. On the other hand, suppose that $T(M_w)$ is separated. Then by Proposition 1 the inequality

$$\| w^{-1}(py')' \|_{w,I}^2 + \| w^{-1}qy \|_{w,I}^2 \le K \| M_w[y] \|_{w,I}^2 + L \| y \|_{w,I}^2\} \tag{1.9}$$

holds on $\mathcal{D}$. Let $N_w[y]$ be defined by $-w^{-1}(py')'$ and $\mathcal{D}_{N_w}$ signify the domain of $T(N_w)$. It is a special case of [15, Theorem 1] that N_w is Dirichlet and *SLP* at ∞. Thus $\lim_{c\to\infty} (py\bar{z}) = 0$ for all $y, z \in \mathcal{D}_{N_w}$. Since (1.9) implies that $\mathcal{D} \subseteq \mathcal{D}_{N_w}$, and M_w and N_w have the same Lagrange billinear form, this is also true for all $y, z \in \mathcal{D}$. Hence M_w is *SLP* at ∞.

In particular if M_w is separated then M_w is *SLP* at ∞. From this we deduce for example that if M_w is limit-circle at ∞ then $T(M_w)$ is not separated. Another consequence is that if M_w is *LP* but not *SLP* at ∞, then $T_0(M_w)$ cannot be separated. For if $T_0(M_w)$ is separated and M_w is *LP* at ∞, then the logic of the first part of the proof of Proposition 2 implies that $T(M_w)$ is separated $\Rightarrow M_w$ is *SLP* at ∞ which contradicts the assumption.

We now sketch the main contributions of the paper. We give several new separation criteria; for technical simplicity they all will be formulated in the left end-point regular, right end-point ∞ setting of this paper, but may be easily modified to include other cases. The first result (Theorem 1) is an extension of Theorem *B* to the setting $w \ne 1$ that also allows nontrivial p satisfying the minimal condition $p^{-1} \in L_{loc}(I)$. The method of proof follows that of Everitt and Giertz [13]. Theorems 6 and 7 comprise (unweighted) extensions of Theorem *C* in that we do not require q to be L^r integrable but only that the L^r norm of q or the L^s norm of p' for certain ranges of r and s be uniformly bounded on the intervals of a sufficiently fine partition of I. There is some overlap, however, between these two theorems. In Theorem 6 we assume that $2 \le r < \infty$ which allows us to prove a separation inequality as well as an extended version of Theorem C. On the other hand in Theorem 7 we show that for $p = 1$ the statements of Theorem 6 except for separation remain true if $1 \le r < \infty$. The remaining results are new criteria which are independent of the results of Everitt and Giertz. Theorem 2 for example allows p to be *any* positive function such that $p^{-1} \in L_{\text{loc}}(I)$ provided q is of slow enough growth. Theorem 3 (we hope) includes the correct version of a test which is stated erroneously in Dunford and Schwartz [7]. All these criteria can fail for $p = 1$ and q unbounded at ∞ but oscillating rapidly. However such q can be allowed if strong enough conditions which guarantee the existence of an appropriate Hardy inequality are put conjointly on p and q (Theorem 4). Theorem 5 derives separation from the existence of either a certain sum or product type inequality while Corollaries 3–4 show that the discreteness of the spectrum of the minimal operator determined by a certain expression $\tilde{M}_{\tilde{w}}$ guarantees separation for $T_0(M_w)$. Thus satisfaction by one operator of any of a number of well-known discrete spectrum criteria can be used to show separation M_w on $\mathscr{D}_0$. The precise statement of all these results together with examples are given in Section 2; their proofs are delayed until Section 3.

We end this section with a few remarks on notation. Constants will be denoted by capital letter such as K, C, etc., and may change their value from line to line. We sometimes distinguish different constants by notation such as $K_1, K_2, \ldots$ Notation such as $K(\varepsilon)$, $K(\alpha, \beta)$ signals dependence of the constant on the indicated parameters. Finally, a local property is indicated by the subscript "loc", e.g., $L^2_{\text{loc}}(I)$.

2. The Separation Criteria

Theorem 1 Suppose $p^{-1} \in L_{\text{loc}}(I)$, w is a positive function in $L_{\text{loc}}(I)$, $pq \ge 0$, and $q \in AC_{\text{loc}}(I)$, where $I = [a, b)$, $-\infty < a < b \le \infty$. Then the separation inequality (1.7) holds for all $y \in \mathscr{D}_0$ in the following cases

(i) with certain constants A, $C < 1$, $B < 2$, $K = 1$ and $L = 0$ if

$$\left| \frac{w p^{1/2} (w^{-1} q)'}{q^{3/2}} \right| \le \theta < 2 \tag{2.1}$$

for all $t \in I$. Moreover, precise (although possibly not sharp values of A, B and C) may be calculated in terms of θ;

(ii) if $A = 1$, $B = 2 - \varepsilon$, $C = 1$, $K = 1 + \varepsilon$, $L = D^4/4\varepsilon^3$ for $\varepsilon \in (0, 2)$, $p, q \ge 0$, and there exists a positive constant D such that

$$\sqrt{wp} \left| \frac{(w^{-1} q)'}{q} \right| \le D. \tag{2.1'}$$

Corollary 1 Under the hypotheses of Theorem 1, the inequality (1.7) is true for $A = 1$, $B = 1 - \varepsilon\theta/2$, $C = 1 - \theta/\varepsilon$, $K = 1$, $L = 0$, $\theta \in (0, \sqrt{2})$, and $\varepsilon \in (\theta, 2/\theta)$. If $\theta = \sqrt{2}$ then (1.7) holds with $A = C = 1$, $B = 0$, $K = 5$, and $L = 0$.

Theorem 2 Suppose $p^{-1} \in L^1_{\text{loc}}$, $pq \ge 0$, q is locally absolutely continuous on I. Then the separation inequality (1.7) hold on $\mathscr{D}_0$ with $A = 1 - 2\sqrt{K_1}$, $B = 2$, $C = 1 - 2\sqrt{K_1}$, $K = 1$, $L = 0$ where and $K_1 < 1/4$ is defined by

$$\sup_{t\in I}\left(\int_a^t w\right)\left(\int_t^\infty w(w^{-1}q)'/q)^2\right) = K_1 \tag{M1}$$

Furthermore, (1.7) is true on $\mathscr{D}_0$ with $A = 1 - \varepsilon$, $B = 2$, $C = 1$, $K = 1$, $L = 4K_2/\varepsilon$ if the condition

$$\sup_{t\in I}\left(\int_a^t w\right)\left(\int_t^\infty w^{-1}((w^{-1}q)')^2\right) = K_2 \tag{M2}$$

for finite K_2 is satisfied.

Example 1 Suppose $w = 1$ and $q(x) = x^\beta$ with $\beta > 0$. Then (M1) holds if and only if $\beta < 1/2$. If $I = (1, \infty)$, $q(x) = K \log(x)$ a calculation shows that (M2) (but not necessarily (M1)) is satisfied. Notice that in either example *no* condition is put on p.

A slight variation of the proof of Theorem 1 gives an improvement of the result probably intended in [7, Chapter XIII, 9.B5, p. 1541][2]

Theorem 3 Suppose $p^{-1} \in L_{\text{loc}}(I)$, $pq \geq 0$, q, p are differentiable, then the separation inequality (1.7) is true on $\mathscr{D}_0$ under the following circumstances

(i). with certain constants $A, C < 1$, $B < 2$, $K = 1$ and $L = 0$

$$\left|\frac{w(p(w^{-1}q)')'}{q^2}\right| \leq \theta \tag{2.3}$$

for some $0 < \theta < 2$. Moreover, in the case $\theta = \sqrt{2}$ the inequality is also true if $A = 2$, $B = 1$, $K = 5$, and $L = 0$;

(ii) with $A = 1$, $B = 2$, $C = 1$, $K = 1 + \varepsilon$, $L = E^2/4\varepsilon$, $p, q \geq 0$ and E is a positive constant such that

$$\left|\frac{(p(w^{-1}q)')'}{q}\right| \leq E. \tag{2.3'}$$

Corollary 2 Under the hypotheses of Theorem 3, the inequality (1.7) is true for $A = 1$, $B = 1 - \varepsilon\theta/2$, $C = 1 - \theta\varepsilon$, $K = 1$, $L = 0$, $\theta \in (0, \sqrt{2})$, and $\varepsilon \in (0, \theta)$. If $\theta = \sqrt{2}$ then (1.7) holds with $A = C = 1$, $B = 0$, $K = 5$, and $L = 0$.

---

[2] The result was misstated as $\limsup_{t\to\infty} |(pq')'|\, q^2 < 1$. As noted by Everitt and Giertz [12] $p(x) = 1$, $q(x) = -x$ for $x \in [0, \infty)$ satisfies the condition and yet separation does not occur.

Example 2 To see that 2 is the best possible upper bound on θ in Theorems 1 and 3, let

$$w = x^{\gamma},\ p = x^{\alpha},\ q = K_0 x^{\beta}$$

where $\beta = \alpha - 2$, and $I = [a, \infty)$, $a > 0$. Suppose $f = x^{\delta}$. A calculation shows that the following is true

(i) $f \in L^2(w; I)$ if and only if

$$\delta < -\left(\frac{\gamma + 1}{2}\right);$$

(ii) $\int_I w \,|\, w^{-1} q f \,|^2 = \infty$ if and only if $2(\alpha + \delta - 1) \geq \gamma + 1$. (Thus this is a sufficient condition for the nonseparation of M_w on $\mathscr{D}$.)

(iii) $M_w[f] = 0$ if and only if $K_0 = \delta(\alpha + \delta - 1)$.

Now let $\alpha = 1$ and $\gamma < -1$. Then (i)–(iii) above will hold for

$$\frac{\gamma + 1}{2} < \delta < -\left(\frac{\gamma + 1}{2}\right) \quad \text{and } K_0 = \delta^2.$$

Consider the condition $\left| \dfrac{w p^{1/2} (w^{-1} q)'}{q^{3/2}} \right| \leq \theta < 2 \Leftrightarrow$

$$\frac{|\beta - \gamma|}{K_0^{1/2}} x^{\frac{\alpha}{2} - \frac{\beta}{2} - 1} = \frac{|\beta - \gamma|}{K_0^{1/2}}$$

$$= \left| \frac{\gamma + 1}{\delta} \right| \leq \theta \text{ (since } \beta = \alpha - 2 \text{ and } \alpha = 1).$$

Let $\varepsilon > 0$. Since $\left| \dfrac{\gamma + 1}{\delta} \right| = 2 + \varepsilon \Leftrightarrow$

$$\delta = -\frac{\gamma + 1}{2} \cdot \frac{1}{(1 + \varepsilon/2)} < -\frac{(\gamma + 1)}{2},$$

the possibility $\theta > 2$ requires that (i) be violated. It follows that $\theta \leq 2$. This example works to show the same upper bound on θ in Theorem 3. It is an open question as to whether $\theta = 2$ in Theorems 1 and 3 is allowed.

Example 3 In (ii) of either Theorems 1 or 3 let $w = q > 0$. Then conditions (2.1′) or (2.3′) are automatically satisfied. If we take $D = 4^{1/4}\,\varepsilon$ or $E = 2\varepsilon$ (1.7) becomes the inequality

$$\int_I q^{-1} \,|\, (py')' \,|^2 + (2 - \varepsilon) \int_I p \,|\, y' \,|^2$$

$$+ (1 - \varepsilon) \int_I q \,|\, y \,|^2 \leq (1 + \varepsilon) \int_I q^{-1} \,|\, M[y] \,|^2.$$

Since $\varepsilon > 0$ is arbitrary, we obtain that

$$\int_I q^{-1} \,|\, (py')' \,|^2 + 2 \int_I p \,|\, y' \,|^2 + \int_I q \,|\, y \,|^2 \leq \int_I q^{-1} \,|\, M[y] \,|^2$$

for all $y \in \mathcal{D}_0'$. (This inequality is also implied immediately for the choice $w = q$ in the proofs of (ii) of Theorems 1 and 3 below.)

Theorem 4 Suppose that p, q are nonnegative and that p^{-1} and q^2 are locally integrable on I. Assume additionally that either

$$\sup_t \left(\int_0^t w^{-1} q^2 \right) \left(\int_t^\infty p^{-1} \right) < \infty \tag{2.4}$$

or

$$\sup_t \left(\int_t^\infty w^{-1} q^2 \right) \left(\int_0^t p^{-1} \right) < \infty. \tag{2.5}$$

Then the inequalities

$$\| w^{-1} qy \|_{w,I} \leq K(\varepsilon) \| y \|_{w,I} + \varepsilon \| M_w[y] \|_{w,I}, \tag{2.6}$$

$$\| w^{-1}(py')' \|_{w,I} \leq K_1(\varepsilon) \| M_w[y] \|_{w,I} + \varepsilon \| y \|_{w,I} \tag{2.7}$$

are valid on $\mathcal{D}_0$ for every $\varepsilon > 0$.

Example 4 If $w^{-1}q^2$ and p^{-1} are integrable, then (2.4) and (2.5) are true and the inequalities (2.6) and (2.7) follow. Similarly (2.6) and (2.7) hold when $p = 1$ if

$$\sup_{t>a} t \int_t^\infty w^{-1} q^2 < \infty.$$

Such a condition could allow significant oscillation in q.

Theorem 5 Suppose $p \geq 0$. Then (1.7) holds on $\mathcal{D}_0$ with $A = 1$, $B = 2$, ε, $\mu \in (0, 1)$, $C = 1 - \varepsilon$, $K = 1 + \varepsilon$, $L = K(\varepsilon)$, if either the sum or multiplicative inequalities

$$\| \sqrt{|\, (p(w^{-1}q)')' \,|}\, y \|_I \leq \varepsilon \| w^{-1/2} qy \|_I + K(\varepsilon) \| p^{1/2} y' \|_I,$$

$$\| \sqrt{|\, (p(w^{-1}q)')' \,|}\, y \|_I \leq K \| w^{-1/2} qy \|_I^\mu \, \| p^{1/2} y' \|_I^{1-\mu}$$

are true for $y \in \mathcal{D}_0'$.

Corollary 3 The conclusions of Theorem 5 hold if the maximal operator in $L^2(\tilde{w}, I)$ generated by

$$\tilde{M}_{\tilde{w}}[y] := \tilde{w}^{-1}[-(py')' + \tilde{q}y]$$

has discrete spectrum where $\tilde{w} = |(p(w^{-1}q)')'|$ and $\tilde{q} = w^{-1}q^2$.

Corollary 4 The conclusions of Theorem 5 hold under any of the following circumstances

(i) $\lim_{t\to\infty} \dfrac{w(p(w^{-1}q)')'}{q^2} = 0.$

(ii) There is a positive constant M such that

$$\sqrt{\frac{p}{\tilde{w}}}\left[\frac{\tilde{w}'}{\tilde{w}} + \frac{p'}{p}\right] \le M,$$

and for each $\theta > 0$

$$\lim_{t\to\infty} W^{-1}(t) \int_t^{t+\theta W(t)} (\tilde{q}/\tilde{w})(t)\, dt = \infty$$

where $W(t) = \sqrt{p/\tilde{w}}$.

(iii) $\lim_{t\to\infty} \dfrac{1}{\tilde{w}}\left(\tilde{q} + \dfrac{1}{4ph^2}\right) = \infty$ where

$$h(t) = \begin{cases} \int_t^\infty p^{-1}, & \text{if } p^{-1} \in L^1(I) \\ \int_a^t p^{-1}, & \text{otherwise.} \end{cases}$$

(iv) $(p(w^{-1}q)')' \in L^r(I)$ for some $r \in (0, \infty)$ and the functions $w^{-1}q$ and p are bounded uniformly below by positive constants.

Remark 2 As we will see (i)–(iii) in Corollary 4 are just different sufficient conditions that $T(\tilde{M}_{\tilde{w}})$ have discrete spectrum. Presumably the more extremely general conditions of Oĭnarov [21] or Curgus and Read [6] could be used as well.

Example 5 If $p = w = 1$. The condition in Corollary 4 (i) for separation becomes

$$\lim_{t\to\infty} \frac{|q''|}{q^2} = 0.$$

As far as separation goes this is a special case of Theorem 3, but we have better control on the left-hand constants of the separation inequality (1.7). Under the same assumptions (ii) requires that

$$\left| \frac{q'''}{(q'')^{3/2}} \right| \leq M,$$

$$\lim_{t \to \infty} \sqrt{|q''|} \int_x^{x+q''} \frac{q^2}{q''} = \infty.$$

An immediate consequence of (iii) shows that when $w = 1$ there will be separation if $(p(q')'ph^2(t) \to 0$ as $t \to \infty$.

Theorem 6 Let $w = 1$, q be possibly complex and in $L^2_{\text{loc}}(I)$ and p be positive, differentiable, and essentially bounded uniformly away from zero. Suppose also that the following two conditions hold.

(H1) For some $2 \leq r < \infty$, and $\theta > 0$

$$\sup_{t \in I} \int_t^{t+\theta} |q|^r = L < \infty. \tag{2.8}$$

(H2) for some $2 \leq s \leq \infty$, and $\phi > 0$

$$\sup_{t \in I} \int_t^{t+\phi} |p'|^s = M < \infty. \tag{2.9}$$

Then

(i) $|q|^{r/2}y$, $|p'|^{s/2}y$, $|q|y$, and $|p'|y$ are in $L^2(I)$ for all $y \in \mathcal{D}$;

(ii) M is separated on $\mathcal{D}$ and the inequality (1.7) holds for $A = C = 1$, $B = 0$, $K = 1 + \varepsilon$, and $L = K(\varepsilon)$ where $K(\varepsilon) \to \infty$ as $\varepsilon \to 0^+$;

(iii) $y, y' \in \mathcal{D}$ are uniformly bounded and

$$\lim_{t \to \infty} |y(t)| = 0, \qquad \lim_{t \to \infty} |y'(t)| = 0;$$

(iv) if $p = 1$ M satisfies the Dirichlet formula

$$\int_I \{y'\bar{z}' + qy\bar{z}\} = -(y\bar{z}')(0) + \int_I M[\bar{z}]y; \tag{2.10}$$

for all $y, z \in \mathcal{D}$, and if q is real, then M is SLP at ∞.
Moreover, if $p = 1$ and

$$\sup_{t \in I} \int_t^{t+\theta} |q'|^2 = K_q < \infty z \tag{2.11}$$

for some constant $\theta > 0$, then (1.7) is true on $\mathcal{D}_0$ for $A = 1 - DK_q\varepsilon^2\theta^2$, $B = 2$,

$C = 1$, $K = 1$, and $L = DK_q(\varepsilon\theta)^{-2} + 1$, where D is the best constant of the interpolation inequality for $y \in \mathscr{D}_0'$

$$\| y' \|_I^2 \le D \{\eta^{-1} \| y \|_I^2 + \eta \| y'' \|_I^2\}, \qquad \eta > 0\cdot$$

Theorem 7 If $p = w = 1$, $(H1)$ holds for $1 \le r < 2$, and the other hypotheses of Theorem 6 are satisfied then

(i) $| q |^{r/2} y$ and $| q |^{1/2} y$ are in $L^2(I)$ for all $y \in \mathscr{D}$;
(ii) $y, y' \in \mathscr{D}$ are uniformly bounded and

$$\lim_{t\to\infty} | y(t) | = 0, \quad \lim_{t\to\infty} | y'(t) | = 0;$$

(iii) for all $y, z \in \mathscr{D}$, M satisfies the Dirichlet formula (2.10), and if q is real, then M is *SLP* at ∞.

Remark 3 When $p = 1$ Theorems 6 and 7 extend Theorem *C*. Note that $(H2)$ is satisfied if $I = [1, \infty)$, $p(t) = t^\tau$, $\tau \le 1$, $p(t) = \log t$, etc. and (H1) holds for similar choices for q.

3. Proof of the Theorems

This section gives the proofs of the results stated in Section 2. We begin with a Lemma required to extend (1.7) from $\mathscr{D}_0'$ to all of $\mathscr{D}_0$. Our proof is similar to that of Everitt and Giertz [11, pp. 313–314] who however assume slightly more restrictive conditions on p, q and prove a different separation inequality.

Lemma 1 Suppose p and q satisfy minimal conditions and if $B \neq 0$ we assume that $pq > 0$ and $w^{-1}q \in L_{\text{loc}}(I)$. Then if the separation inequality (1.7) is true on $\mathscr{D}_0'$ it is true on $\mathscr{D}_0$ with constants $K_1 := 2C(1 + K/A) + K$ and $L_1 := L(2C/A + 1)$ replacing K and L in the original inequality. If, however, we suppose that either $q \in L^2_{\text{loc}}(I)$ or that $q \ge c > 0$ a.e. for some constant c, $w = 1$, and $q \in L_{\text{loc}}$ then the constants of (1.7) on D_0 are the same as they are on $\mathscr{D}_0'$.

Proof Let $y \in \mathscr{D}_0$. There is a sequence $\{y_n\} \subset \mathscr{D}_0'$ such that $y_n \to y$ and $M_w[y_n] \to M_w[y]$ in $L^2(w; I)$. By hypothesis (1.7) is true for each y_n and for functions $y_n - y_m$. Therefore if $A, B, C > 0$ each of the sequences $\{w^{-1}(py_n')'\}$, $\{w^{-1}\sqrt{pq}\, y_n'\}$, and $\{w^{-1}qy_n\}$ is Cauchy in $L^2(w; I)$. From [22, §17] or [11, *p.* 313], the maximal operator N generated by $w^{-1}(py')'$ is closed.[3] Hence $w^{-1}(py_n')' \to w^{-1}(py')'$ in $L^2(w; I)$. Since

$$A \| w^{-1}(py_n')' \|_{w,I}^2 \le K \| M_w[y_n] \|_{w,I}^2 + L \| y_n \|_{w,I}^2,$$

[3]The argument is given only in the case $w = 1$; but the extension to general w is immediate.

this inequality holds also for y. However by the triangle inequality and the elementary estimate $(X + Y)^2 \le 2(X^2 + Y^2)$

$$\| w^{-1}qy \|_{w,I}^2 \le 2\,(\| M_w[y] \|_{w,I}^2 + \| w^{-1}(py')' \|_{w,I}^2).$$

Therefore a calculation shows that

$$A \| w^{-1}(py')' \|_{w,I}^2 + C \| w^{-1}qy \|_{w,I}^2$$
$$\le [2C(1 + K/A) + K] \| M_w[y] \|_{w,I}^2 + L(2C/A + 1) \| y \|_{w,I}^2. \qquad (3.1)$$

This proves the Lemma when $B = 0$ in (1.7).

If $B > 0$ we have by integration and the Cauchy-Schwartz inequality that

$$|(y_n' - y')(t)| = |p^{-1}(t)| \left| \int_a^t (p(y_n - y)')' \right|$$
$$\le |p^{-1}(t)| \left(\int_a^t w \right)^{1/2} \left(\int_I w^{-1} |(p(y_n - y)')'|^2 \right)^{1/2}$$

Hence on any finite subinterval $\Delta \subset I$ (since $w^{-1}q$ is locally integrable)

$$\int_\Delta w^{-1} |pq|\, |y_n' - y'|^2 = O(\| w^{-1}(p(y_n - y)')' \|_{w,I}^2).$$

Substitution of $w^{-1}(p(y_n - y)')'$ into (3.9) shows that $\| w^{-1}(p(y_n - y)')' \|_{w,I} \to 0$. Thus $w^{-1}\sqrt{pq}\,y_n' \to w^{-1}\sqrt{pq}\,y$ in $L^2(w; \Delta)$. However we already know that $w^{-1}\sqrt{pq}\,y_n'$ converges in $L^2(w; I)$ to a function g. By uniqueness of limits $g = w^{-1}\sqrt{pq}\,y'$ on any Δ and hence

$$B \int_I w^{-1}pq\, |y_n'|^2 \le K \| M_w[y_n] \|_{w,I}^2 + L \| y_n \|_{w,I}^2,$$

the same inequality is satisfied by y. Addition of this inequality with (3.1) gives (1.7) with the constants K_1 and K_2.

If $w^{-1/2}q$ is locally L_w integrable, on any compact interval Δ we have

$$\int_\Delta |q|\, |y_n - y| \le \| w^{-1}q \|_{w,\Delta} \| y_n - y \|_{w,\Delta}.$$

This means that $qy_n \to qy$ pointwise a.e. Since $w^{-1}qy_n$ converges to a function f in $L^2(w; I)$, f must equal $w^{-1}qy$ a.e. If on the other hand q is locally integrable, uniformly bounded below by $c > 0$, and $w = 1$, the fact that $\{(py_n')'\}$ is Cauchy in $L^2(I)$ implies that $\{q^{-2}(py_n')'\}$ is Cauchy in $L^2(q^2; I)$. Because $T[q^{-2}(py')']$ is a closed operator in $L^2(q^2)$, $\{y_n\}$ converges to a function $\tilde{y} \in L^2(q^2; I)$ such that $p\tilde{y}'$ is locally absolutely continuous and $q^{-2}(p\tilde{y}')' \in L^2(q^2; I)$. This means that $\{q^{-2}(py')'\}$ converges pointwise a.e. to $q^{-2}(p\tilde{y}')'$. However we have seen that $(py_n')' \to (py')'$ and $y_n \to y$ in $L^2(I)$ where py' is locally absolutely continous and $(py')') \in L^2(I)$. Since this convergence is also pointwise a.e., it follows that

$$q^{-2}(p\tilde{y}')' = q^{-2}(py')' \text{ a.e.} \Rightarrow \tilde{y} = y.$$

In either case since (1.7) is satisfied by $\{y_n\}$, and we have shown that

$$w^{-1}(py_n')' \to w^{-1}(py')'$$

$$w^{-1}\sqrt{pq}\,y_n' \to w^{-1}\sqrt{pq}\,y'$$

$$w^{-1}qy_n \to w^{-1}qy,$$

(where $w = 1$ if $q \geq c > 0$, etc.) the same inequality with the same constants must be satisfied by y.

Lemma 1 will allow us in all cases to prove the particular separation inequality only on $\mathscr{D}_0'$. Moreover, since we are directly or indirectly assuming in every Theorem that $q \in L^2_{\text{loc}}(I)$ the stated constants of inequality (1.7) will not change in this extension.

Proof of Theorem 1 There is no loss of generality in restricting ourselves to real functions y in $\mathscr{D}_0'$. Let γ be a real parameter in (0, 2). We begin with the basic identity

$$w\,|\,M_w[y]\,|^2 + \gamma w^{-1}(py')'\,qy = w^{-1}(py')'^2 + (\gamma - 2)\,w^{-1}(py')'\,qy + w^{-1}q^2y^2.$$

We then integrate both sides over I and apply integration by parts to the term $w^{-1}(py')'(qy)$ on the right. This gives that

$$\begin{aligned}\|M_w[y]\|^2_{w,I} + \gamma \int_I w^{-1}(py')'\,qy &= \|w^{-1}(py')'\|^2_{w,I} \\ &+ (2-\gamma)\,\|w^{-1}\sqrt{pq}\,y'\|^2_{w,I} + (2-\gamma)\int_I py'(w^{-1}q)'\,y \\ &+ \|w^{-1}qy\|^2_{w,I}. \end{aligned} \tag{3.2}$$

By Hölder's inequality, the condition (2.1), and the arithmetic-geometric mean inequalities we obtain the estimates

$$\left|\int_I w^{-1}(py')'\,q\,y\right| \leq (1/2)\,\{\mu\,\|w^{-1}(py')'\|^2_{w,I} + \mu^{-1}\,\|w^{-1}qy\|^2_{w,I}\}$$

$$\left|\int_I p(w^{-1}q)'\,y\,y'\right| \leq (\theta/2)\,\{\varepsilon\,\|w^{-1}\sqrt{pq}\,y'\|^2_{w,I} + \varepsilon^{-1}\,\|w^{-1}qy\|^2_{w,I}\}.$$

Substituting these into (3.2) and rearranging gives finally that

$$\| M_w[y] \|_{w,I}^2 \geq (1 - \gamma\mu/2) \| w^{-1}(py')' \|_{w,I}^2$$

$$+ (2 - \gamma)[1 - \varepsilon\theta/2] \| w^{-1}\sqrt{pq}y' \|_{w,I}^2$$

$$+ (1 - \gamma/2\mu - \theta(2 - \gamma)/2\varepsilon) \| w^{-1}qy \|_{w,I}^2. \qquad (3.3)$$

In order that this inequality holds with positive coefficients we must have $\gamma\mu \in (0, 2)$ and

$$\theta < \min\left\{\frac{2}{\varepsilon}, \left(\frac{2\mu - \gamma}{2 - \gamma}\right)\left(\frac{\varepsilon}{\mu}\right)\right\}. \qquad (3.4)$$

Let $\varepsilon = 1$, $\mu = 2/\gamma - \delta$ where $0 < \delta < 2/\gamma$. Then from (3.4) it should be the case that

$$\theta < \frac{4 - 2\gamma\delta - \gamma^2}{(2 - \gamma\delta)(2 - \gamma)}.$$

Suppose $\theta = 2 - \eta$. If we choose $\gamma = 2 - \eta$ and $\delta = \eta^2/(2\gamma)$ we will have that

$$\frac{4 - 2\gamma\delta - \gamma^2}{(2 - \gamma\delta)(2 - \gamma)} > \frac{4 - \eta^2 - \gamma^2}{2(2 - \gamma)}$$

$$= \frac{2 + \gamma}{2} - \frac{\eta^2}{2(2 - \gamma)} = 2 - \eta/2 - \eta/2 = \theta.$$

The inequality will follow. From the form of (3.3) it is obvious finally that $A < 1$, $B < 2$, and $C < 1$.

Turning to (ii), we take $\gamma = 0$ and get by rearrangement and obvious estimates in (3.2) that

$$\| w^{-1}(py')' \|_{w,I}^2 + 2 \| w^{-1}\sqrt{pq}y' \|_{w,I}^2 + \| w^{-1}qy \|_{w,I}^2 \leq \| M_w[y] \|_{w,I}^2$$

$$+ 2\int_I \sqrt{wp}\left|\frac{(qw^{-1})'}{q}\right| |q^{1/2}y|\,|w^{-1/2}\sqrt{pq}y'|.$$

The hypothesis, Hölder's inequality, and the arithmetic-geometric mean inequality then gives

$$\| w^{-1}(py')' \|_{w,I}^2 + 2 \| w^{-1}\sqrt{pq}y' \|_{w,I}^2 + \| w^{-1}qy \|_{w,I}^2 \leq \| M_w[y] \|_{w,I}^2$$

$$+ (D^2/\varepsilon)\left(\int_I q|y|^2 + \varepsilon \| w^{-1}\sqrt{pq}y' \|_{w,I}^2\right). \qquad (3.5)$$

Now

$$\int_I q\,|y|^2 \le \int_I wM_w[y]\,y \le \|M_w[y]\|_{w,I}\,\|y\|_{w,I}$$

$$\le (\delta/2)\,\|M_w[y]\|^2 + (1/2\delta)\,\|y\|^2_{w,I}.$$

Hence substituting this into (3.5) we get

$$\|w^{-1}(py')'\|^2_{w,I} + (2-\varepsilon)\,\|w^{-1}\sqrt{pq}\,y'\|^2_{w,I} + \|w^{-1}qy\|^2_{w,I}$$

$$\le \left(1 + \frac{D^2\delta}{2\varepsilon}\right)\|M_w[y]\|^2_{w,I} + \left(\frac{D^2}{2\varepsilon\delta}\right)\|y\|^2_{w,I}.$$

The choice $\varepsilon < 2$ and $\delta = 2\varepsilon^2/D^2$ gives the stated version of (1.7).

Proof of Corollary 1 We take $\gamma = 0$. Then (3.3) becomes

$$\|M_w[y]\|^2_{w,I} \ge \|w^{-1}(py')'\|^2_{w,I} + 2(1 - \varepsilon\theta/2)\,\|w^{-1}\sqrt{pq}\,y'\|^2_{w,I}$$

$$+ (1 - \theta/\varepsilon)\,\|w^{-1}qy\|^2_{w,I}.$$

The coefficients will be positive if $\theta \in (0, \sqrt{2})$ and $\varepsilon \in (\theta, 2/\theta)$.

If $\gamma = 0$ and $\theta = \varepsilon = \sqrt{2}$ we have the inequality $\|w^{-1}(py')'\|_{w,I} \le M_w[y]\|_{w,I}$ in (3.3). The triangle inequality gives

$$\|qy\|_{w,I} = \|M_w[y] + w^{-1}(py')'\|_{w,I}$$

$$\le \|M_w[y]\|_{w,I} + \|w^{-1}(py')'\|_{w,I}$$

$$\le 2\,\|M_w[y]\|_{w,I}.$$

Hence by addition we have that

$$\|w^{-1}(py')'\|^2_{w,I} + \|qy\|^2_{w,I} \le 5\,\|M_w[y]\|^2_{w,I}.$$

We now require a Lemma concerning the weighted Hardy inequality. The proof may be found either in Muckenhoupt [21] or Opic and Kufner [24].

Lemma 2 Let u, v be positive a.e. measurable functions such that u, v, and v^{-1} are locally integrable on I. If either

$$\sup_{t>a}\left(\int_t^\infty u\right)\left(\int_a^t v^{-1}\right) = K < \infty \tag{3.6}$$

or

$$\sup_{t>a}\left(\int_a^t u\right)\left(\int_t^\infty v^{-1}\right) = K < \infty \tag{3.6$'$}$$

is true, then the Hardy-type inequality

$$\int_I u\,|f|^2 \le D\int_I v\,|f'|^2 \tag{3.7}$$

holds for absolutely continuous functions f satisfying the condition $f(a) = 0$ in (3.6) and $\lim_{t \to \infty} f(t) = 0$ in (3.6′). Also in (3.7) the constant $D \le 4K$.

Proof of Theorem 2 The proof in part follows that of Theorem 1 with $\gamma = 0$. Suppose (M1) holds. Consider the identity

$$\| w^{-1}(py')' \|_{w,I}^2 - 2\int_I w^{-1}(py')' qy + \| w^{-1}qy \|_{w,I}^2 = \| M_w[y] \|_{w,I}^2.$$

Integrating by parts gives

$$-2\int_I w^{-1}(py')' qy = 2\| w^{-1}\sqrt{pq}y' \|_{w,I}^2 + 2\int_I p(w^{-1}q)' yy'.$$

Hence by substitution, rearrangement, and integration of the final term by parts, and Hölder's inequality it follows that

$$\| w^{-1}(py')' \|_{w,I}^2 + 2\| w^{-1}\sqrt{pq}y' \|_{w,I}^2 + \| qy \|_{w,I}^2 - \| M_w[y] \|_{w,I}^2$$

$$\le 2\left|\int_I p(w^{-1}q)' yy'\right| \le 2\| w^{-1}(py')' \|_{w,I} \left(\int_I w \left|\int_t^\infty (w^{-1}q)' y\right|^2\right)^{1/2} \tag{3.8}$$

Let $Y = \int_t^\infty (w^{-1}q)'y$. Then Y has compact support and so $\lim_{t \to \infty} Y = 0$. By (3.6′) of Lemma 2 with $u = w$ and $v = w^{-1}(q/(w^{-1}q)')^2$ the inequality

$$\int_I w|Y|^2 \le D\int_I w(q/(w^{-1}q)')^2 |Y'|^2 = \| w^{-1}qy \|_{w,I}^2$$

holds with $D \le 4K_1$ if and only ($M1$) is true. Hence by ($M1$) and the arithmetic-geometric mean inequality we get that

$$\| w^{-1}(py')' \|_{w,I}^2 + 2\| w^{-1}\sqrt{pq}y' \|_{w,I}^2 + \| w^{-1}qy \|_{w,I}^2$$

$$- \| M_w[y] \|_{w,I}^2 \le 4\sqrt{K_1}\| w^{-1}(py')' \|_{w,I} \| w^{-1}qy \|_{w,I}$$

$$\le 2\sqrt{K_1}\left(\| w^{-1}(py')' \|_{w,I}^2 + \| w^{-1}qy \|_{w,I}^2\right).$$

Since $K_1 < 1/4$ the result follows if the final right-hand side of this sequence of inequalities is subtracted from the left-hand side.

Alternatively, with the same definition of Y by (3.6′) of Lemma 2 again with $u = w$ and $v = w((w^{-1}q)')^{-2}$ ($M2$) gives the Hardy inequality

$$\int_I w|Y|^2 \le D\int_I w((w^{-1}q)')^{-2}|Y'|^2 = C\int_I w|y|^2.$$

We obtain that

$$\| w^{-1}(py')' \|_{w,I}^2 + 2 \| w^{-1}\sqrt{pq}y' \|_{w,I}^2 + \| w^{-1} qy \|_{w,I}^2$$

$$\leq \| M_w[y] \|_{w,I}^2 + 4\sqrt{K_2} \| w^{-1}(py')' \|_{w,I} \| y \|_{w,I}$$

$$\leq \| M_w[y] \|_{w,I}^2 + \varepsilon \| w^{-1}(py')' \|_{w,I}^2 + (4K_2/\varepsilon) \| y \|_{w,I}^2;$$

the proof is completed as before.

Proof of Theorem 3 In regard to (i), let γ be a parameter in (0, 2) we begin with the integration by parts identity

$$(2-\gamma)\int_I p(w^{-1}q)'yy' = -(1/2)(2-\gamma)\int_I \left[\frac{w(p(w^{-1}q)')'}{q^2}\right](w^{-1}q^2y^2),$$

and substitute this into the first line of (3.2). This gives

$$\| M_w[y] \|_{w,I}^2 + \gamma \int_I w^{-1}(py')'qy = \| w^{-1}(py')' \|_{w,I}^2$$

$$+ (2-\gamma) \| w^{-1}\sqrt{pq}y' \|_{w,I}^2$$

$$+ (1/2)(2-\gamma)\int_I \left[\frac{w(p(w^{-1}q)')'}{q^2}\right](w^{-1}q^2y^2) + \| w^{-1} qy \|_{w,I}^2.$$

Hölder's inequality, the arithmetic-geometric mean equality applied to the left side, the hypothesis, and obvious rearrangement gives us as in the proof of Theorem 1 that

$$\| M_w[y] \|_{w,I}^2 \geq (1-\gamma\mu/2) \| w^{-1}(py')' \|_{w,I}^2$$

$$+ (2-\gamma) \| w^{-1}\sqrt{pq}y' \|_{w,I}^2 + (1-\gamma/2\mu - \theta(2-\gamma)/2) \| w^{-1} qy \|_{w,I}^2. \tag{3.9}$$

In order that the first and last terms on the right be positive we must have

$$\mu = 2/\gamma - \delta,\ 0 < \delta < 2/\gamma$$

$$\theta < \left(\frac{2\mu-\gamma}{2-\gamma}\right)\left(\frac{1}{\mu}\right).$$

The argument that the parameters may be chosen so that these inequalities are satisfied for any $\theta \in (0, 2)$ will be omitted since it is the same as that given in the proof of Theorem 1.

(ii) is also similar to the proof of the analogous statement for Theorem 1.

$$\| w^{-1}(py')' \|_{w,I}^2 + 2 \| w^{-1}\sqrt{pq}y' \|_{w,I}^2 + \| w^{-1}qy \|_{w,I}^2 \le \| M_w[y] \|_{w,I}^2$$

$$+ \int_I |(p(w^{-1}q)')'/q \,|\, q \,|\, y|^2$$

$$\le \| M_w[y] \|_{w,I}^2 + E \| q^{1/2} y \|^2$$

$$\le \| M_w[y] \|_{w,I}^2 + E \left| \int_I wM_w[y]\bar{y} \right|$$

$$\le \| M_w[y] \|_{w,I}^2 + E \| M_w[y] \|_{w,I} \| y \|_{w,I}$$

$$\le \| M_w[y] \|_{w,I}^2 + \varepsilon \| M_w[y] \|_{w,I}^2 + (E^2/4\varepsilon) \| y \|_{w,I} \text{ etc.}$$

Proof of Corollary 2 The proof is the same as that of Corollary 1.

Proof of Theorem 4 By the triangle inequality and the local L^2 integrability of q we have that

$$\| w^{-1}(py')' \|_{w,I} \le \| M_w[y] \|_{w,I} + \| w^{-1} qy \|_{w,I} \tag{3.10}$$

for $y \in \mathscr{D}_0'$. However again by Lemma 2 the conditions (3.6) or (3.6′) imply that $\| w^{-1}qy \|_{w,I} \le K \| p^{1/2}y' \|$ for all $y \in \mathscr{D}_0'$. Hence

$$\| w^{-1}qy \|_{w,I} < K(\| p^{1/2}y' \|_I^2 + \| q^{1/2} y \|_I^2)^{1/2}$$

$$\le K\left(\int_I wM_w[y]\bar{y} \right)^{1/2}$$

$$\le K \| M_w[y] \|_{w,I} \| y \|_{w,I}$$

$$\le \varepsilon \| M_w[y] \|_{w,I} + (K^2/4\varepsilon) \| y \|_{w,I})^{1/2}. \tag{3.11}$$

(3.1) is equivalent to (2.6); further (3.10) and (3.11) yield (2.7).

Proof of Theorem 5 Integrating by parts, we can derive the inequality

$$\left| 2 \int_I p(w^{-1}q)' yy' \right| \le \| \sqrt{|(p(w^{-1}q)')'|}\, y \|_I^2 . \tag{3.12}$$

If the multiplicative inequality is satisfied, substitution of (3.12) into (3.2) with $\gamma = 0$ and the general arithmetic-geometric mean inequality $A^\mu B^{1-\mu} \le \mu A + (1 - \mu) B$ gives that

$$\| w^{-1}(py')' \|_{w,I}^2 + 2 \| w^{-1}\sqrt{pq}y' \|_{w,I}^2 + \| w^{-1} qy \|_{w,I}^2$$

$$\le \| M_w[y] \|_{w,I}^2 + \| \sqrt{|(p(w^{-1}q)')'|}\, y \|_I^2$$

$$\leq \| M_w[y] \|_{w,I}^2 + K \| w^{-1} qy \|_{w,I}^{\mu} \; \| p^{1/2} y' \|_I^{(1-\mu)}$$

$$\leq \| M_w[y] \|_{w,I}^2 + \varepsilon \| w^{-1} qy \|_{w,I}^2 + K(\varepsilon) \| p^{1/2} y' \|_I^2$$

$$\leq \| M_w[y] \|_{w,I}^2 + \varepsilon \| w^{-1} qy \|_{w,I}^2 + K(\varepsilon) \| (M_w[y] \|_{w,I} \| y \|_{w,I})$$

$$\leq \| M_w[y] \|_{w,I}^2 + \varepsilon \| w^{-1} qy \|_{w,I}^2 + \varepsilon \| M_w[y] \|_{w,I}^2 + K(\varepsilon) \| y \|_{w,I}^2$$

which what is to be proved. The argument using the additive inequality is similar and will be omitted.

Proof of Corollary 3 Let $W^{1,2}(I; \tilde{q}, p)$ denote the space defined appropriate locally absolutely continuous functions y such that the norm

$$\| y \|_{I, \tilde{q}, p} := \| w^{-1/2} qy \|_I + \| p^{1/2} y' \|_I$$

is finite. If $T(M_{\tilde{w},\tilde{q}})$ has discrete spectrum it follows by a version of Rellich's Theorem that there is a compact mapping of $W^{1,2}(I; \tilde{q}, p)$ into $L^2(\tilde{w}; I)$. Now suppose that the additive inequality of Theorem 5 is not true on $\mathscr{D}_0'$. Then for every $\varepsilon > 0$ there exists $y_n \in \mathscr{D}_0' \subset W^{1,2}(I; \tilde{q}, p)$ such that

$$\| \tilde{w}^{1/2} y_n \|_I > \varepsilon \| \tilde{q}^{1/2} y_n \|_I - n \| p^{1/2} y_n' \|_I . \tag{3.13}$$

Let $g_n := y_n / \| y_n \|_{I,\tilde{q},p}$. Then (3.13) is true if y_n is replaced by g_n. Further by compactness a subsequence $\{g_{n_i}\}$ converges to a function g in $L^2(\tilde{w}; I)$. But by (3.13) this forces $g_{n_i}' \to 0$ in $L^2(p; I)$. By Hölder's inequality g_{n_i} converges uniformly on any subinterval interval $\Delta_t = [a, t]$ of I. Therefore $g = 0$ a.e. on Δ_t. But then $g_{n_i} \to 0$ in $L^2(\tilde{q}; I)$ and

$$1 = \| g_{n_i} \|_{I,\tilde{q},p} = \| g_{n_i} \|_{\tilde{q},I} + \| g_{n_i}' \|_{p,I} \to 0$$

which is impossible. Hence we can satisfy the hypothesis of Theorem 5, etc.

Proof of Corollary 4 (i) By a weighted version of Weyl's Theorem $T(M_{\tilde{w},\tilde{q}})$ has discrete spectrum if $\tilde{q}/\tilde{w} \to \infty$. This is equivalent to the condition (i). The result follows by Corollary 3. (ii) The conditions are a weighted generalization of the Molchanov condition due to Hinton [18] and imply that $T(M_{\tilde{w},\tilde{q}})$ has discrete spectrum. (iii) Again this is a discrete spectrum criterion due to Friedrich [16] (iv) Let $s = 2r/(r-1)$. Then by the hypothesis and Hölder's inequality

$$\int_I |(p(w^{-1} q)')' \| y |^2 \leq \left(\int_I | (p(w^{-1} q)')' |^r \right)^{1/r} \left(\int_I | y |^s \right)^{1/s} .$$

But by Theorem 1 of Brown and Hinton [3] we have the multiplicative inequality

$$\left(\int_I |(p(w^{-1} q)')'|^r\right)^{1/r} \left(\int_I |y|^s\right)^{1/s} \leq K \,||\, w^{-1/2} q y \,\|_I^{\mu}\, \| p^{1/2} y' \|_I^{1-\mu}$$

where $\mu = (1 - 1/2 + 1/s)$. It follows that Theorem 5 may be applied.

Proof of Theorem 6 Let $y \in \mathscr{D}$. We note first that since p is both positive and locally absolutely continuous, $y' \in AC_{\text{loc}}$ since

$$y'(t) \equiv p(t)^{-1} \int_a^t (py')' + p(t)^{-1} p(a) y'(a).$$

Let $\delta \leq \min\{\theta, \phi\}$ and be sufficiently small. Given an arbitrary positive integer n set $J_{\delta,n} := [a, a + n\delta]$. The triangle inequality and the fact that $\| M[y] \|_I \geq \| M[y] \|_{J_{\delta,n}}$ gives the estimates

$$\| (py')' \|_{J_{\delta,n}} \leq \| M[y] \|_I + \| qy \|_{J_{\delta,n}}$$

$$\| py'' \|_{J_{\delta,n}} \leq \| M[y] \|_I + \| p'y' \|_{J_{\delta,n}} + \| qy \|_{J_{\delta,n}}. \tag{3.14}$$

A Hölder's inequality argument shows that the right-hand side of this is finite since $p', q \in L^2_{\text{loc}}(I)$ and $y, y' \in AC_{\text{loc}}(I)$.

We now partition I so that it is the union of mutually disjoint intervals $\{\Delta_i\}$ of constant length δ. Then both (H1) and (H2) are satisfied with δ replacing θ or ϕ. Next a standard interpolation lemma (see Brown and Hinton [2, Lemma 2.1]), the fact that $p \geq c > 0$ on I, and Hölder's inequality yield successively that

$$|y(s)| \leq K \left\{\delta^{-1} \int_{\Delta_i} |y| + \delta \int_{\Delta_i} |y''|\right\}$$

$$|y(s)|^2 \leq K_1 \left\{\delta^{-1} \int_{\Delta_i} |y|^2 + \delta^3 c^{-2} \int_{\Delta_i} p^2 |y''|^2\right\} \tag{3.15}$$

$$\| q^{r/2} y \|^2_{\Delta_i} \leq K_1 \left(\int_{\Delta_i} q^r\right) \left\{\delta^{-1} \int_{\Delta_i} |y|^2 + \delta^3 c^{-2} \int_{\Delta_i} p^2 |y''|^2\right\}$$

$$\| qy \|^2_{\Delta_i} \leq K_1 \left(\int_{\Delta_i} q^r\right)^{2/r} \left\{\delta^{-2/r} \int_{\Delta_i} |y|^2 + \delta^{(4r-2)/r} c^{-2} \int_{\Delta_i} p^2 |y''|^2\right\}.$$

Addition, the hypothesis (H1) on q, and replacement of $\| y \|_{J_{\delta,n}}$ by $\| y \|_I$ gives

$$\| q^{r/2} y \|^2_{J_{\delta,n}} \leq K_1 L \left\{\delta^{-1} \int_I |y|^2 + \delta^3 c^{-2} \int_{J_{\delta,n}} p^2 |y''|^2\right\} \tag{3.16}$$

$$\| qy \|^2_{J_{\delta,n}} \le K_1 L^{2/r} \{\delta^{-2/r} \int_I | y |^2 + \delta^{(4r-2)/r} c^{-2} \int_{J_{\delta,n}} p^2 | y'' |^2\}. \quad (3.17)$$

The same partitioning method, addition, and using (H2) in place of (H1) gives us first that

$$| y'(s) | \le K_2 \left\{\delta^{-1} \int_{\Delta_i} | y | + \int_{\Delta_i} | y'' |\right\}$$

$$| y'(s) |^2 \le K_3 \left\{\delta^{-3} \int_{\Delta_i} | y |^2 + \delta c^{-1} \int_{\Delta_i} p^2 | y'' |^2\right\}$$

$$\| p'^{s/2} y' \|^2_{\Delta_i} \le K_3 \left(\int_{\Delta_i} p^s\right) \left\{\delta^{-3} \int_{\Delta_i} | y |^2 + \delta c^{-2} \int_{\Delta_i} p^2 | y'' |^2\right\} \quad (3.18)$$

$$\| p' y' \|^2_{\Delta_i} \le K_1 \left(\int_{\Delta_i} p^s\right)^{2/s} \left\{\delta^{-2(1+1/s)} \int_{\Delta_i} | y |^2 + \delta^{2(1-1/s} c^{-2} \int_{\Delta_i} p^2 | y'' |^2\right\},$$

and then that

$$\| p'^{s/2} y' \|^2_{J_{\delta,n}} \le K_3 M \left\{\delta^{-3} \int_I | y |^2 + \delta \int_{J_{\delta,n}} | y'' |^2\right\} \quad (3.19)$$

$$\| p' y' \|^2_{J_{\delta,n}} \le K_1 M^{2/s} \left\{\delta^{-(2+1/s)} \int_I | y |^2 + \delta^{(2-1/s} c^{-2} \int_{J_{\delta,n}} p^2 | y'' |^2\right\}. \quad (3.20)$$

Obvious square root estimates in (3.17), (3.20), substitution into (3.14), followed by subtraction (for small enough δ) gives the inequality

$$\| py'' \|_{J_{\delta,n}} \le K(\delta)\{ \| y \|_I + \| M [y] \|_I\}. \quad (3.21)$$

Taking δ sufficiently small in relation to a parameter $\varepsilon > 0$ and substitution into (3.16) and (3.17) or (3.19) and (3.20) yields the inequalities

$$\| q^{r/2} y \|_{J_{\delta,n}} \le K_1(\varepsilon) \| y \|_I + \varepsilon \| M[y] \|_I$$

$$\| qy \|_{J_{\delta,n}} \le K_2(\varepsilon) \| y \|_I + \varepsilon \| M[y] \|_I$$

$$\| p'^{s/2} y' \|_{J_{\delta,n}} \le K_3(\varepsilon) \| y \|_I + \varepsilon \| M[y] \|_I \quad (3.22)$$

$$\| p' y' \|_{J_{\delta,n}} \le K_4(\varepsilon) \| y \|_I + \varepsilon \| M[y] \|_I.$$

Since both (3.21) and the above inequalities are independent of n, they hold with the same constants on all of I. This proves (i).

Since $\| (py')' \|_I \le \| M[y] \|_I + \| qy \|_I$, (3.22) gives the separation inequality

(1.7) with $A = C = 1$, $B = 0$, $L = K_2(\varepsilon)$ and $K = \varepsilon$. Since the $K_i(\varepsilon)$, $i = 1, \ldots, 4$ involve negative powers of δ it is evident that $K_i(\varepsilon) \to \infty$ as $\varepsilon \to 0$. This completes the proof of (ii).

(iii) follows the same pattern. But this time we replace I by the interval $I_n := [n, \infty)$ and choose $n > a$ so large that for a particular y, $\| M[y] \|_{I_n} < \varepsilon$ and $\| y \|_{I_n} < \varepsilon$. If I_n is partitioned into intervals of unit length, (3.15) and (3.18) added over these intervals, and we apply (3.21) (rederived on I_n) we get that for $t > n$

$$\begin{aligned} | y(t) | &\leq K \{ \| y \|_{I_n} + \| py'' \|_{I_n} \\ &\leq K_1 \| y \|_{I_n} + \| M[y] \|_{I_n} \\ &= O(\varepsilon) \end{aligned}$$

and

$$\begin{aligned} | y'(t) | &\leq K_2 \{ \| y \|_{I_n} + \| py'' \|_{I_n} \\ &\leq K_3 \| y \|_{I_n} + \| M[y] \|_{I_n} \\ &= O(\varepsilon). \end{aligned}$$

These arguments show that y, y' are both essentially bounded on I and vanish at ∞, proving (iii).

If $p = 1$ and q is real the fact that $y\bar{z}'(t) \to 0$ shows that M is *SLP* at ∞. That M satisfies the Dirichlet condition for real or complex q when $r \geq 2$ is easy to prove. First $\int_I | q || y |^2 < \infty$ since (we follow here [15, Lemma 1])

$$\int_I | q || y |^2 \leq \int_{E_1} | y |^2 + \int_{E_2} q^2 \; | y |^2$$

where $E_1 := \{t : | q(t) | \leq 1\}$ and $E_2 := \{t : | q(t) > 1\}$. Hence as a consequence of the Dirichlet formula (1.1) with $y = z$

$$\int_a^t | y' |^2 \leq | (y\bar{y}')(t) | + | (y\bar{y}')(a) | + \int_I | q | | y |^2 + \int_I | M[y] \bar{y} |. \quad (3.23)$$

Since the integrals on the right are all finite and $(y\bar{y}')(t) \to 0$ as $t \to \infty$, it follows that $\int_a^\infty | y' |^2 < \infty$. Application of the Cauchy-Schwartz inequality and the fact that $\lim_{t \to \infty} y\bar{z}' = 0$ for $y, z \in \mathscr{D}$ shows that the Dirichlet condition (2.10) is satisfied.

Suppose now that $p = 1$ and q' satisfies (2.11). Using (C_3) of [2, Theorem 1] (with $f(t) = \theta$) and noting that

$$\sup_{t\in I}\int_t^{t+\varepsilon\theta} |q'|^2 \le \sup_{t\in I}\int_t^{t+\theta} |q'|^2 = K_q$$

for $0 < \varepsilon < 1$ we obtain the inequality

$$\int_I |q'|^2 |y'|^2 \le DK_q\left\{(\varepsilon\theta)^{-2}\int_I |y|^2 + (\varepsilon\theta)^2 \int_I |y''|\right\}. \qquad (3.24)$$

(Here we must assume that $y \in \mathscr{D}_0'$ in order to guarantee the existence of $\int_I |y''|^2$.) By the Hölder inequality, arithmetic-geometric mean inequality, and substitution of (3.24) we get that

$$2\int_I |q'y'\|y| \le \int_I |q'|^2 \, |y'|^2 + \int_I |y|^2$$

$$\le (DK_q(\varepsilon\theta)^{-2}) \int_I |y|^2 + (\varepsilon\theta)^2 \int_I |y''|^2.$$

We now substitute the last inequality for the first right-hand term of the basic identity (3.2) (with $p = w = 1$ and $\gamma = 0$) to obtain that

$$\|y''\|_I^2 + 2\|\sqrt{q}y'\|_I^2 + \|qy\|_I^2 - \|M[y]\|_I^2$$

$$\le (DK_q(\varepsilon\theta)^{-2} + 1)\int_I |y|^2 + DK_q(\varepsilon\theta)^2\int_I |y''|^2.$$

Subtraction and rearrangement yields the separation inequality.

Proof of Theorem 7 The essential ideas are similar to those of [15]. As before we partition $J_{\delta,n}$ into n intervals Δ_i of length δ. Then [2, Lemma 2.1] and Hölder's inequality gives the inequalities for $s \in \Delta_i$

$$|y(s)| \le K\left\{\delta^{-1}\int_{\Delta_i} |y| + \int_{\Delta_i} |y'|\right\}$$

$$\le K\left\{\delta^{-1/2}\left(\int_{\Delta_i} |y|^2\right)^{1/2} + \delta^{1/2}\left(\int_{\Delta_i} |y'|^2\right)^{1/2}\right\}. \qquad (3.25)$$

By squaring integrating both sides of the result over Δ_i, using the condition on q, Hölder's inequality, and finally adding over $J_{\delta,n}$ we get

$$\int_{\Delta_i} |q|^r \, |y|^2 \le KM\left\{\delta^{-1}\int_{\Delta_i} |y|^2 + \delta\int_{\Delta_i} |y'|^2\right\}$$

$$\int_{\Delta_i} |q||y|^2 \le KM^{1/r}\left\{\delta^{-1/r}\int_{\Delta_i}|y|^2 + \delta^{(2-1/r)}\int_{\Delta_i}|y'|^2\right\}$$

$$\int_{J_{\delta,n}} |q|^r\,|y|^2 \le KM\left\{\delta^{-1}\int_{J_{\delta,n}}|y|^2 + \delta\int_{J_{\delta,n}}|y'|^2\right\}$$

$$\int_{J_{\delta,n}} |q||y|^2 \le KM^{1/r}\left\{\delta^{-1/r}\int_{J_{\delta,n}}|y|^2 + \delta^{(2-1/r)}\int_{J_{\delta,n}}|y'|^2\right\} \quad (3.26)$$

But as in (3.23)

$$\int_{J_{\delta,n}} |y'|^2 \le |(y\bar{y}')(a+n\delta)| + |(y\bar{y}')(a)| + \int_{J_{\delta,n}}|q||y|^2 + \int_I |M[y]\bar{y}|.$$

Combining this with the last line of (3.26), replacing $\| y \|^2_{J_{\delta,n}}$ by $\| y \|^2_I$ and rearranging gives us the estimate

$$(1 - KM^{1/r}\delta)\int_{J_{\delta,n}}|y'|^2 \le |(y\bar{y}')(a+n\delta)| + |(y\bar{y}')(a)|$$

$$+ KM^{1/r}\delta^{-1/r}\int_I |y|^2 + \int_I |M[y]\bar{y}|.$$

For small enough (but fixed) δ the left side is positive. If now $n \to \infty$ and $\int_{J_{\delta,n}} |y'|^2 \to \infty$ then $|(y\bar{y}')(a+n\delta)| \to \infty$. This means that

$$\left|\frac{d|y|^2}{dt}\right| = 2\mathrm{Re}\,(\bar{y}y') \to \infty,$$

contradicting the fact that $y \in L^2(I)$. It follows that $y' \in L^2$. Using this fact in (3.26) we see that

$$\int_{J_{\delta,n}} |q|^r|y|^2 \le KM\{\delta^{-1}\int_I |y|^2 + \delta\int_I |y'|^2\}$$

$$\int_{J_{\delta,n}} |q||y|^2 \le KM^{1/r}\{\delta^{-1/r}\int_I |y|^2 + \delta^{(2-1/r)}\int_I |y'|^2\}.$$

Consequently both $|q|^{1/2}y$ and $|q|^{r/2}$ are in $L^2(I)$. This completes the proof of (i). From (3.25) we readily get that

$$|y(t)| \le K\left\{\delta^{-1/2}\left(\int_t^\infty |y|^2\right)^{1/2} + \delta^{1/2}\left(\int_t^\infty |y'|^2\right)^{1/2}\right\}.$$

Since for large enough t, the integrals on the right are uniformly bounded and become as small as we please, we have that y is uniformly bounded and that $\lim_{t\to\infty} y(t) \to 0$. We now consider a nonnegative real C_0^∞ function z_δ such that (i) $z(a) = 1, z'(a) = 0$, (ii) z_δ has support in $[a, a + \delta)$, (iii) $z_\delta \leq 1$. Let $t > a$ and set $z_{t,\delta} := z_\delta(s - t)$ which we trivially extend to $[t, \infty]$ by defining it to be zero off of $[t, t + \delta)$. Integration by parts gives

$$\int_t^\infty M[y]\, z_\delta - \left(\int_t^\infty y' z_\delta' + q y z_\delta \right) = y'(t).$$

By Cauchy-Schwartz and the Hölder inequality this leads to the estimate

$$| y'(t) | \leq \| M[y] \|_{[t,\infty)} \, \delta^{1/2} + \left(\int_t^{t+\delta} | y' |^2 \right)^{1/2} \| z_{t,\delta}' \|_{[t,t+\delta]}$$

$$+ M^{1/r} \delta^{1-1/r} \| y \|_{\infty,[t,t+\delta]}.$$

Since the quantities on the right are all uniformly bounded and

$$\lim_{t\to\infty} \| M[y] \|_{[t,\infty)} = 0, \lim_{t\to\infty} \int_t^{t+\delta} | y' |^2 = 0, \quad \lim_{t\to\infty} \int_t^{t+\delta} | y |^2 = 0,$$

it follows that $\lim_{t\to\infty} y'(t) = 0$ which completes the proof of (ii).

The proofs of (iii) and (iv) are the same as in Theorem 6.

References

1. F.V. Atkinson, *On some results of Everitt and Giertz,* Proc. Royal Soc. Edinburg **71A** (1972/3), 151–58.
2. R.C. Brown and D.B. Hinton, *Sufficient conditions for weighted inequalities of sum form.,* J. Math. Anal. Appl. **112** (1985), 563–578.
3. R.C. Brown and D.B. Hinton, *Sufficient conditions for weighted Gabushin inequalities,* Časopis Pěst. Mat. pp. 113–122. **111** (1986), 113–121.
4. R.S. Chisholm and W.N. Everitt, *On bounded integral operators in the space of integrable square functions,* Proc. R.S.E. (A) **69** (1970/71), 199–204.
5. E.A. Coddington and N. Levinson, *Theory of ordinary differential equations,* McGraw-Hill Book Company, New York, 1955.
6. B. Curgus and T.T. Read, *Discreteness of the spectrum of second order differential operators and associated embedding theorems,* J. Differential Equations (to appear).
7. N. Dunford and J.T. Schwartz, *Linear operators, part II: spectral theory,* Interscience, 1963.
8. D.E. Edmunds and W.D. Evans, *Spectral theory and differential operators,* Oxford University Press, Oxford, UK, 1987.
9. W.N. Everitt, *A Note on the Dirichlet conditions for second-order Differential Expressions,* Can. J. Math. **28(2)** (1976), 312–320.

10. W.N. Everitt and M. Giertz, *Some properties of the domains of certain differential operators,* Proc. London Math. Soc. (3) **23** (1971), 301–24.
11. W.N. Everitt *Some inequalities associated with the domains of ordinary differential operators,* Math. Zeit. **126** (1972), 308–328.
12. W.N. Everitt *On limit-point and separation criteria for linear differential expressions,* Proceedings of the 1972 Equadiff Conference, Brno, 1972, pp. 31–41.
13. W.N. Everitt *Inequalities and separation for certain ordinary differential operators,* Proc. London Math. Soc **28(3)**(1974), 352–372.
14. W.N. Everitt *Inequalities and separation for Schrodinger type operators in L_2R^n,* Proc. Royal Soc. Edinburgh **79A** (1977), 257–265.
15. W.N. Everitt and Weidmann, *Some remarks on a separation and limit-point criterion of second- order ordinary differential expressions,* Math. Ann **200** (1973), 335–346.
16. K. Friedrichs, *Criteria for discrete spectra,* Comm. Pure. Appl. Math. **3** (1950), 439–449.
17. M. Glazman, *Direct methods of qualitative spectral analysis of singular differential operators,* Israel Program for Scientific Translation, Jerusalem, 1965.
18. D. Hinton, *Molchanov's discrete spectra criterion for a weighted operator,* Can. Math. Bull. **22** (1979), 425–431.
19. R. Kauffman, *On the limit-n classification of ordinary differential operators with positive coefficients,* Proc. Lond. Math. Soc. (3) **35** (1977), 496–526.
20. R. Kauffman T. Read, and A. Zettl, *The deficiency index problem for powers of differential operators, "Lecture notes in mathematics"*, vol. 621, Springer-Verlag, New York, Berlin, Heidelberg, and Tokyo, 1977.
21. B. Muchkenhoupt, *Hardy's inequality with weights,* Studia Math. **44** (1972), 31–38.
22. M.A. Naimark, *Linear differential operators, part II,* Frederick Unger, New york, 1968.
23. R. Oinarov and M. Otelbaev, *A. criterion for a general Sturm-Liouville operator to have a discrete spectrum,* Differential Eqns. **24** (1988), 402–408.
24. B. Opic and A. Kufner, *Hardy-type inequalities,* Longman Scientific and Technical, Harlow, Essex, UK, 1990.
25. J. Weidmann, *Spectral theory of ordinary differential operators,* lecture notes in mathematics, Vol 1258, Springer-Verlag, Berlin.

Function Spaces and Applications
D.E. Edmunds et al (Eds)

3. Weak Type Estimates for Averaging Operators

María J. Carro*

Departament de Matemàtica Aplicada i Anàlisi. Facultat de Matemàtiques.
Universitat de Barcelona, 08071 Barcelona, Spain

1. Introduction

Let E be a function space on a σ-finite measure space $(\mathscr{M}, \mu)$ and let $\|\cdot\|_E$ be a quasinorm on E. Let (N, P) be a probability measure space and consider a measurable function

$$f : N \times M \to \mathbb{R}.$$

Define the average operator

$$Tf(x) = \int_N f(\theta, x)\, dP(\theta).$$

This note deals with the following problem: if there exists a positive constant C so that $\| f(\theta, \cdot) \|_E \le C$ uniformly in θ, can we give an estimate for the "norm" of Tf in the same or another function space F?

Obviously, since the probability measure space (N, P) can be $(\{0\}, \delta_0)$, we get that the space F, we are looking for, has to satisfy that $E \subset F$. Also, if E is a Banach space, Minkowski's integral inequality implies that $\| Tf \|_E \le C$ and, hence, $F = E$. Therefore, it suffices to consider the case of non normable function spaces, for which in general F will be strictly bigger.

In the category of Lorentz spaces $L^{p,q}$, the most significant non normable space is $L^{1,\infty}$. For this, nothing reasonable can be said in general since, if we take, for example, $N = (0, 1)$ with the Lebesgue measure and $f(\theta, x) = \dfrac{1}{|x - \theta|}$, for $x \in (0, 1)$, then $Tf \equiv \infty$ while $\| f(\theta, \cdot) \|_{L^{1,\infty}} \le 1$. However, if the measure space P is completely atomic, the following result due to Stein and Weiss holds [5]:

1991 *Mathematics Subject Classification.* 46M35, 46E30.

*This work has been partly supported by the DGICYT PB97-0986 and 1997SGR 00185.

Theorem Let $E = L^{1,\infty}$, $N = \mathbb{N}$ and $P = \sum_{n \in \mathrm{N}} c_n \delta_n$. Then, $Tf \in L^{1,\infty}$ whenever $\sum_n c_n \mid \log c_n \mid < \infty$.

Section 2 presents the main results for a general probability measure space N and Section 3 is restricted to the atomic case.

As usual, constants such as C may change from one occurrence to the next.

2. Main Results

Let $(\mathcal{M}, \mu)$ be a σ-finite measure space. Let W be a positive function on $(0, \infty)$ so that $W(0) = 0$ and let us define

$$L_{W,\infty}(\mu) = \{f \text{ measurable; } \| f \|_{L_{W,\infty}(\mu)} = \sup_{y>0} W(y)\, \lambda_f^{\mu}(y) < \infty\},$$

where $\lambda_f^{\mu}(y) = \mu(\{x; \mid f(x) \mid > y\})$ is the distribution function of f.

The first trivial observation is that we can always assume that W is an increasing function since, by using that λ_f^{μ} is a right-continuous decreasing function, we obtain

$$\sup_{y>0} W(y)\, \lambda_f^{\mu}(y) = \sup_{y>0} W(y) \sup_{x \geq y} \lambda_f^{\mu}(x)$$

$$= \sup_{x>0} \lambda_f^{\mu}(x) \sup_{y \leq x} W(y) = \sup_{y>0} \lambda_f^{\mu}(x) \overline{W}(x),$$

with $\overline{W}$ the least increasing majorant of W. From now on, we shall assume that W is an increasing function.

These spaces can be considered as a weak version of modular spaces (see for example [4]) in the following sense: If

$$L_W(\mu) = \{f \text{ measurable; } \| f \|_{L_W(\mu)} = \int_{\mathcal{M}} W(\mid f(x) \mid)\, d\mu(x) < \infty\}$$

is the modular space associated to W, then, obviously $L_W(\mu) \subset L_{W,\infty}(\mu)$ and $\| f \|_{L_W(\mu)}$. If $W(y) = y$ then $L_w(\mu) = L^1(\mu)$ and $L_{W,\infty}(\mu) = L^{1,\infty}(\mu)$.

If W satisfies the Δ_2-condition, that is, there exists $C > 0$ so that, for every $t > 0$, $W(2t) \leq CW(t)$ then, one can easily see that $L_W(\mu)$ and $L_{W,\infty}(\mu)$ are linear spaces and the functional $\| \cdot \|_W$, that it will denote both $\| \cdot \|_{L_W(\mu)}$ and $\| \cdot \|_{L_{W,\infty}(\mu)}$, satisfies the following properties [1]:

(1) $\| f \|_W \geq 0$.

(2) There exists $C > 0$ so that $\| f + g \|_W \leq C (\| f \|_W + \| g \|_W)$, for every f, g.

(3) $\| -f \|_W = \| f \|_W$ and $\| \lambda f \|_W \leq \| f \|_W$ for every $\mid \lambda \mid \leq 1$.

(4) If $\mid f \mid \leq \mid g \mid$, then $\| f \|_W \leq \| g \|_W$.

(5) $\| f \|_{L_W(\mu)} = \int_0^{\infty} W(f_{\mu}^*(t))\, dt$, where f_{μ}^* denotes the decreasing

rearrangement of f with respect to the measure μ and $\| f \|_{L_{W,\infty}(\mu)} = \| f_\mu^* \|_{L_{W,\infty}(m)}$, with m the Lebesgue measure in $\mathbb{R}^+$.

From now on we shall simply write λ_f and $L_{W,\infty}$ instead of f_μ^* and $L_{W,\infty}(\mu)$. To prove our main result, we need the following lemma.

Lemma 2.1

$$\sup_{f\uparrow} \frac{\int_0^R f}{\int_0^\infty \frac{f(t)}{W(t)}\,dt} = \sup_{x\le R} \frac{R-x}{\int_x^\infty \frac{1}{W(u)}\,du}.$$

Proof Using Theorem 3.2 in [2], we obtain

$$\sup_{f\uparrow} \frac{\int_0^R f}{\int_0^\infty \frac{f(t)}{W(t)}\,dt} = \sup_{g\downarrow} \frac{\int_0^R g(1/s)\,ds}{\int_0^\infty \frac{g(1/t)}{W(t)}\,dt}$$

$$= \sup_{g\downarrow} \frac{\int_{1/R}^\infty g(u)\frac{du}{u^2}}{\int_0^\infty \frac{g(u)}{W(1/u)}\frac{du}{u^2}} = \sup_{x>1/R} \frac{\int_{1/R}^x \frac{du}{u^2}}{\int_0^x \frac{1}{W(1/u)}\frac{du}{u^2}}$$

$$= \sup_{x>1/R} \frac{R-\frac{1}{x}}{\int_0^\infty \frac{1}{W(1/u)}\frac{du}{u^2}} = \sup_{x\le R} \frac{R-x}{\int_x^\infty \frac{1}{W(u)}\,du}. \qquad \square$$

Let us denote
$$\tilde{W}(R) = \sup_{x\le R} \frac{R-x}{\int_x^\infty \frac{1}{W(u)}\,du}.$$

Observe that since W is increasing

$$\tilde{W}(R) = \sup_{x\le R} \frac{R-x}{\int_x^R \frac{1}{W}} \le \sup_{x\le R} \frac{R-x}{(R-x)\frac{1}{W(R)}} = W(R),$$

and hence, $L_{W,\infty} \subset L_{\tilde{W},\infty}$. Our main result shows that the space $F = L_{\tilde{W},\infty}$ is the one we are looking for.

Theorem 2.2 If there exists $C > 0$ so that $\| f(\theta,\cdot) \|_{L_{W,\infty}} \le C$, then $Tf \in L_{\tilde{W},\infty}$ and $\| Tf \|_{L_{\tilde{W},\infty}} \le C$.

Proof Let $\phi(t) = \int_0^t h(s)\, ds$, with h a positive and increasing function. Then ϕ is a convex function and, by Jensen's inequality,

$$\phi(\,|\,Tf(x)\,|\,) \le \int_N \phi(\,|\,f(\theta, x)\,|\,)\, dP(\theta).$$

Hence, for every $R > 0$, $\phi(R)\chi_{\{x;|Tf(x)|>R\}}(x) \le \int_N \phi(\,|\,f(\theta, x)\,|\,)\, dP(\theta)$ and integrating over $\mathcal{M}$, we obtain

$$\phi(R)\lambda_{Tf}(R) \le \int_N \int_{\mathcal{M}} \phi(\,|\,f(\theta, x)\,|\,)\, d\mu(x)\, dP(\theta).$$

Now,

$$\int_{\mathcal{M}} \phi(\,|\,f(\theta,x)\,|\,)\, d\mu(x) = \int_0^\infty \lambda_{f(\theta,\cdot)}(y)\, d\phi(y) = \int_0^\infty \lambda_{f(\theta,\cdot)}(y)\, h(y)\, dy$$

and hence, we deduce from (1) that

$$\phi(R)\lambda_{Tf}(R) \le \int_N \int_0^\infty \lambda_{f(\theta,\cdot)}(y)\, h(y)\, dy\, dP(\theta) \le C\int_0^\infty \frac{h(y)}{W(y)}\, dy,$$

and therefore

$$\left(\sup_{h\uparrow} \frac{\int_0^R h}{\int_0^\infty \frac{h(y)}{W(y)}\, dy} \right) \lambda_{Tf}(R) \le C.$$

Applying the previous lemma, we obtain the result. □

Remark 2.3 Observe that for $W(u) = u$ or, in general for every function W so that $1/W$ is not integrable at infinity, we get $\tilde{W} \equiv 0$ and hence, we do not obtain any information from Theorem 2.2.

Example 2.4
(1) If $W(u) = u(1 + |\log u\,|)^2$, then one can easily check that

$$\tilde{W}(R) \approx \begin{cases} \dfrac{R}{2} & \text{if } R \le 1, \\ R \log R & \text{if } R > 1. \end{cases}$$

(2) If
$$W(u) = \begin{cases} u & \text{if } u \le 1, \\ u(1 + |\log u\,|)^2 & \text{if } u \ge 1. \end{cases}$$

then
$$\tilde{W}(R) \approx \begin{cases} \dfrac{R}{1+\log R} & \text{if } R \le 1, \\ R\,(\log R + 1) & \text{if } R > 1. \end{cases}$$

Sometimes, we can obtain some more information for Tf if we assume some extra condition on the function f, namely $\| f(\theta, \cdot) \|_\infty \le C$. In this case, we can proceed as follows: Let us take $\phi(x) = \int_0^x h(t)\, dt$ where now, h is a positive function so that h is increasing in $[0, S]$ for some S fixed and let us denote by C_S the collection of these functions (Observe that every function in C_S is an increasing function so that it is convex in $[0, S]$). Then, similarly to Lemma 2.1, we have the following:

Lemma 2.5 For every $R \le S$

$$\sup_{\phi \in C_S} \frac{\phi(R)}{\displaystyle\int_0^\infty \frac{1}{W(t)}\, d\phi(t)} = \sup_{x \le R} \frac{R - x}{\displaystyle\int_x^S \frac{1}{W(u)}\, du}.$$

Proof As in Lemma 2.1, if $R \le S$,

$$\sup_{\phi \in C_S} \frac{\phi(R)}{\displaystyle\int_0^\infty \frac{1}{W(t)}\, d\phi(t)} = \sup_{h\uparrow}\ \sup_{d\phi \ge 0} \frac{\displaystyle\int_0^R h(s)\, ds}{\displaystyle\int_0^S \frac{h(t)}{W(t)}\, dt + \int_S^\infty \frac{1}{W(t)}\, d\phi(t)}$$

$$= \sup_{h\uparrow} \frac{\displaystyle\int_0^R h(s)\, ds}{\displaystyle\int_0^S \frac{h(t)}{W(t)}\, dt} = \sup_{g\downarrow} \frac{\displaystyle\int_0^R g(1/s)\, ds}{\displaystyle\int_0^S \frac{g(1/t)}{W(t)}\, dt} = \sup_{g\downarrow} \frac{\displaystyle\int_{1/R}^\infty g(u) \frac{du}{u^2}}{\displaystyle\int_{1/S}^\infty \frac{g(u)}{W(1/u)} \frac{du}{u^2}}$$

$$= \sup_{x > 1/R} \frac{\displaystyle\int_{1/R}^x \frac{du}{u^2}}{\displaystyle\int_{1/S}^x \frac{1}{W(1/u)} \frac{du}{u^2}} = \sup_{x > 1/R} \frac{R - \frac{1}{x}}{\displaystyle\int_{1/S}^x \frac{1}{W(1/u)} \frac{du}{u^2}} = \sup_{x \le R} \frac{R - x}{\displaystyle\int_x^S \frac{1}{W(u)}\, du}. \qquad \square$$

Observe, that if $W(u) = u$, then, for every $R \le S$,

$$\sup_{x \le R} \frac{R - x}{\log \frac{S}{x}} \ge \frac{R}{2 \log \frac{2S}{R}} \ne 0$$

and hence, we can get some information in the case $E = L^{1,\infty}$ (see Example 2.7).

Let us denote

$$\tilde{W}_S(R) = \begin{cases} \sup\limits_{x \le R} \dfrac{R - x}{\displaystyle\int_x^S \frac{1}{W(u)}\, du} & \text{if } R \le S \\ 0 & \text{if } R > S \end{cases}$$

Then, similarly, to Theorem 2.2, we obtain the following result.

Theorem 2.6 If there exist $C > 0$ and $S > 0$ so that $\| f(\theta, \cdot) \|_{Lw,\infty} \le C$ and $\| f(\theta, \cdot) \|_\infty \le S$ then $Tf \in L_{\tilde{W}_{S,\infty}}$ and $\| Tf \|_{L_{\tilde{W}_{S,\infty}}} \le C$.

Example 2.7 Let M_θ be the Hardy-Littlewood operator in the direction $\theta \in \sum\limits_{n-1}$; that is

$$M_\theta f(x) = \sup_{r>0} \frac{1}{2r} \int_{-r}^{r} | f(x - s\theta) |\, ds$$

and let

$$Tf(x) = \frac{1}{|\sum\limits_{n-1}|} \int_{\sum\limits_{n-1}} M_\theta f(x)\, d\theta.$$

Then, it is known [3] that T is not of weak type (1, 1). However, if we take $f \in L^1 \cap L^\infty$, then we have that $\| M_\theta f \|_{L^{1,\infty}} \le \| f \|_1$ and $\| M_\theta f \|_{L^\infty} \le \| f \|_\infty$ and therefore, as a consequence of Theorem 2.6, we get that

$$\sup_{R>0} \frac{R}{\log \dfrac{\| f \|_\infty}{R}} \lambda_{Tf}(R) \le C \| f \|_1.$$

In particular, for every measurable set,

$$\sup_{R>0} \frac{R}{|\log R|} \lambda T\chi_E(R) \le C |E|.$$

The above result can also be obtained by interpolation.

3. The Atomic Case

In this section, we shall assume that the measure space N is completely atomic; that is $N = \mathbb{N}$ and $P(\{n\}) = c_n$ so that $\sum\limits_n c_n = 1$. Therefore, we shall be dealing with operators of the form

$$Tf(x) = \sum_n c_n f_n(x).$$

As in the previous section, we shall assume that $\| f_n \|_{Lw,\infty} \le C$. Modifying slightly the proof of Stein and Weiss Lemma, we get the following result:

Theorem 3.1 If there exists $C > 0$ so that, for every n, $\| f_n \|_{L_{W,\infty}} \leq C$, then $Tf \in L_{V,\infty}$ with

$$V(3y) = \min\left(W(y), \frac{1}{\sum_n \frac{1}{W\left(\frac{y}{c_n}\right)}}, \frac{y}{\sum_n c_n \int_y^{y/c_n} \frac{1}{W}} \right).$$

Proof Let us fix y and let us write

$$f_n(x) = f_n(x)\chi_{\{x;\,|f_n(x)| \leq y\}} + f_n(x)\chi_{\{x;\, y \leq |f_n(x)| \leq y/c_n\}} + f_n(x)\chi_{\{x;\,|f_n(x)| > y/c_n\}}$$

$$= f_n^1(x) + f_n^2(x) + f_n^3(x).$$

Then $Tf = \sum_n c_n f_n = \sum_n c_n f_n^1 + \sum_n c_n f_n^2 + \sum_n c_n f_n^3 = g_1 + g_2 + g_3$. Therefore, $\lambda_{Tf}(3y) \leq \lambda_{g_1}(y) + \lambda_{g_2}(y) + \lambda_{g_3}(y)$.

Now, $\lambda_{g_1}(y) = 0$ and

$$\lambda_{g_3}(y) \leq \sum_n \lambda_{f_n}\left(\frac{y}{c_n}\right) \leq C \sum_n \frac{1}{W\left(\frac{y}{c_n}\right)}.$$

On the other hand,

$$V(3y)\lambda_{g_2}(y) \leq \frac{V(3y)}{y} \| g_2 \|_1$$

$$= \frac{V(3y)}{y} \left(\sum_n c_n y \lambda_{f_n^2}(y) + \sum_n c_n \int_y^{y/c_n} \lambda_{f_n^2}(z)\, dz\right)$$

$$\leq C \frac{V(3y)}{y} \left(\frac{y}{W(y)} + \sum_n c_n \int_y^{y/c_n} \frac{1}{W(z)}\, dz \right) \leq C,$$

and therefore, $\sup_y \lambda_{Tf}(3y)V(3y) \leq C$; that is $Tf \in L_{V,\infty}$. □

Finally, observe that if $W(x) = x$, that is $L_{W,\infty} = L^{1,\infty}$ then $V(3y) = \min\left(y, y, \frac{y}{\sum_n c_n |\log c_n|} \right)$ and therefore $L_{V,\infty} = L^{1,\infty}$ and we recover the result of Stein and Weiss.

References

1. C. Bennett and R. Sharpley, *Interpolation of Operators,* Academic Press, 1988.
2. M.J. Carro and J. Soria, *Boundedness of some integral operators,* Can. J. Math. **45** (1993), 1155–1166.
3. R. Fefferman, *A theory of Entropy in Fourier Analysis,* Adv. In Math. **30** (1978), 171–201.
4. J. Musielak, *Orlicz spaces and modular spaces,* Lecture Notes in Math., Springer-Verlag **1034** (1983).
5. E.M. Stein and N.J. Weiss, *On the convergence of Poisson integrals,* Trans. Amer. Math. Soc. **140** (1969), 34–54.

Function Spaces and Applications
D.E. Edmunds et al (Eds)

4. Norms of Interpolation Operators Controlled by the Dicesar Function

María J. Carro,*[1] Ludmila Nikolova†[2] and Lars-Erik Persson[3]

[1]Departament de Matemàtica Aplicada i Anàlisi. Facultat de Matemàtiques. Universitat de Barcelona, 08071 Barcelona, Spain

[2]Department of Mathematics, Sofia University, 1126 Sofia, Bulgaria

[3]Department of Mathematics, Luleå University, S-971 87 Luleå, Sweden

1. Introduction

In the theory of interpolation (see [1]) one usually considers Banach couples, i.e. pairs (A_0, A_1) such that A_0 and A_1 are Banach spaces embedded in a common topological vector space $\mathcal{U}$. The most important among the various constructions of interpolation with respect to a given couple is the complex method leading to the spaces $[A_0, A_1]_\theta$ (where $0 < \theta < 1$) and the real method leading to the spaces $(A_0, A_1)_{\theta,q}$ (where $0 < \theta < 1$ and $0 < q \leq \infty$).

Parts of the theory concerning interpolation between two Banach spaces can be generalized to cover also cases where one interpolates between finitely many Banach spaces and even between general families of (infinite many) Banach spaces. In this direction let us mention that a theory of complex interpolation between families of Banach spaces was developed by Coifman, Cwikel, Rochberg, Sagher and Weiss (see [8]) and independently by Krein and Nikolova (see [15], [16]).

A new nontrivial application of this theory was recently found by Ferenczi [12] as a further development of some important works of Maurey and Govers.

A theory of real interpolation between n-tuples of Banach spaces was worked out by Sparr [25]. A parallel theory of interpolation between 2^k-tuples of Banach spaces was developed by Fernández [13]. Lately Cobos and Peetre [7] have developed a theory which, in particular, covers both of the constructions of Sparr and Fernández with $n = 3$ and $n = 4$ respectively. On the other hand,

---

1991 *Mathematics Subject Classification.* 46M35.

*This work has been partly supported by the DGICYT PB97-0986 and by 1997SGR 00185.

†The research was supported by Bulgarian Ministry of Education and Science by contract MM-703-97.

even earlier, the construction of Sparr had been extended by Cwikel and Janson [9] to the case of interpolation between a fairly general family $\overline{A} = (A_t)_{t\in\Gamma}$, where the A_t are Banach spaces and Γ is a general probability space. Moreover, some new real interpolations methods for families of Banach spaces were introduced and studied in [2], [3] (continuous method) and in [22] and [23] (discrete method). These methods were compared and some new (limit construction) methods were introduced in [5].

The theory developed in this paper will deal with the discrete method although most of the results holds (with the obvious changes) for the continuous one.

The paper is organized as follows: in Section 2 we, very briefly, present some necessary notations and basic facts. In section 3, we study the Dicesar function and in Section 4 we prove the interpolation results for C-subadditive operators with applications to weak-type and weakened-type operators. Finally, in section 5, we give some other new interpolation results for summing and correct operators in terms of the Dicesar function.

We shall write $f \sim g$ to indicate the existence of two positive constants A and B so that $Af \leq g \leq Bf$. Also, C denotes a positive constant, not the same in different occurrences.

2. Preliminaries

Here we recall some necessary definitions from [5] and for more details we refer the reader to this paper.

Let D denote the unit disc $\{z \in \mathbf{C} : |z| < 1\}$ and Γ its boundary and let $\overline{A} = \{A(\gamma) : \gamma \in \Gamma; \mathscr{A}, \mathscr{U}\}$ be an interpolation family (i.f.) on Γ with $\mathscr{U}$ as the containing Banach space and $\mathscr{A}$ as the log-intersection space, in the sense of [8]. For simplicity, we shall work, in this paper, with bounded families, that is, we assume that $\| a \|_{\mathscr{U}} \leq \| a \|_\gamma$ for every $\gamma \in \Gamma$.

For each $a \in \mathscr{U}$, we define the discrete K-functional

$$K(\alpha, a) = \inf \{\sum_j \alpha(\gamma_j) \| a_{\gamma_j} \|_{\gamma_j}\},$$

where the infimum is taken over all representations of the element $a = \sum_j a_{\gamma_j}$ with convergence in $\mathscr{U}$ and $a_{\gamma_j} \in A(\gamma_j)$ (see [19] and also [22], [23]).

For each $a \in \mathscr{A}$, we also define the J-functional by

$$J(\alpha, a) = \sup_{\gamma\in\Gamma} \alpha(\gamma) \| a \|_\gamma .$$

Let

$$\mathscr{L} = \{\alpha : \Gamma \to \mathbf{R}^+; \ \alpha \text{ is measurable and } \log \alpha \in L^1(\Gamma)\},$$

and, for each $\alpha \in \mathscr{L}$ and $z \in D$, let us define

$$\alpha(z) = \exp\left(\int_\Gamma \log \alpha(\gamma) \, P_z(\gamma) \, d\gamma\right),$$

where P_z is the Poisson kernel. Let $S \subset \mathscr{L}$ and $0 < p \leq \infty$.

(a) The space $(\overline{A})^S_{z0,p;K}$ consists of all $a \in \mathscr{U}$ for which

$$\left(\frac{K(\alpha, a)}{\alpha(z_0)}\right)_{\alpha \in S} \in l^p(S),$$

endowed with the quasi-seminorm

$$\| a \|_{(\overline{A})^S_{z0,p;K}} = \left(\sum_{\alpha \in S} \left(\frac{K(\alpha, a)}{\alpha(z_0)}\right)^p\right)^{1/p}$$

(b) The space $(\overline{A})^S_{z0,p;J}$ is the set of all elements $a \in \mathscr{U}$ such that there exists $\{u(\alpha)\}_{\alpha \in S}$ in $\mathscr{A}$ satisfying $a = \sum_{\alpha \in S} u(\alpha)$ (in the $\mathscr{U}$-norm) and

$$\left(\sum_{\alpha} \left(\frac{J(\alpha, u(\alpha))}{\alpha(z_0)}\right)^p\right)^{1/p} < +\infty.$$

This space will be endowed with the quasi-seminorm

$$\| a \|_{(\overline{A})^S_{z0,p;J}} = \inf \left\{\left(\sum_{\alpha} \left(\frac{J(\alpha, u(\alpha))}{\alpha(z_0)}\right)^p\right)^{1/p}\right\}$$

where the infimum extends to all possible representations of a.

Also, it is known (see [2]) that if we want to have good properties on the interpolation spaces we need to have a "size condition" on the set S, namely

$$\sum_{\alpha \in S} \left(\frac{\inf_{\gamma \in \Gamma} \alpha(\gamma)}{\alpha(z_0)}\right)^p < +\infty, \tag{1}$$

and therefore we shall assume throughout this paper that this condition holds.

As usual

$$\Sigma(\overline{A}) = \{\alpha \in \mathscr{U}; K(1, a) < \infty\},$$

and

$$\Delta(\overline{A}) = \{a \in A(\gamma), \forall \gamma \in \Gamma; \sup_{\gamma \in \Gamma} \| a \|_\gamma < \infty\},$$

and we say that a Banach space B belongs to the class $K^S_{z_0}(\overline{A})$ if $B \subset \Sigma(\overline{A})$ and

$$\frac{K(\alpha, a)}{\alpha(z_0)} \leq C \| a \|_B,$$

for every $a \in B$ and all $\alpha \in S$.

Also, a Banach space B belongs to the class $J_{z_0}^S(\overline{A})$ if $\Delta(\overline{A}) \subset B$ and

$$\| a \|_B \leq C \frac{J(\alpha, a)}{\alpha(z_0)},$$

for every $a \in B$ and all $\alpha \in S$.

Finally, a linear operator $T : \overline{A} \to \overline{B}$ is an interpolation operator if $T : \mathscr{U} \to \mathscr{V}$ and $T : A(\gamma) \to B(\gamma)$ are bounded for every $\gamma \in \Gamma$.

Several examples of these interpolation spaces were studied in [2] and [5]. Let us just mention that this method recovers, by choosing S appropriately, Sparr, Fernández and Cobos-Peetre methods.

3. The Dicesar Function

Let us recall that if $\overline{A}$ and $\overline{B}$ are two i.f. and $T : \overline{A} \to \overline{B}$ is an interpolation operator such that $\| T \|_{A(\gamma) \to B(\gamma)} \leq M(\gamma)$ a.e. $\gamma \in \Gamma$ with $M \in \mathscr{L}$, then (see [3], [5]) we have that

$$\| T \|_{F(\overline{A}) \to F(\overline{B})} \leq D_{z_0}^S(M),$$

where F is either $(\cdot)_{z_0;K}^S$ or $(\cdot)_{z_0;J}^S$ and if S is a multiplicative group

$$D_{z_0}^S(M) = \inf_{\alpha \in S} \left\{ \frac{\sup \alpha(\gamma) M(\gamma)}{\alpha(z_0)} \right\} \tag{2}$$

was introduced in [5] as the Dicesar function, since the authors found it for the first time in the paper of Dicesar Fernández [13].

This function is a generalization of the function $M_0^{1-\theta} M_1^{\theta}$ for the classical case of two Banach spaces. Other special cases of the Dicesar function are $D_\theta(M)$ in [22] and $D_{(\alpha,\beta)}(M_1, \ldots, M_N)$ in [7].

Since $1 \in S$, $M_{z_0}^S(M) \leq \| M \|_\infty$ and obviously $M_{z_0}^S(M) \geq M(z_0)$. Moreover, if $M \in S$, $M_{z_0}^S(M) = M(z_0)$. Finally, the connection between the function $M_{z_0}^S(M)$ and the classes $K_{z_0}^S$ and $J_{z_0}^S$ was studied in [5] where also some concrete examples of the Dicesar function were computed.

The Dicesar function will play a fundamental role in this paper but here we will also consider the case when S is not a group and also a new function $D_{z_0,p;J \to K}^S(M)$ (see Definition 3.5) that will give us an upper estimate for the norm of the interpolation operator when acting from $(\overline{A})_{z_0,p;J}^S$ into $(\overline{B})_{z_0,p;K}^S$.

Definition 3.1 Let S be a subset of $\mathscr{L}$, let $z_0 \in D$ and let $M \in \mathscr{L}$. The Dicesar function is then defined by

$$D_{z_0}^S(M) = \inf_{\alpha \in \tilde{S}} \left\{ \frac{\sup \alpha(\gamma) M(\gamma)}{\alpha(z_0)} \right\},$$

where $\tilde{S}$ is the set of all $\beta \in \mathscr{L}$ so that, for every $\alpha \in S$, $\alpha\beta^{-1} \in S$ and the mapping $\alpha \to \alpha\beta^{-1}$ is injective.

Obviously if S is a multiplicative group $\tilde{S} = S$ and this is just the Dicesar function we have talked about in the introduction of this section. Also, $\tilde{S}$ is always non-empty since $1 \in \tilde{S}$, for every S.

As a new example (when S is not a group), let us take for some $\alpha \in \mathscr{L}$, $S^+ = \{\alpha^n;\, n \in \mathbb{N}\}$ and $S^- = \{\alpha^n;\, n \in \mathbb{Z}^-\}$, then $\widetilde{S^+} = S^-$. In particular, if

$$S = \left\{ \alpha_n(\gamma) = \begin{cases} 1 & \text{if } \gamma \in \Gamma_0 \\ 2^n & \text{if } \gamma \in \Gamma_1 \end{cases}, n \in \mathbb{Z} \right\}$$

with $\{\Gamma_0, \Gamma_1\}$ a partition of Γ and $M(\gamma) = M_j$ if $\gamma \in \Gamma_j$, one can easily check that

$$D_{z_0}^{S^+}(M) = \inf_{n\le 0} \frac{\max(M_0, 2^n M_1)}{2^{n|\Gamma_1|_{z_0}}} \sim M_0^{1-\theta} M_1^{\theta} \inf_{n\le 0} \frac{\max(1, 2^{n+k})}{2^{(n+k)|\Gamma_1|_{z_0}}},$$

where $2^k \le M_0/M_1 \le 2^{k+1}$. Hence, if $k \le 0$, we get that $D_{z_0}^{S^+}(M) \sim M_0$, while if $k > 0$, $D_{z_0}^{S^+}(M) \sim M_0^{1-\theta} M_1^{\theta}$.

As it was seen in [3] for the case when S is a group, one can easily see that the following result holds.

Proposition 3.2 If $T : \overline{A} \to \overline{B}$ is an interpolation operator such that

$$\| T \|_{A(\gamma)\to B(\gamma)} \le M(\gamma)$$

for every $\gamma \in \Gamma$ with $M \in \mathscr{L}$ and $S \subset \mathscr{L}$, then

$$\| T \|_{(\overline{A})^S_{z_0,F} \to (\overline{B})^S_{z_0,F}} \le D^S_{z_0}(M),$$

where F is either K or J.

Sometimes, it is very useful to mix the J and K methods for an interpolation operator. In this line, we have the following estimates:

Proposition 3.3 Let S_1 and S_2 be two sets in $\mathscr{L}$. Then, for any i.f $\overline{A}$ and $\overline{B}$ and any interpolation operator T such that $\| T \|_{A(\gamma)\to B(\gamma)} \le M(\gamma)$, for every $\gamma \in \Gamma$, we have that

(a) If $p \ge 1$, then

$$\| T \|_{(\overline{A})^{S_1}_{z_0,p;J} \to (\overline{A})^{S_2}_{z_0,p;K}}$$

$$\le \left(\sup_{\beta\in S_1} \sum_{\alpha\in S_2} \left(\sum_{\lambda\in S_1} \frac{\inf_{\gamma\in\Gamma} (\alpha\lambda^{-1}M)(\gamma)}{\alpha(z_0)\lambda^{-1}(z_0)} \right)^{p-1} \frac{\inf_{\gamma\in\Gamma} (\alpha\beta^{-1}M)(\gamma)}{\alpha(z_0)\beta^{-1}(z_0)} \right)^{1/p}$$

(b) If $p < 1$, then

$$\| T \|_{(\bar{A})^{S_1}_{z_0,p;J} \to (\bar{A})^{S_2}_{z_0,p;K}} \leq \left(\sup_{\beta \in S_1} \sum_{\alpha \in S_2} \left(\frac{\inf_{\gamma \in \Gamma} (\alpha \beta^{-1} M)(\gamma)}{\alpha(z_0)\beta^{-1}(z_0)} \right)^p \right)^{1/p}.$$

Proof The case $p \geq 1$ was proved in [4]. Hence, we shall only give a proof for the case $p < 1$. Let $a \in (\bar{A})^{S_1}_{z_0,p;J}$. Given $\varepsilon > 0$, let $a = \sum_{\beta \in S_1} a_\beta$, such that $a_\beta \in \mathscr{A}$ and

$$\left(\sum_{\beta \in S_1} \left(\frac{J(\beta, a_\beta)}{\beta(z_0)} \right)^p \right)^{1/p} \leq \| a \|_{(\bar{A})^{S_1}_{z_0,p;J}} + \varepsilon.$$

Let us write $a_\beta = a_\beta \sum_j \varphi(\gamma_j)$ with $\varphi \in L^\infty$ so that $\sum_j \varphi(\gamma_j) = 1$. Then,

$$\begin{aligned} K(\alpha, Ta) &\leq K(\alpha M, a) \leq \sum_{\beta \in S_1} K(\alpha M, a_\beta) \\ &\leq \sum_{\beta \in S_1} \inf \Big\{ \sum \alpha(\gamma_j) M(\gamma_j) \beta^{-1}(\gamma_j) \beta(\gamma_j) \\ &\qquad \times \| a_\beta \|_{\gamma_j} \varphi(\gamma_j) : \sum_j \varphi(\gamma_j) = 1 \Big\} \\ &\leq \sum_{\beta \in S_1} \inf_{\Gamma} (\alpha M \beta^{-1})(\gamma) J(\beta, a_\beta). \end{aligned}$$

Then,

$$\begin{aligned} \| Ta \|^p_{(\bar{A})^{S_2}_{z_0,p;K}} &= \sum_{\alpha \in S_2} \left(\frac{K(\alpha, Ta)}{\alpha(z_0)} \right)^p \\ &\leq \sum_{\alpha \in S_2} \sum_{\beta \in S_1} \left(\frac{\inf_\Gamma (\alpha \beta^{-1} M)(\gamma)}{\beta^{-1}(z_0)\alpha(z_0)} \right)^p \left(\frac{J(\beta, a_\beta)}{\beta(z_0)} \right)^p \\ &\leq \sup_{\beta \in S_1} \sum_{\alpha \in S_2} \left(\frac{\inf_\Gamma (\alpha \beta^{-1} M)(\gamma)}{\beta^{-1}(z_0)\alpha(z_0)} \right)^p \sum_{\beta \in S_1} \left(\frac{J(\beta, a_\beta)}{\beta(z_0)} \right)^p \\ &\leq \sup_{\beta \in S_1} \sum_{\alpha \in S_2} \left(\frac{\inf_\Gamma (\alpha \beta^{-1} M)(\gamma)}{\beta^{-1}(z_0)\alpha(z_0)} \right)^p (\| a \|_{(\bar{A})^{S_1}_{z_0,p;J}} + \varepsilon)^p. \end{aligned}$$

From this we get the result by letting ε tends to zero. □

Remark 3.4 The above general estimate give us a sufficient condition to have the embedding

$$(\bar{A})^{S_1}_{z_0,p;J} \subset (\bar{A})^{S_2}_{z_0,p;K}$$

when applying it to the identity operator.

In particular, if $S_1 = S_2 = S$ and S is a multiplicative group, the above embedding holds if S satisfies the size condition (1) for p fixed. This was proved in [2].

Let us now define the announced "mixed" Dicesar function $D^S_{z_0,p;J\to K}(M)$:

Definition 3.5
(a) If $p \geq 1$, then

$$D^S_{z_0,p;J\to K}(M) = \left(\sup_{\lambda\in S} \sum_{\alpha\in S} \left(\sum_{\beta\in S} \frac{\inf_\Gamma(\alpha\beta^{-1}M)}{\alpha(z_0)\beta^{-1}(z_0)} \right)^{p-1} \frac{\inf_\Gamma(\alpha\lambda^{-1}M)}{\alpha(z_0)\lambda^{-1}(z_0)} \right)^{1/p}$$

(b) If $p < 1$, then

$$D^S_{z_0,p;J\to K}(M) = \left(\sup_{\alpha\in S} \sum_{\beta\in S} \left(\frac{\inf_\Gamma(\alpha\beta^{-1}M)}{\alpha(z_0)\beta^{-1}(z_0)} \right)^{p} \right)^{1/p}$$

Proposition 3.6 It holds that

$$D^S_{z_0,p;J\to K}(M) \leq D^S_{z_0}(M)\, D^S_{z_0,p;J\to K}(1).$$

Proof Let $\mu \in \tilde{S}$ and $p \geq 1$.
Then

$$\sum_{\beta\in S} \frac{\inf_\Gamma(\alpha\beta^{-1}\mu^{-1}\mu M)}{\alpha(z_0)(\beta^{-1}\mu^{-1})(z_0)\mu(z_0)} \leq \frac{\sup_\Gamma(\mu(\gamma)M(\gamma))}{\mu(z_0)} \sum_{\beta\in S} \frac{\inf_\Gamma(\alpha\beta^{-1}\mu^{-1})}{(\alpha\beta^{-1}\mu^{-1})(z_0)}$$

$$\leq \frac{\sup_\Gamma(\mu(\gamma)M(\gamma))}{\mu(z_0)} \sum_{\beta\in S} \frac{\inf_\Gamma(\alpha\beta^{-1})}{(\alpha\beta^{-1})(z_0)},$$

since $\alpha\mu^{-1} \in S$. Taking now the infimum over all $\mu \in \tilde{S}$ we have that

$$\sum_{\beta\in S} \frac{\inf_\Gamma(\alpha\beta^{-1}M)}{(\alpha(z_0)\beta^{-1})} \leq D^S_{z_0}(M) \sum_{\beta\in S} \frac{\inf_\Gamma(\alpha\beta^{-1})}{(\alpha\beta^{-1})(z_0)}.$$

From this and the analogous estimate for the other term in $D^S_{z_0,p;J\to K}(M)$ the claimed estimate follows.

The proof for the case $p < 1$ is similar. □

4. Interpolation Results for Some Nonlinear Operators

We say that the i.f. A is an i.f. of Banach lattices in $L^0(\mu)$ if, for every $\gamma \in \Gamma$,

$A(\gamma)$ is a Banach lattice and the containing space of the family is contained in $L^0(\mu)$.

Also, the operator T is said to be C-subadditive from the family $\overline{A}$ into $L^0(\mu)$ if $T : \mathcal{U} \to L^0(\mu)$ and if $a = \sum_j a_{\gamma_j}$ (convergence in $\mathcal{U}$) with $a_{\gamma_j} \in A(\gamma_j)$ and $A(\gamma_j) \neq A(\gamma_k)$ for every $j \neq k$, then

$$\left| T\left(\sum_j a_{\gamma_j} \right)(t) \right| \leq C \sum_j | T(a_{\gamma_j})(t) |$$

almost everywhere $[\mu]$ (see [23]).

Before going further, let us give some examples of C-subadditive operators:

(1) If $T_j : \overline{A} \to \overline{B}$ is a collection of linear interpolation operator with $\mathcal{U}_{\overline{B}} \subset L^0(\mu)$, then the maximal operator $\sup_j | T_j |$ is clearly 1-subadditive.

(2) If $\overline{A}$ is finite, then any subadditive operator is a C-subadditive operator.

(3) If T is a C-subadditive operator and $S_T a = (Ta)^{**}$ then S_T is also C-subadditive.

Theorem 4.1 Let T be *a* C-subadditive operator from the i.f. $\overline{A}$ into $L^0(\mu)$ and let $\overline{B}$ be an i.f. of Banach lattices with the containing space $\mathcal{V}$ in $L^0(\mu)$.

Let us assume that $T : A(\gamma) \to B(\gamma)$ is such that $Ta \in B(\gamma)$ and $\| Ta \|_{B(\gamma)} \leq M(\gamma) \| a \|_{A(\gamma)}$, where $M \in \mathcal{L}$. Then, if $S^{-1} = \{\alpha^{-1}; \alpha \in S\} \subset L^\infty$, we have that

$$\| Ta \|_{(\overline{B})^S_{z_0,p;K}} \leq C D^S_{z_0}(M) \| a \|_{(\overline{A})^S_{z_0,p;K}}$$

Remark 4.2 A corresponding result for the Sparr-method was proved in [23, Corollary 6]. For the case of a finite family, Theorem 4.1 implies also a corresponding result both for the Cobos-Peetre and the Fernández spaces. For the case of only two spaces this is a result of Maligranda [18, Theorem 7].

Proof Let $a \in (\overline{A})^S_{z_0,p;K}$ and let $\alpha \in S$. Then, for every $\varepsilon > 0$ there exists a representation $a = \sum_{j=1}^{\infty} a_{\gamma_j}$ (convergence in $\mathcal{U}$) so that $a_{\gamma_j} \in A(\gamma_j)$ and

$$\sum_j M(\gamma_j)\, \alpha(\gamma_j) \| a_{\gamma_j} \|_{\gamma_j} \leq (1 + \varepsilon)\, K(\alpha M, a).$$

Now, we can assume that $D^S_{z_0}(M) < \infty$ and hence there exists $\lambda \in \tilde{S}$ so that $M_\lambda = \sup_\Gamma(\lambda(\gamma) M(\gamma)) < \infty$.

Hence, $K(\alpha M, a) = K(\alpha\lambda^{-1} M\lambda, a) \leq M_\lambda K(\alpha\lambda^{-1}, a) < \infty$ since $\alpha\lambda^{-1} \in S$ and thus $K(\alpha M, a) < \infty$. From this and the fact that $S^{-1} \subset L^\infty$, we can deduce that $\sum_j \| a_{\gamma_j} \|_{A(\gamma_j)} M(\gamma_j)$ converges and therefore, since

$$\| \sum_N^{N'} | Ta_{\gamma_j} | \|_V \leq \sum_N^{N'} \| a_{\gamma_j} \|_{A(\gamma_j)} M(\gamma_j),$$

we get that the left hand side converges to zero whenever N and N' goes to infinity. That is, $\sum_{1}^{N} | Ta_{\gamma_j} |$ converges in $\mathscr{V}$ to a function which is finite a.e. (μ).

Hence we can define

$$b_j = \frac{T\left(\sum_j a_{\gamma_j}\right)}{\sum_j | Ta_{\gamma_j} |} | Ta_{\gamma_j} |,$$

on the support of $\sum_j | Ta_{\gamma_j} |$ and $b_j = 0$ elsewhere.

Since T is C-subadditive we have the estimate $| b_j | \leq C | Ta_{\gamma_j} |$ and thus, since $B(\gamma)$ are lattices we have that

$$\| b_j \|_{B(\gamma_j)} \leq C \| Ta_j \|_{B(\gamma_j)} \leq CM(\gamma_j) \| a_j \|_{A(\gamma_j)}.$$

Now, since $\sum_j b_j = Ta$, we find that,

$$K(\alpha, Ta) \leq \sum_j \alpha(\gamma_j) \| b_j \|_{B(\gamma_j)}$$

$$\leq C \sum_j \alpha(\gamma_j) M(\gamma_j) \| a_j \|_{A(\gamma_j)} \leq (1+\varepsilon) K(\alpha M, a),$$

and therefore, by letting ε goes to zero, we have that, for every $\beta \in \tilde{S}$,

$$\frac{K(\alpha, Ta)}{\alpha(z_0)} \leq \frac{K(\beta M \alpha \beta^{-1}, a)}{(\alpha\beta^{-1})(z_0)\beta(z_0)} \leq C \sup_{\gamma \in \Gamma} \frac{\beta(\gamma)M(\gamma)}{\beta(z_0)} \frac{K(\alpha\beta^{-1}, a)}{(\alpha\beta^{-1})(z_0)}. \quad (3)$$

It follows that

$$\| Ta \|_{(\bar{B})^S_{z_0,p;K}}$$

$$\leq \left(\sum_{\alpha \in S} \left(\frac{K(\alpha, Ta)}{\alpha(z_0)} \right)^p \right)^{1/p} \leq C \sup_{\gamma \in \Gamma} \frac{\beta(\gamma)M(\gamma)}{\beta(z_0)} \left(\sum_{\alpha \in S} \left(\frac{K(\alpha\beta^{-1}, a)}{(\alpha\beta^{-1})(z_0)} \right)^p \right)^{1/p}$$

$$\leq C \sup_{\gamma \in \Gamma} \frac{\beta(\gamma)M(\gamma)}{\beta(z_0)} \left(\sum_{\alpha \in S} \left(\frac{K(\alpha, a)}{\alpha(z_0)} \right)^p \right)^{1/p}$$

$$= C \sup_{\gamma \in \Gamma} \frac{\beta(\gamma)M(\gamma)}{\beta(z_0)} \| a \|_{(\bar{A})^S_{z_0,p;K}}$$

and taking the infimum in $\beta \in \tilde{S}$ we are done. □

An analogue of inequality (3) was proved in [23, Theorem 5].

Theorem 4.3 Let T be *a* C-subadditive operator from the i.f. $\overline{A}$ into $L^0(\mu)$ and let $\overline{B}$ be an i.f. of Banach lattices with the containing space $\mathscr{V}$ in $L^0(\mu)$.

Let us assume that $T : A(\gamma) \to B(\gamma)$ is such that $Ta \in B(\gamma)$ and $\| Ta \|_{B(\gamma)} \leq M(\gamma) \| a \|_{A(\gamma)}$, where $M \in \mathscr{L}$.

(a) If $0 < p \leq 1$, then

$$\| Ta \|_{(\overline{B})^S_{z_0,p;J}} \leq C D^S_{z_0}(M) \| a \|_{(\overline{A})^S_{z_0,p;J}} \tag{4}$$

(b) If $p > 1$, then (4) still holds if, in addition,

$$\left(\left(\sum_{\alpha \in S} \frac{\inf_\Gamma (\alpha^{-1}(\gamma) M(\gamma))}{\alpha^{-1}(z_0)} \right)^{p'} \right)^{1/p'} < \infty.$$

Proof We shall only prove it for $p \geq 1$ since the case $p < 1$ is quite similar.

Let $a \in (\overline{A})^S_{z_0,p;J}$. Then, for every $\varepsilon > 0$ there exists a representation $a = \sum_{j=1}^{\infty} a_{\alpha_j}$ (convergence in $\mathscr{U}$) so that $a_{\alpha_j} \in \mathscr{A}$ and

$$\left(\sum_j \left(\frac{J(\alpha_j, a_{\alpha_j})}{\alpha_j(z_0)} \right)^p \right)^{1/p} \leq (1 + \varepsilon) \| a \|_{(\overline{A})^S_{z_0,p;J}}$$

Let us take N and N' in $\mathbb{N}$. Then, by elementary estimates and Hölder's inequality,

$$\begin{aligned}
\| \sum_N^{N'} | Ta_{\alpha_j} | \|_{\mathscr{V}} &\leq \sum_N^{N'} \inf_\Gamma (\| a_{\alpha_j} \|_{A(\gamma)} M(\gamma)) \\
&\leq \sum_N^{N'} \inf_\Gamma \alpha_j(\gamma) \| a_{\alpha_j} \|_{A(\gamma)} M(\gamma)\, \alpha_j^{-1}(\gamma) \\
&\leq \sum_N^{N'} \frac{J(\alpha_j, a_{\alpha_j})}{\alpha_j(z_0)} \frac{\inf_\Gamma \alpha_j^{-1}(\gamma)\, M(\gamma)}{\alpha_j^{-1}(z_0)} \\
&\leq C \left(\sum_1^{N'} \left(\frac{J(\alpha_j, a_{\alpha_j})}{\alpha_j(z_0)} \right)^p \right)^{1/p} \left(\sum_j \left(\frac{\inf_\Gamma \alpha_j^{-1}(\gamma)\, M(\gamma)}{\alpha_j^{-1}(z_0)} \right)^{p'} \right)^{1/p'}
\end{aligned}$$

Now, the sum on the right is finite and we have that it converges to zero whenever N and N' goes to infinity. Thus, we have that $\sum_1^N | Ta_{\alpha_j} |$ converges in $\mathscr{V}$ to a measurable function which is finite a.e. (μ) and hence we can define, as in the previous theorem,

$$b_j = \frac{T\left(\sum_j a_{\alpha_j}\right)}{\sum_j |Ta_{\alpha_j}|} |Ta_{\alpha_j}|$$

on the support of $\sum_j |Ta_{\alpha_j}|$ an $b_j = 0$ elsewhere.

Then, we get that $b_j \in \mathscr{B}$, $Ta = \sum_j b_j$ (convergence in $\mathscr{V}$) and, for every $\beta \in \tilde{S}$

$$\frac{J(\alpha_j, b_j)}{\alpha_j(z_0)} \leq C \frac{J(\beta M \alpha_j \beta^{-1}, a_{\alpha_j})}{(\alpha_j \beta^{-1})(z_0)\beta(z_0)} \leq C \sup_{\gamma \in \Gamma} \frac{\beta(\gamma)M(\gamma)}{\beta(z_0)} \frac{J(\alpha_j \beta^{-1}, a_{\alpha_j})}{(\alpha_j \beta^{-1})(z_0)}.$$

From this we have, by first taking the l^p-norm in j and after that taking the infimum in $\beta \in \tilde{S}$, that

$$\| Ta \|_{(\bar{B})^S_{z_0,p;J}} \leq C D^S_{z_0}(M) \| a \|_{(\bar{A})^S_{z_0,p;J}}. \quad \square$$

Theorem 4.4 Under the same hypothesis as those in Theorem 4.3 it yields that

$$\| Ta \|_{(\bar{B})^S_{z_0,p;K}} \leq C D^S_{z_0,p;J\to K}(M) \| a \|_{(\bar{A})^S_{z_0,p;J}}.$$

Proof Again we shall only prove the statement for the case $p \geq 1$. Let $a \in (\bar{A})^S_{z_0,p;J}$. Then, for every $\varepsilon > 0$, there exists a representation $a = \sum_{j=1}^{\infty} a_{\alpha_j}$ (convergence in $\mathscr{U}$) so that $a_{\alpha_j} \in \mathscr{A}$ and

$$\left(\sum_j \left(\frac{J(\alpha_j, a_{\alpha_j})}{\alpha_j(z_0)} \right)^p \right)^{1/p} \leq (1 + \varepsilon) \| a \|_{(\bar{A})^S_{z_0,p;J}}$$

Then $Ta_{\alpha_j} \in \mathscr{B}$ and, as it was done in the previous theorem, we can define

$$b_j = \frac{T\left(\sum_j a_{\alpha_j}\right)}{\sum_j |Ta_{\alpha_j}|} |Ta_{\alpha_j}|,$$

on the support of $\sum_j |Ta_{\gamma_j}|$ and $b_j = 0$ elsewhere.

Then, we get that $b_j \in \mathscr{B}$, $\sum_j b_j = Ta$ and, for every $\alpha \in S$,

$$K(\alpha, Ta) \leq \sum_j K(\alpha, b_j) \leq C \sum_j \inf_{\gamma \in \Gamma} (\alpha M \alpha_j^{-1})(\gamma)\, J(\alpha_j, a_{\alpha_j}).$$

From this, we get the result by doing the same type of computations as those in the proof of Proposition 3.3. □

We will point out some applications of the results above but first we need the following definitions:

Let $f^*(t)$, $0 < t < \infty$, denote the nonincreasing rearrangement of the function f defined on a σ-finite measure space (Ω, μ). Set $f^{**}(t) = \frac{1}{t}\int_0^t f^*(s)\, ds$ and let $\Psi(t)$ denote a measurable function (usually quasiconcave). Let now T be a C-subadditive operator from a Banach space A into a space of μ-measurable and essentially bounded functions. We say that T is of weak-type (A, Ψ) (resp. weakened-type (A, Ψ)) with constant C if, for every $f \in A$ and $0 < t < \infty$,

$$(Tf)^*(t) \leq C(\, \| f \|_A / \Psi(t)),$$

(resp. $(Tf)^{**}(t) \leq C(\, \| f \|_A / \Psi(t))$), see [10], [11].

Observe that, in the case that Ψ is quasiconcave, T is of weak-type (A, Ψ) if and only if

$$T : A \to \Lambda^{1,\infty}(w),$$

where $\Lambda^{1,\infty}(w)$ is the weak weighted Lorentz space (see, for example, [6]) and w is such that $\Psi(t) \sim \int_0^t w(s)\, ds$. Similarly, T is of weakened-type (A, Ψ) if and only if

$$T : A \to \Gamma^{1,\infty}(w)$$

where $\Gamma^{1,\infty}(w) = \left\{ f; f \text{ is measurable and } \sup_{t>0} f^{**}(t) \int_0^t w < \infty \right\}.$

While the space $\Gamma^{1,\infty}(w)$ is a Banach space, this is not, in general, the case of the space $\Lambda^{1,\infty}(w)$. However, these two spaces coincides if and only if w satisfies a B_1 condition (see [24]).

Also, by defining

$$L^\infty(\Psi) = \{ f; f \text{ is measurable and } \sup_{t>0} \Psi(t)\, | f(t) | < \infty \},$$

we have that if $S_T f = (Tf)^{**}$, then T is of weakened type (A, Ψ) whenever $S_T : A \to L^\infty(\Psi)$.

Theorem 4.5 Let $\overline{A}$ be an i.f., let B belong to the class $K^S_{z_0}$ and let T be the C-subadditive operator as above.

(a) If T is of weakened-type $(A(\gamma), \Psi_\gamma)$ with constants $c(\gamma) \in \mathcal{L}$, then $T_{|B}$ is of the weakened-type (B, Φ), where

$$\Phi(t) = \inf_{\alpha \in S} \frac{D^S_{z_0}(\alpha^{-1} c(\Psi_{\cdot}(t))^{-1})}{\alpha^{-1}(z_0)}. \tag{5}$$

(b) If the i.f. is finite and T is of weak-type $(A(\gamma),\ \Psi_\gamma)$ with constants $c(\gamma) \in \mathscr{L}$ where the functions Ψ_γ are quasiconcave, the $T_{|B}$ is of the weak-type $(B,\ \Phi)$ with Φ as in (5).

Proof (a) The proof of (a) can be easily obtained as a corollary of Theorem 4.2 and example 5.5 in [5], however we shall give a direct proof that can be used to also prove (b).

Let $a \in B$ and choose an arbitrary $\varepsilon > 0$. We can find a representation $a = \sum a_j$, $a_j \in A(\gamma_j)$, such that

$$\sum_j \alpha(\gamma_j) \| a_j \|_{\gamma_j} \le (1 + \varepsilon) K(\alpha, a).$$

By using this fact, our assumptions and the subadditive property

$$(f + g)^{**}(t) \le f^{**}(t) + g^{**}(t),$$

we can make, for every $\beta \in \tilde{S}$, the following estimates:

$$(Ta)^{**}(t) = (T(\sum_j a_j))^{**}(t) \le C(\sum_j | Ta_j |)^{**}(t)$$

$$\le C \sum_j c(\gamma_j) \frac{\| a_j \|_{\gamma_j}}{\Psi_{\gamma_j}(t)}$$

$$\le C\alpha^{-1}(z_0) \sum_j \frac{(c(\gamma_j)\alpha^{-1}(\gamma_j)\beta(\gamma_j)}{\Psi_{\gamma_j}(t)\alpha^{-1}(z_0)} \| a_j \|_{\gamma_j} \alpha(\gamma_j)\beta^{-1}(\gamma_j)$$

$$\le C\alpha^{-1}(z_0)\beta(z_0) \sup_{\gamma\in\Gamma} \frac{\alpha(\gamma)^{-1}\beta(\gamma)c(\gamma)}{\Psi_\gamma(t)\alpha^{-1}(z_0)\beta(z_0)}$$

$$\times \sum_j \| a_j \|_{\gamma_j} \alpha(\gamma_j)\beta^{-1}(\gamma_j)$$

$$\le C(1 + \varepsilon) \frac{K(\alpha\beta^{-1}, a)}{\alpha(z_0)\beta^{-1}(z_0)} \sup_{\gamma\in\Gamma} \frac{\alpha(\gamma)^{-1}\beta(\gamma)c(\gamma)}{\Psi_\gamma(t)\alpha^{-1}(z_0)\beta(z_0)}$$

$$\le C(1 + \varepsilon) \| a \|_A \sup_{\gamma\in\Gamma} \frac{\alpha(\gamma)^{-1}\beta(\gamma)c(\gamma)}{\Psi_\gamma(t)\alpha^{-1}(z_0)\beta(z_0)}.$$

Thus, by taking infimum over all $\beta \in \tilde{S}$ first and then over all $\alpha \in S$, we find that

$$(Ta)^{**}(t) \le C \| a \|_A \inf_{\alpha\in S} \frac{D^S_{z_0} \alpha^{-1}c(\Psi\cdot(t))^{-1}}{\alpha^{-1}(z_0)}.$$

(b) The proof is similar to that of (a). In fact, we only need to replace the corresponding estimates in the proof of (a) by the following estimates:

$$(\sum_{j=1}^{N} | Ta_j(t) |)^* \leq \sum_{j=1}^{N} (Ta_j)^*(t/n) \leq \sum_{j}^{n} c(\gamma_j) \| a_j \|_{A(\gamma_j)} / \Psi_{\gamma_j}(t/n)$$

$$\leq n \sum_{j}^{n} c(\gamma_j) \| a_j \|_{A(\gamma_j)} / \Psi_{\gamma_j}(t).$$

These estimates hold because $(f+g)^*(t_1+t_2) \leq f^*(t_1) + g^*(t_2)$ and $\Psi_{\gamma_j}(t)/t$ are nonincreasing so that, in particular, $\Psi_{\gamma_j}(t/n) \geq \Psi_{\gamma_j}(t)/n$. □

By using Theorem 4.5 (and some recent results in [5]) we obtain the following result for the complex (St. Louis) spaces $A[z_0]$ (c.f. also [20]):

Corollary 4.7 Let $A = A[z_0]$ ($z_0 \in D$) be the complex interpolation space for the i.f. $\overline{A}$ and let $\Psi_\gamma(t)$ be measurable functions in $\mathcal{L}$, for every fixed t. If $T_{|A(\gamma)}$ is of weakened type $(A(\gamma), \Psi_\gamma)$ with constants $c(\gamma) \in \mathcal{L}$, then $T_{|A[z_0]}$ is of weakened type $(A[z_0], \Psi_{z_0})$.

Proof According to Theorem 2.1 in [5] and well-known interpolation theorems for the complex method (see [8]) we have that $A[z_0] \in K_{z_0}^S$ for any group S suitable for the constructions of the real method. Therefore, the proof follows by applying Theorem 4.5. □

5. Further Interpolation Results with Operator Norms Controlled by the Dicesar function

A Banach lattice A is a KB-space if the condition that $0 < x_n$, $x_n \leq x_{n+1}$, $n = 1, 2, \ldots$, $\sup \| x_n \| < \infty$ implies that there exists $x \in A$ such that $\| x_n - x \| \to 0$ (see e.g. [14]).

An operator acting from a Banach space A into a Banach lattice B is called correct and we write $T \in \Pi(A, B)$ if $T \in L(A, B)$ and the set $T(U_A)$ is order bounded in B (see e.g. [14]; here, as usual, U_A denotes the unit ball in A). Moreover,

$$\| T \|_{\Pi(A,B)} = \inf \{ \| b \| : | Ta | \leq b, \text{ for all } a \in U_A \}.$$

For the special case when B is a KB-space

$$\| T \|_{\Pi(A,B)} = \sup \{ \| \, | Ta_1 | \vee | Ta_2 | \vee \ldots | Ta_n | \, \| \},$$

where supremum is taken over all finite sets $\{a_1, a_2, \ldots, a_n\} \subset U_A$.

Theorem 5.1 Let A be a Banach space, let $\overline{B} = (B(\gamma), \gamma \in \Gamma\}$ be an i.f. of *KB*-spaces, let $T : A \to \Delta(\overline{B})$ and $T \in \Pi(A, B(\gamma))$ for all $\gamma \in \Gamma$ and let $\| T \|_{\Pi(A,B(\gamma))} \leq M(\gamma)$ with $M \in \mathcal{L}$. If the *KB*-space B belongs to $J_{z_0}^S(\overline{B})$, then T is a correct operator from A into B and

$$\| T \|_{\Pi(A,B)} \leq C \inf_{\alpha \in S} \frac{D_{z_0}^{S^{-1}}(\alpha M)}{\alpha(z_0)}.$$

The analogous result for the class $J_\theta(\overline{B})$ was proved in [21].

Proof Take $a_1, a_2, \ldots, a_n \in U_A$ and consider $y = |Ta_1| \vee |Ta_2| \vee \ldots \vee |Ta_n|$. Since $T \in \Pi(A, B(\gamma))$ and $B(\gamma)$ is a KB-space we have that $\| y \|_{B(\gamma)} \leq M(\gamma)$. This implies that $y \in \mathscr{B}$ and since $\| y \|_B \leq C \dfrac{J(\alpha, y)}{\alpha(z_0)}$ for every $\alpha \in S$, we get that, for $\alpha \in S$ and $\beta \in \tilde{S}$, $\alpha\beta^{-1} \in S$, and hence, since $\widetilde{S^{-1}} = \tilde{S}^{-1}$, we have that

$$\| y \|_B \leq C \inf_{\alpha \in S} \inf_{\beta \in \tilde{S}} \frac{\sup_\gamma (\alpha(\gamma)\beta^{-1}(\gamma) \| y \|_\gamma)}{(\alpha\beta^{-1})(z_0)}$$

$$\leq C \inf_{\alpha \in S} \inf_{\beta \in \tilde{S}} \frac{\sup_\gamma (\alpha(\gamma)\beta^{-1}(\gamma) M(\gamma))}{(\alpha\beta^{-1})(z_0)} = \inf_{\alpha \in S} \frac{D_{z_0}^{S^{-1}}(\alpha M)}{\alpha(z_0)}.$$

Now the proof follows by taking supremum over all finite sets $\{a_1, a_2, \ldots, a_n\} \subset U_A$. □

In particular, if S is a group we get $\| T \|_{\Pi(A,B)} \leq C D_{z_0}^S(M)$.

An operator T acting from a Banach lattice A into a Banach space B is called summing ($T \in S(A, B)$) if $T \in L(A, B)$ and

$$\| T \|_{S(A,B)} = \sup \frac{\sum_1^n \| Ta_k \|_B}{\| \sum_1^n | a_k | \|_A} < \infty,$$

where the supremum is taken over all finite sets $\{a_1, a_2, \ldots, a_n\} \subset A$. For these operators the following result follows easily.

Theorem 5.2 Let $\overline{A} = \{A(\gamma), \gamma \in G\}$ be an i.f. of Banach lattices such that the dual spaces $A'(\gamma)$ are KB-spaces and let B be a Banach space. Let $T \in L(\Sigma(\overline{A}), B)$ be a summing operator from $A(\gamma)$ into B a.e. $\gamma \in \Gamma$ and $\| T \|_{S(A(\gamma),B)} \leq M(\gamma)$, with $M \in \mathscr{L}$ for $j = 2$ or $M \in L^\infty$ if $j = 1$. If $A \in K_{z_0}^S(\overline{A})$ and $S, S^{-1} \subset L^\infty$, then T is a summing operator from A into B and $\| T \|_{S(A,B)}$ $\leq C \inf_{\alpha \in S} \dfrac{D_{z_0}^S(M\alpha^{-1})}{\alpha^{-1}(z_0)}$.

The corresponding result for the class $K_\theta(\overline{B})$ was proved in [21].

In particular, if S contains the function 1 we get that $\| T \|_{S(A,B)} \leq C D_{z_0}^S(M)$.

References

1. J. Bergh and J. Löfström, *Interpolation Spaces. An Introduction.*, vol. 101, Grundlehren der Mathematischen Wissenschaften 223, Springer Verlag, Berlin-Heidelberg-New York, 1976.

2. M.J. Carro, *Real interpolation for families of Banach spaces,* Stud. Math. **109** (1994), 1–21.
3. M.J. Carro, *Real interpolation for families of Banach spaces, II,* Collect. Math. **72** (1993) 47–60
4. M.J. Carro and L.I. Nikolova, *Real interpolation for families of Banach spaces and the compactness property,* Preprint (1996).
5. M.J. Carro, L.I. Nikolova, J. Peetre and L.E. Persson, *Some real interpolation methods for families of Banach spaces-A comparison,* J. Approx. Theory **89** (1997), 26–57.
6. M.J. Carro and J. Soria, *Weighted Lorentz Spaces and the Hardy Operator,* J. Funct. Anal. **112** (1993), 480–494.
7. F. Cobos and J. Peetre, *Interpolation of compact operators: the multidimensional case,* Proc. London Math. Soc. **63** (1991), 371–400.
8. R.R. Coifman, M. Cwikel, R.R. Rochberg, Y. Sagher and G. Weiss, *Complex interpolation for families of Banach spaces,* In; Proc. Symp. Pure Math. 35, Part 2, American Mathematical Society, Providence, R.I. (1979), 269–282.
9. M. Cwikel and S. Janson, *Real and complex interpolation methods for tinite and infinite families of Banach spaces,* Adv. Math. **66** (1987), 234–290.
10. V.I. Dmitriev and S.G. Krein, *Interpolation of operators of weak type,* Anal. Math. **4** (1978), 83–99.
11. V.I. Dmitriev, S.G. Krein and V.I. Ovchinnikov, *Fundamentals of the theory of interpolation of linear operators,* In: Geometry of Linear Spaces and Operator Theory, Yaroslavl (1977), 31–74 (in Russian).
12. V. Ferenci, *A uniformly convex heriditarily indecomposable Banach space,* Israel J. Math. **102** (1997), 199–225.
13. D.L. Fernández, *Interpolation of 2^n Banach spaces,* Studia Math. **65** (1979), 175–201.
14. L.V. Kantorovitsch and G.P. Akilov, *Functional Analysis,* Nauka,Moscow (1977).
15. S.G. Krein and L.I. Nikolova, *Holomorphic functions in a family of Banach spaces and interpolation,* Dokl. Akad. Nauk SSSR **250** (1980), 547–550; Englishtranslation: Soviet Math. dokl. **21** (1980),131–134.
16. S.G. Krein and L.I. Nikolova,*Complex interpolation offamilies of Banach spaces,* Ukr. Mat. Zh. 34 (1982), 31–42; English translation: Ukr. Math J. **34** (1982), 26–36.
17. V.I. Levin, *About two classes of linear operators acting between Banach spaces and Banach lattices,* (in Russian), Sib. Mat. J. 10, Nr. **4** (1969), 903–909.
18. L. Maligranda, *On interpolation of nonlinear operators, Comment. Math. Prace Mat.* **28** (1989), 253–275.
19. L.I. Nikolova, *Measure of noncompactness of operators acting in interpolation spaces-the multidimensional case,* C.R. Acad. Bulg. Sci. **44** (1991), 5–8.
20. L.I. Nikolova, *On the interpolation of operators of weakened type,* Complex Analysis and Applications (Varna 1985), Sofia 1986, 478–481.
21. L.I. Nikolova, *On the interpolation of some classes of operators acting in families of Banach spaces,* Constructive Theory of Functions 87, Sofia (1988), 352–359.
22. L.I. Nikolova and L.E. Persson, *Real interpolation methods for families of Banach spaces,* manuscript, Luleå University (1992).
23. L.I. Nikolova and L.E. Persson, *Interpolation of nonlinear operators betwen families of Banach spaces,* Math Scand. **72** (1993), 253–275.
24. J. Soria, *Lorentz spaces of weak-type,* Quarterly J. Math **49** (1998), 93–103.
25. G. Sparr, *Interpolation of several Banach spaces,* Ann. Mat. Pura et Appl. **99** (1974), 247–316.

Function Spaces and Applications
D.E. Edmunds et al (Eds)

5. On the García-Falset Coefficient in Orlicz Sequence Spaces Equipped with the Orlicz Norm

Yunan Cui[1*] and Henryk Hudzik[2†]

[1]Department of Mathematics, Harbin University of Science and Technology, Harbin, 150080, P.R. China

[2]Faculty of Mathematics and Computer Science, Adam Mickiewicz University, Matejki 48/49, 60-769 Poznań, Poland

[2]Institute of Mathematics, Poznań University of Technology, Piotrowo 3a, 60-965 Poznań, Poland

1. Introduction

Throughout this paper X is a *Banach space* which is assumed not to have the Schur property, that is, X has a weakly convergent sequence that is not norm convergent. $S(X)$ and $B(X)$ denote the unit sphere and the unit ball of X, respectively, and l^0 denotes the space of all real sequences.

To obtain the weak fixed point property for some Banach spaces, García-Falset introduced (see [5]) the following coefficient

$$R(X) = \sup \{ \liminf_{n \to \infty} \| x_n - x \| : \{x_n\} \subset B(X), x_n \xrightarrow{w} 0, x \in B(X)\}.$$

He proved that a Banach space X with $R(X) < 2$ has the weak fixed point property (see [6]).

A Banach space X is said to be nearly uniformly smooth (NUS) provided that for every $\varepsilon > 0$ there is $\eta > 0$ such that if $t \in (0, \eta)$ and (z_n) is a basic sequence in $B(X)$, then there exists $k > 1$ such that $\| x_1 + tx_k \| \le 1 + t\varepsilon$ (see [11]).

A natural generalization of this notion is the weak near uniform smoothness (WNUS for short).

A Banach space X is said to be WNUS whenever it satisfies the above condition from the definition of NUS with "for some $\varepsilon \in (0, 1)$" in place of "for every $\varepsilon > 0$" (see [6]).

It is well known that a Banach space X is WNUS if and only if it is reflexive and $R(X) < 2$ (see [5]) and that WNUS Banach spaces have the weak

*Supported by Chinses National Science Foundation Grant.
†Supported by KBN Grant 2PO3A 031 10.

point property. For the definition of the fixed point property and the weak fixed point property see [4], [7] [13] and [14].

A mapping $\Phi : \mathscr{R} \to \mathscr{R}_+$ is said to be an *Orlicz* function if Φ vanishes at 0, Φ is even and convex on the whole line R and $\lim_{u \to \infty} \frac{\Phi(u)}{u} = \infty$ and $\lim_{u \to 0} \frac{\Phi(u)}{u} = 0$.

For every *Orlicz* function Φ we define its complementary function $\Psi : \mathscr{R} \to [0, \infty)$ by the formula

$$\Psi(v) = \sup_{u>0} \{u \mid v \mid - \Phi(u)\}$$

for every $v \in \mathscr{R}$. The complementary function Ψ is also an *Orlicz* function.

The Orlicz sequence space l_Φ is defined to be the set $\{x \in l_0 : I_\Phi(\lambda x) = \sum_{i=1}^{\infty} \Phi(\lambda x(i)) < \infty$ for some $\lambda > 0\}$ equipped with the *Luxemburg* norm

$$\| x \| = \inf \left\{k > 0 : I_\Phi\left(\frac{x}{k}\right) \leq 1\right\}$$

or the *Amemiya-Orlicz* norm

$$\| x \|^0 = \inf \left\{\frac{1}{k}(1 + I_\Phi(kx)) : k > 0\right\}$$

To simplify notations, we write shortly l_Φ for $(l_\Phi, \| \cdot \|_\Phi)$ and l_Φ^0 for $(l_\Phi, \| \cdot \|_\Phi^0)$. Put $h_\Phi = \{x \in l_0 : I_\Phi(\lambda x) = \sum_{i=1}^{\infty} \Phi(\lambda x(i)) < \infty$ for any $\lambda > 0\}$. Then $x \in h_\Phi$ if and only if x has the absolutely continuous norm, i.e. $\| \sum_{i=n+1}^{\infty} x(i)\, e_i \| \to 0$ as $n \to \infty$, where $e_i = (0, 0, \ldots, 1^{\text{ith}}, 0, \ldots)$.

An *Orlicz* function Φ is said to satisfy the δ_2-condition ($\Phi \in \delta_2$) if there exist $u_0 > 0$ and $K \geq 2$ such that

$$\Phi(2u) \leq K\Phi(u)$$

whenever $| u | \leq u_0$.

We say that an *Orlicz* function Φ satisfies the $\overline{\delta}_2$-condition if its complementary function Ψ satisfies the δ_2-condition.

The basic facts on Orlicz spaces can be found in [1], [10] and [12].

2. Results

Theorem 1 If $\Phi \notin \delta_2$, then $R(l_\Phi^0) = 2$.

Proof Since $\Phi \notin \delta_2$, for any $\varepsilon > 0$ there exists $x \in S(l_\Phi^0)$ such that

$$1 - \varepsilon \leq \| \sum_{i=n+1}^{\infty} x(i)\, e_i \|^0 \leq 1 \text{ for all } n \in \mathscr{N} \text{ (see [2]).}$$

Take $0 = i_0 < i_1 < i_2 < i_3 < \ldots$ such that

$$1 - 2\varepsilon \leq \| \sum_{i=i_{n-1}+1}^{i_n} x(i)\, e_i \|^0 \leq 1 \text{ for all } n \in \mathscr{N}.$$

Such a sequence $\{i_n\}$ exists because the space l_Φ^0 has the semi-*Fatou* property, i.e. $\| x_n \|^0 \to \| x \|^0$ whenever $0 \leq x_n \leq x \in l_\Phi^0$ and $x_n \to x$ coordinatewise (see [1] and [9]).

Put $x_n = \sum_{i=i_{n-1}+1}^{i_n} x(i)\, e_i$ for all $n \in \mathscr{N}$. Then the condition

$$\limsup \frac{I_\Phi(\lambda x_n)}{\lambda} \leq \limsup_{\lambda \to 0} \frac{I_\Phi(\lambda x)}{\lambda} = 0$$

and the *Young* inequality and fact that any singular functional over l_Φ^0 vanishes on $x_n (n = 1, 2, \ldots)$ yield that $x_n \xrightarrow{w} 0$.

By $\liminf_{n\to\infty} \| x_n + x \|^0 \geq \liminf_{n\to\infty} 2 \| x_n \|^0 \geq 2(1 - 2\varepsilon)$ and the arbitrariness of $\varepsilon > 0$, we get $R(l_\Phi^0) = 2$. □

Theorem 2 The equality $R(h_\Phi^0) = \alpha(h_\Phi^0)$ holds for any *Orlicz* function Φ.

Proof First, we will show that $R(h_\Phi^0) = R_1(h_\Phi^0)$, where

$$R_1(l_\Phi) = \sup \{\liminf_{n\to\infty} \| x_n - x \|^0 : x_n = \sum_{i=i_{n-1}+1}^{I_n} x_n(i)\, e_i \in S(h_\Phi^0),\, x_n \xrightarrow{w} 0,$$

$$x = \sum_{i=1}^{I_1} x(i)\, e_i \in S(h_\Phi^0),\, I_1 < I_2 < \ldots,\, n = 2, 3, \ldots\}.$$

We need only to show the inequality $R(h_\Phi^0) \leq R_1(h_\Phi^0)$.

For any $\varepsilon > 0$ there exist a weakly null sequence $\{x_n\} \subset B(h_\Phi^0)$ and $x_0 \in B(h_\Phi^0)$ such that

$$\liminf_{n\to\infty} \| x_n - x_0 \|^0 \geq R(h_\Phi^0) - \varepsilon.$$

For the sake of convenience, we may assume that $\| x_n - x_0 \|^0 \geq R(h_\Phi^0) - 2\varepsilon$ for all $n \in \mathscr{N}$. There exists $i_0 \in \mathscr{N}$ such that

$$\| \sum_{i=i_0+1}^{\infty} x_0(i)\, e_i \|^0 < \varepsilon.$$

By $x_n \xrightarrow{w} 0$, we have $x_n(i) \to 0$ as $n \to \infty$ for all $i \in \mathcal{N}$. So there exists $n_1 \in \mathcal{N}$ such that

$$\| \sum_{i=1}^{i_0} x_n(i)\, e_i \|^0 < \varepsilon \ \text{ whenever } n \ge n_1.$$

Fix $z_1 = \sum_{i=i_0+1}^{\infty} x_{n_1}(i)\, e_i$. Then there exists $i_1 > i_0$ such that

$$\| \sum_{i=i_1+1}^{\infty} z_1(i)\, e_i \|^0 < \varepsilon.$$

By coordinatewise convergence of x_n to 0, there exists $n_2 > n_1$ such that

$$\| \sum_{i=1}^{i_1} x_n(i)\, e_i \|^0 < \varepsilon \ \text{ whenever } n \ge n_2.$$

Set $z_2 = \sum_{i=i_1+1}^{\infty} x_{n_2}(i)\, e_i$. Proceeding in such a way by induction, we get a sequence $\{z_k\} \subset B(h_\Phi^0)$ such that $z_k = \sum_{i=i_{k-1}+1}^{\infty} x_{n_k}(i)\, e_i$ and

$$\| x_{n_k} \|^0 - \varepsilon \le \| z_k \|^0 \le \| x_{n_k} \|^0$$

for each $k \in \mathcal{N}$. Put $w_0 = \sum_{i=1}^{i_0} x_0(i)\, e_i$ and $w_k = \sum_{i=i_{k-1}+1}^{i_k} z_k(i)\, e_i$ for all $k \in \mathcal{N}$. Then $\{w_k\} \subset B(h_\Phi^0)$ and

$$\| x_{n_k} \|^0 - 2\varepsilon \le \| w_k \|^0 \le \| x_{n_k} \|^0$$

for each $k \in \mathcal{N}$. It is easy to see that the condition

$$\limsup_{\lambda \to 0} \frac{I_\Phi(\lambda w_k)}{\lambda} = \limsup_{\lambda \to 0} \frac{I_\Phi(\lambda x_{n_k})}{\lambda} = 0$$

and the *Young* inequality imply that $x_n \xrightarrow{w} 0$.

Moreover,

$$\limsup_{k\to\infty} \| w_k - w_0 \|^0$$

$$= \liminf_{k\to\infty} \| x_{n_k} - x_0 - \sum_{i=1}^{i_{k-1}} x_{n_k}(i)\, e_i - \sum_{i=i_k+1}^{\infty} x_{n_k}(i)\, e_i + \sum_{i=i_k+1}^{\infty} x_{n_k}(i)\, e_i \|^0$$

$$\ge \liminf_{k\to\infty} \| x_{n_k} - x_0 \|^0 - 3\varepsilon \ge R(X) - 4\varepsilon.$$

By the arbitrariness of $\varepsilon > 0$, we get $R(h_\Phi^0) = R_1(h_\Phi^0)$, whence $\alpha(h_\Phi^0) \ge R(h_\Phi^0)$.

On the other hand, for any $\varepsilon > 0$ there exist $x, y \in S(h_\Phi^0)$ with $|x| \wedge |y| = 0$ such that $\alpha(h_\Phi^0) - \varepsilon < \| x + y \|^0$. For the sake of convenience, we assume that supp x is contained in the set of odd natural numbers and supp y is contained in the set of even natural numbers. Since $y \in h_\Phi^0$, there exists $i_0 \in \mathcal{N}$ such that

$$\alpha(h_\Phi^0) - \varepsilon \le \| x + \sum_{i=1}^{i_0} y(i)\, e_i \|^0.$$

Put $y_n = \sum_{i=ni_0}^{(n+1)i_0} y(i)\, e_i$. Then by supp $x \cap$ supp $y = \phi$ and symmetry of l_Φ^0, we get $\| y_n + x \|^0 = \| x + \sum_{i=1}^{i_0} y(i)\, e_i \|^0 \ge \alpha(h_\Phi^0) - \varepsilon$. Moreover, we can easily get that $y_n \xrightarrow{w} 0$. By the arbitrariness of $\varepsilon > 0$, we have $\alpha(h_\Phi^0) \le R(h_\Phi^0)$, whence $\alpha(h_\Phi^0) \le R(h_\Phi^0)$. □

For any $x, y \in S(h_\Phi^0)$ with $|x| \wedge |y| = 0$ and $k > 1$ there exist $c_{x,y,k} > 0$ such that

$$k - 1 = I_\Phi\left(\frac{kx}{c_{x,y,k}}\right) + I_\Phi\left(\frac{ky}{c_{x,y,k}}\right),$$

which follows by the continuity of the function $f(c) = I_\Phi\left(\frac{kx}{c}\right) + I_\Phi\left(\frac{ky}{c}\right)$ for every $x, y \in h_\Phi^0$.

Put $c_{x,y} = \inf \{c_{x,y,k} > 0 : k > 1\}$. Then we will show that for any $x, y \in S(h_\Phi^0)$ with $|x| \wedge |y| = 0$ there exists $k_0 > 1$ such that $c_{x,y} = c_{x,y,k_0}$

Take $k_n > 1$ $(n = 1, 2, \ldots)$ such that $\lim_{n\to\infty} c_{x,y,k_n} = c_{x,y}$. Consider first the case that $\{k_n\}$ is not bounded. Then we can assume without loss of generality that $k_n \to \infty$. By

$$k_n - 1 = I_\Phi\left(\frac{k_n x}{c_{x,y,k_n}}\right) + I_\Phi\left(\frac{k_n y}{c_{x,y,k_n}}\right), \tag{*}$$

we get

$$\frac{I_\Phi\left(\frac{k_n x}{c_{x,y,k_n}}\right) + I_\Phi\left(\frac{k_n y}{c_{x,y,k_n}}\right)}{k_n} \to 1 \text{ as } n \to \infty.$$

However, the equality (*) and $k_n \to \infty$ yield $\frac{k_n}{c_{x,y,k_n}} \to \infty$. Since $(\Phi(u)/u) \to \infty$ as $u \to \infty$, this yields

$$\frac{I_\Phi\left(\frac{k_n x}{c_{x,y,k_n}}\right) + I_\Phi\left(\frac{k_n y}{c_{x,y,k_n}}\right)}{k_n} \to \infty \text{ as } n \to \infty,$$

a contradiction. Therefore the sequence $\{k_n\}$ is bounded and we may assume without loss of generality that it is convergent, say $k_n \to k_0$. We have

$$I_\Phi\left(\frac{k_0 x}{c_{x,y,k_0}}\right) + I_\Phi\left(\frac{k_0 y}{c_{x,y,k_0}}\right) = k_0 - 1. \qquad (**)$$

Equalities (*) and (**) and $k_n \to k_0$ yield

$$I_\Phi\left(\frac{k_n x}{c_{x,y,k_n}}\right) + I_\Phi\left(\frac{k_n y}{c_{x,y,k_n}}\right) \to I_\Phi\left(\frac{k_0 x}{c_{x,y,k_0}}\right) + I_\Phi\left(\frac{k_0 y}{c_{x,y,k_0}}\right).$$

Moreover, by $k_n \to k_0$ and $c_{x,y,k_n} \to c_{x,y}$, there holds

$$I_\Phi\left(\frac{k_n x}{c_{x,y,k_n}}\right) + I_\Phi\left(\frac{k_n y}{c_{x,y,k_n}}\right) \to I_\Phi\left(\frac{k_0 x}{c_{x,y}}\right) + I_\Phi\left(\frac{k_0 y}{c_{x,y}}\right),$$

which implies that $c_{x,y,k_0} = c_{x,y}$, i.e. the infimum is attained at k_0. □

Theorem 3 There holds the equality

$$\alpha(h_\Phi^0) = \sup\{c_{x,y} : x = \sum_{i=1}^{m} x(i)\, e_i \in S(h_\Phi^0),$$

$$y = \sum_{i=m+1}^{n} y(i)\, e_i \in S(h_\Phi^0) \text{ with } n > m\}.$$

Proof Put $d = \sup\{c_{x,y} : x = \sum_{i=1}^{m} x(i)\, e_i,\ y = \sum_{i=m+1}^{n} y(i)\, e_i \in S(l_\Phi^0)$, where $n > m\}$. Then for any such a couple of x and y there exists $k > 1$ such that $c_{x,y} = c_{x,y,k}$ (see the considerations before Theorem 3). Hence

$$\left\| \frac{x+y}{d} \right\|^0 \le \frac{1}{k}\left(1 + \sum_{i=1}^{m} \Phi\left(\frac{kx(i)}{d}\right) + \sum_{i=m+1}^{n} \Phi\left(\frac{ky(i)}{d}\right)\right)$$

$$\le \frac{1}{k}\left(1 + \sum_{i=1}^{m} \Phi\left(\frac{kx(i)}{c_{x,y}}\right) + \sum_{i=m+1}^{n} \Phi\left(\frac{ky(i)}{c_{x,y}}\right)\right)$$

$$= \frac{1}{k}\left(1 + \sum_{i=1}^{m} \Phi\left(\frac{kx(i)}{c_{x,y,k}}\right) + \sum_{i=m+1}^{n} \Phi\left(\frac{ky(i)}{c_{x,y,k}}\right)\right) = \frac{1}{k}(1 + k - 1) = 1,$$

i.e. $\| x + y \|^0 \le d$, whence $\alpha(h_\Phi) \le d$.

On the other hand, for any $\varepsilon > 0$ there exists $x = \sum_{i=1}^{m} x(i)\, e_i$ and $y = \sum_{i=m+1}^{n} y(i)\, e_i$ in $S(X)$ with $n > m$ such that $c_{x,y} + \varepsilon > d$. So, we have for all $k > 1$

$$\frac{1}{k}\left(1 + I_\Phi\left(\frac{k(x+y)}{d-\varepsilon}\right)\right) = \frac{1}{k}\left(1 + I_\Phi\left(\frac{kx}{d-\varepsilon}\right) + I_\Phi\left(\frac{ky}{d-\varepsilon}\right)\right)$$

$$\geq \frac{1}{k}\left(1 + I_\Phi\left(\frac{kx}{c_{x,y}}\right) + I_\Phi\left(\frac{ky}{c_{x,y}}\right)\right)$$

$$\geq \frac{1}{k}\left(1 + I_\Phi\left(\frac{kx}{c_{x,y,k}}\right) + I_\Phi\left(\frac{ky}{c_{x,y,k}}\right)\right) = 1.$$

Hence we get $\| x + y \|^0 > d - \varepsilon$, whence by the arbitrariness of $\varepsilon > 0$, it follows that $\alpha(l_\Phi^0) \geq d$. Consequently, $\alpha(h_\Phi^0) = d$. □

Since l_Φ^0 has the semi-*Fatou* property, we can easily get the following result.

Corollary 1 For any Orlicz function Φ there holds the equality $\alpha(h_\Phi^0) = \alpha(h_\Phi^0)$.

Corollary 2 If $\| x \|^0 = c\Phi^{-1}(I_\Phi(x))$ for any $x \in h_\Phi^0$, then

$$\alpha(h_\Phi^0) = \frac{1}{c} \inf_{k>1} \left(\frac{k}{\Phi^{-1}\left(\frac{k-1}{2}\right)} \right).$$

Proof For any $x, y \in S(h_\Phi^0)$, $| x | \wedge | y | = 0$ and $k > 1$, there holds

$$k - 1 = I_\Phi\left(\frac{kx}{c_{x,y,k}}\right) + I_\Phi\left(\frac{ky}{c_{x,y,k}}\right)$$

$$= \Phi\left(\frac{1}{c}\left\| \frac{kx}{c_{x,y}} \right\|^0\right) + \Phi\left(\frac{1}{c}\left\| \frac{ky}{c_{x,y}} \right\|^0\right) = 2\Phi\left(\frac{1}{c}\frac{k}{c_{x,y}}\right).$$

Hence

$$c_{x,y,k} = \frac{k}{c\Phi^{-1}\left(\frac{k-1}{2}\right)}.$$

Therefore

$$c_{x,y} = \inf_{k>1} \left(\frac{k}{\Phi^{-1}\left(\frac{k-1}{2}\right)} \right) = \alpha(h_\Phi^0). \quad \square$$

Corollary 3 For *Lebesgue* space $l_p (1 < p < \infty)$, we have $R(l_p) = \alpha(l_p) = 2^{1/p}$.

Proof For any $x \in S(l_p)$ we have $\| x \|^0 = q^{1/q} \Phi^{-1}(I_\Phi(x))$, where $\Phi(u) = \frac{|u|^p}{p}$ and $q > 0$ satisfies the equality $\frac{1}{p} + \frac{1}{q} = 1$. Hence

$$\alpha(l_p) = \inf_{k>1} \left(\frac{k}{\Phi^{-1}\left(\frac{k-1}{2}\right)} \right) = \frac{1}{q^{1/q} p^{1/p}} \inf_{k>1} k \left(\frac{k-1}{2} \right)^{1/p} = 2^{1/p}. \quad \square$$

Corollary 4 For *Orlicz* sequence spaces the inequality $R(l_\Phi^0) < 2$ holds if and only if $\Phi \in \delta_2$ and $\Phi \in \bar{\delta}_2$.

Proof Necessity. If $\Phi \notin \delta_2$, then $R(l_\Phi^0) = 2$ (see Theorem 1), which is a contradiction. Suppose that $\Phi \notin \bar{\delta}_2$. Then l_Φ^0 contains an almost isometric copy of l_1. Since $\alpha(l_1) = 2$, we get $\alpha(l_\Phi^0) = 2$, a contradiction.

Sufficiency. For any $x, y \in S(h_\Phi^0)$ with $|x| \wedge |y| = 0$ there are $k > 1$ and $h > 1$ such that

$$\| x \|^0 = \frac{1}{k}(1 + I_\Phi(kx)), \ \| y \|^0 = \frac{1}{h}(1 + I_\Phi(hy))$$

(see [1] and [12]). Let for any $x \in l_\Phi^0$, $K_\Phi(x) = \{k > 0 : \| x \|^0 = \frac{1}{k}(1 + I_\Phi(kx))\}$. Since $\Phi \in \bar{\delta}_2$, $M = \sup\{K_\Phi(x) : x \in S(l_\Phi^0)\} < \infty$ (see [1]). Using again $\Phi \in \bar{\delta}_2$ there is $\theta \in (0, 1)$ such that $\Phi(\lambda u) \le (1 - \theta)\lambda\Phi(u)$, whenever $\lambda \in \left[0, \frac{M}{M+1}\right]$ and $|u| \le \Phi(M)$ (see [1] and [8]). Since $h > 1$ and $k > 1$, we have $\frac{h}{k+h} < \frac{M}{M+1}$ and $\frac{k}{k+h} < \frac{M}{M+1}$, and consequently

$$\begin{aligned}
\| x + y \|^0 &\le \frac{k+h}{kh}\left(1 + I_\Phi\left(\frac{kh}{k+h}(x+y)\right)\right) \\
&= \frac{k+h}{kh}\left(1 + I_\Phi\left(\frac{h}{k+h}kx\right) + I_\Phi\left(\frac{k}{k+h}hy\right)\right) \\
&\le \frac{k+h}{kh}\left(1 + \frac{h}{k+h}(1-\theta)I_\Phi(kx) + \frac{h}{k+h}(1-\theta)I_\Phi(hy)\right) \\
&= \frac{1}{k}(1 + I_\Phi(kx)) + \frac{1}{h}(1 + I_\Phi(hy)) - \frac{\theta}{k}I_\Phi(kx) - \frac{\theta}{h}I_\Phi(hy) \\
&\le 2 - \theta I_\Phi(x) - \theta I_\Phi(y).
\end{aligned}$$

By $\Phi \in \delta_2$, there exists $\delta > 0$ such that $I_\Phi(x) \ge \delta$. Hence $\| x + y \|^0 \le 2 - 2\delta\theta$.

$\square$

Corollary 5 An Orlicz sequence space l_{Φ}^{0} is WNUS if and only if it is reflexive.

Proof Since a Banach space X is WNUS if and only if it is reflexive and $R(X) < 2$ (see [6]) and $R(l_{\Phi}^{0}) < 2$ if and only if l_{Φ} is reflexive (see Corollary 4) the result follows.

Corollary 6 An Orlicz sequence space l_{Φ}^{0} has the fixed point property if and only if it is reflexive.

Proof By Corollary 2, we know that if l_{Φ}^{0} is reflexive, then l_{Φ}^{0} has the fixed point property. On the other hand, we can use the same method to that it was used in the proof of Theorem 5 in [4] to finish the proof.

Acknowledgement

This work was carried out when the first author visited the Department of Mathematics of the University of Newcastle in Australia. He wish to thank the Department for their very good hospitality.

References

1. S.T. Chen, *Geometry of Orlicz Spaces,* Dissertationes Mathematicae, **356**, Warszawa, 1996.
2. S.T. Chen, H. Hudzik and H.Y. Sun, *Complemented copies of l^1 in Orlicz spaces,* Math. Nachr. **158** (1997), 299–309.
3. J. Diestel, *Sequences and Series in Banach Spaces,* Graduate Texts in Mathematics, Springer-Verlag, 1984.
4. P.R. Dowling, C.J. Lennard and B. Turett, *Reflexivity and the fixed-point property for nonexpansive maps*, J. Math. Anal. Appl., **200** (1996), 653–662.
5. J. García-Falset, *Stability and fixed points for nonexpansive mappings,* Houston J. Math. **20** (1994), 495–505.
6. J. García-Falset, *The fixed point property in Banach spaces with NUS property,* Nonlinear Anal., (to appear).
7. R. Goebel and W.A. Kirk, *Topics in Metric Fixed Point Theory,* Cambridge University Press, 1990.
8. H. Hudzik, *Uniformly non-$l_n^{(l)}$ Orlicz spaces with Luxemburg norm,* Studia Math. **81** (3) (1985), 271–284.
9. L.V. Kantorovich and G.P. Akilov, *Functional Analysis,* Nauka Moscow, 1977 (in Russian).
10. J. Musielak, *Orlicz Spaces and Modular Spaces,* Lecture Notes in Math. **1034**, 1983.
11. S. Prus, *Nearly uniformly smooth Banach spaces,* Boll. U.M.I. **7** 3-B (1989), 507–521.
12. M.M. Rao and Z.D. Ren, *Theory of Orlicz Spaces,* Marcel Dekker Inc., New York, Basel, Hong Kong 1991.
13. B. Sims and M. Smith, *On non-uniform conditions giving weak normal structure,* Questiones Math., **18** (1995), 9–19.
14. K. Tan and H. Xu, *On fixed point theorems of nonexpansive mappings in product spaces,* Proc. Amer. Math. Soc., **113** (1991), 983–989.

Function Spaces and Applications
D.E. Edmunds et al (Eds)

6. On Some Fundamental Properties of the Maximal Operator

Alberto Fiorenza*[1] and Miroslav Krbec†[2]

[1]Dipartmento di Matematica e Applicazioni "R. Caccioppoli" via Cintia, 80126 Napoli, Italy.

[2]Institute of Mathematics, Academy of Sciences of the Czech Republic, Žitná 25, 115 67 Prague 1, Czech Republic

1. Background Remarks

The maximal function, or, in other terms, the maximal operator and its various clones has become one of the essential tools in many areas of real analysis. In view of fact that the maximal function estimates many important integral operators and it is relatively easier to be handled with at the same time it is natural to study its behaviour in various function spaces of integrable functions. As is well known the systematic study of the maximal function and related operators has been pursued since the celebrated works by Hardy and Littlewood, and by Wiener in the 1930s, great effort has been done to understand fully its behaviour in weighted spaces of integrable functions, with the well known milestones in the 1970s, when the theory of A_p weights has been developed (Muckenhoupt etc.)

Here we shall turn our attention to basic properties of the maximal operator having in mind conditions ensuring that a function belongs to its domain or to its range. To be more specific, we shall deal with the problem of the strength of various assumptions on the finiteness of the maximal function at a single point to L_r integrability, where $0 < r < 1$, and questions about the range of the maximal operator under minimal necessary hypothesis about the space on which it acts, employing spaces obtained by various extrapolation procedures, in particular, $L^{p)}$ spaces (Iwaniec and Sbordone [17]) and the logarithmic Lebesgue spaces (Edmunds and Triebel [7]).

We shall use standard notation L^p for the Lebesgue spaces; we shall denote by $| E |$ the Lebesgue measure of a set $E \subset \mathbb{R}^N$ and by $B_R(x)$ the open ball centered at x with radius R.

---

*This work has been performed as a part of a National Research Project and partly supported by G.N.A.F.A.

†The research of the second author was partly supported by Grant No. 201/96/0431 of GA ČR.

For $f \in L^1_{\mathrm{loc}}(\mathbb{R}^N)$, $N \geq 1$ we consider the *(global, centered) maximal function*

$$Mf(x) = \sup_{R>0} \frac{1}{|B_R(x)|} \int_{B_R(x)} |f(y)|\, dy, \qquad x \in \mathbb{R}^N,$$

and we denote

$$\mathbb{D} = \{f \in L^1_{\mathrm{loc}}(\mathbb{R}^N);\ Mf \neq \infty\}.$$

In view of Theorem 2.2 and the sublinearity of the maximal operator, $\mathbb{D}$ is a linear subspace of $L^1_{\mathrm{loc}}(\mathbb{R}^N)$. We shall refer to it as the *domain* of the maximal operator.

Let us note that the local integrability of a function is not sufficient for its memberaship in $\mathbb{D}$; a standard example is, for instance, the function $x \to \| x \|$ in $\mathbb{R}^N$.

It is elementary to verify that if one uses balls or cubes containing x (centered or off-centered with respect to x), the cubes having their sides parallel to coordinate axes, or, more generally, even families of regular sets instead of balls in the definition of the maximal function, then the result is an equivalent operator [6].

For simplicity and without loss of generality we shall work only with non-negative functions so that we shall omit the absolute values everywhere.

The relations of $\mathbb{D}$ to the Lebesgue spaces is by no means straightforward; it is immediate that $L^1(\mathbb{R}^N) \subset \mathbb{D}$ but we shall give examples showing that

$$L^1_{\mathrm{loc}}(\mathbb{R}^N) \cap \bigcup_{0<r<1} L^r(\mathbb{R}^N) \not\subset \mathbb{D} \not\subset L^1(\mathbb{R}^N) + L^\infty(\mathbb{R}^N).$$

Let us recall some known properties of Mf when $f \in \mathbb{D}$. We shall first survey known properties of the maximal operator. Let $f \in \mathbb{D}$. Then Mf is measurable lower semicontinuous function. By the Lebesgue Differentiation Theorem, $f \leq Mf$ and if $N > 2$ an equality can hold even without f being a constant (this result is true for the *spherical* maximal operator; see [8]). In terms of membership in various function spaces we recall that if $f \not\equiv 0$ a.e. in $\mathbb{R}^N$ then $Mf \notin L^r(\mathbb{R}^N)$ $\forall\, 0 < r \leq 1$ and $\log(1 + Mf) \notin L^1(\mathbb{R}^N)$ (since there is $c > 0$ such that $Mf(x) \geq c/|x|^N$ for all $|x| > 1$, cf. e.g. [4], p. 117). The behaviour of M in Lebesgue spaces has been studied thoroughlyu: M acts continuously from $L^p(\mathbb{R}^N)$ into $L^p(\mathbb{R}^N)$ if $1 < p \leq \infty$ and from $L^1(\mathbb{R}^N)$ into the weak-$L^1(\mathbb{R}^N)$. Passing to finer scales of spaces, it is possible to find criteria for the boundedness of M in Lorentz and Orlicz spaces, in particular then in Lorentz-Zygmund spaces. A curious phenomenon occurs, namely, a deterioration of the behaviour of M near L^1, expressed in the classical form of the Zygmund inequality, a feature common to other classical operators. This is now well understood also in a more abstract setting as a general property of operators of weak type (1, 1) or its generalizations. From numerous references let us recall at lest [3, 4, 13, 16, 19, 29]. To refine the situation near L^1 we shall make employ the so called grand L^p spaces (introduced in [17]) and logarithmic Lebesgue spaces (see [7]).

After that we shall turn our attention to the properties of the Mf when $f \in L^1(\mathbb{R}^N) + L^\infty(\mathbb{R}^N)$. Among the consequences, we get for instance that for every $0 < r, s < 1$, $\alpha > 0$ we have

$$f \in L^r(\mathbb{R}^N) \cap L^1_{\text{loc}}(\mathbb{R}^N) \text{ and } Mf \in L^s(\{Mf > \alpha\}) \Rightarrow f \in L^1(\mathbb{R}^N).$$

The concluding part is devoted to the local maximal operator $M_\Omega f$ on open bounded sets Ω in $\mathbb{R}^N$. This operator turns out to be extremely useful in many areas, for instance, in the theory of quasiconformal mappings ([2]), when proving the regularity of solutions of P.D.E. or studying minimizers of functionals of the Calculus of Variations (for instance, see [11, 12]), or, recently, when studying homogenization problems without regularity in the coefficients [5]. We shall present a characterization of the domain $\mathbb{D}_\Omega$ of the local maximal operator and we shall see some examples, for instance, functions in $\mathbb{D}_\Omega \setminus L^1_{\text{loc}}(\Omega)$.

2. Domain of the Maximal Operator

In 1939 Wiener [31] found a subspace of $L^1_{\text{loc}}(\mathbb{R}^N)$ contained in $\mathbb{D}$. He proved the following:

Theorem 2.1 If $f \in L^1(\mathbb{R}^N)$, then $Mf < \infty$ *a.e.* in $\mathbb{R}^N$.

Plainly $L^\infty(\mathbb{R}^N) \subset \mathbb{D}$ so that by the sublinearity of the maximal operator $L^1(\mathbb{R}^N) + L^\infty(\mathbb{R}^N) \subset \mathbb{D}$. Therefore the Lebesgue $L^p(\mathbb{R}^N)$ $(1 \le p \le \infty)$, Orlicz $L_A(\mathbb{R}^N)$ (A Young function) and Lorentz spaces are subsets of $\mathbb{D}$. We shall show that $\mathbb{D}$ is effectively larger than $L^1(\mathbb{R}^N) + L^\infty(\mathbb{R}^N)$, but it does not contain any space of the type $L^1_{\text{loc}}(\mathbb{R}^N) \cap (L^r(\mathbb{R}^N) + L^\infty(\mathbb{R}^N))$ where $0 < r < 1$.

We begin with the following characterization of $\mathbb{D}$, whose proof can be found in [10].

Theorem 2.2 Let $f \in L^1_{\text{loc}}(\mathbb{R}^N)$. Then the following statements are equivalent:

(i) there exists $x_0 \in \mathbb{R}^N$ such that $Mf(x_0) < \infty$;

(ii) there exists $x_0 \in \mathbb{R}^N$ such that

$$\limsup_{R\to\infty} \frac{1}{|B_R(x_0)|} \int_{B_R(x_0)} f(y)\,dy < \infty;$$

(iii) there exists $K > 0$ such that

$$\limsup_{R\to\infty} \frac{1}{|B_R(x_0)|} \int_{B_R(x_0)} f(y)\,dy = K < \infty$$

for every $x_0 \in \mathbb{R}^N$;

(iv) $Mf(x) < \infty$ a.e. in $\mathbb{R}^N$;

(v) $f \in \mathbb{D}$.

Remark 2.3 Let Ω be an open set in $\mathbb{R}^N$. The John-Nirenberg space BMO(Ω) is defined as the space of the measurable functions f such that

$$\|f\|_{\text{BMO}(\Omega)} = \sup_{Q\subset\Omega} \frac{1}{|Q|} \int_Q \left| f(x) - \frac{1}{|Q|} \int_Q f(y)\,dy \right| dx < +\infty,$$

where the supremum is taken over all cubes Q with sides parallel to the coordinate axes.

We observe that the implication (i) $\Rightarrow$ (iv) from the previous theorem has been shown to hold for $f \in \mathrm{BMO}(\mathbb{R}^N) \cap \mathbb{D}$ in the book by Bennett-Sharpley (cf. [4], p. 399–400), by using the inequality

$$\frac{1}{|Q|}\int_Q Mf(x)\,dx \le c\,\|f\|_{\mathrm{BMO}(\mathbb{R}^N)} + \inf_{x\in Q} Mf(x), \tag{2.1}$$

true for every Q cube in $\mathbb{R}^N$, $f \in \mathrm{BMO}(\mathbb{R}^N)$. The proof in [10] states this interesting property generally for $f \in L^1_{\mathrm{loc}}(\mathbb{R}^N)$ and it is also by far simpler than that using (2.1). Notice that $\mathbb{D} \not\subset \mathrm{BMO}(\mathbb{R}^N)$ (see next Example 2.8). For another (not elementary) proof of (i) $\Rightarrow$ (iv) for $f \in L^1_{\mathrm{loc}}(\mathbb{R}^N)$ we rever to Wik [30].

It is also $\mathrm{BMO}(\mathbb{R}^N) \not\subset \mathbb{D}$; it suffices to consider $f(x) = |\log|x||$ (cf. [4], p. 400).

Let us return to the promise from Section 1, concerning relations of $\mathbb{D}$ and the Lebesgue spaces: We present two examples showing that if $0 < r < 1$, then we have $L^1_{\mathrm{loc}}(\mathbb{R}^N) \cap L^r(\mathbb{R}^N) \not\subset \mathbb{D} \not\subset L^1(\mathbb{R}^N) + L^\infty(\mathbb{R}^N)$. Notice that the functions f we are going to consider are supported on sets of finite measure, therefore, every level set $\{f(x) > \alpha\}$, $\alpha \ge 0$, is of finite measure.

Example 2.4 Let $A_n = \{n-1 < |x| < n\}$, $n \in \mathbb{N}$ and let F_n be any measurable subset of A_n such that $|F_n| = 2^{-n}$. Put $f = \sum_{n=1}^{\infty} a_n \chi_{F_n}$ with $a_n = 2^n$. Then $f \in L^1_{\mathrm{loc}}(\mathbb{R}^N)$, $f \notin L^1(\mathbb{R}^N) + L^\infty(\mathbb{R}^N)$. At the same time $f \in \mathbb{D}$, in fact,

$$\frac{1}{|B_\rho(0)|}\int_{B_\rho(0)} f(y)\,dy \le \frac{1}{\omega_N \rho^N}\sum_{n=1}^{[\rho+1]} a_n |F_n|, \qquad \rho > 0,$$

where $[\cdot]$ denotes here the integer part and therefore we have

$$\limsup_{\rho\to\infty} \frac{1}{|B_\rho(0)|}\int_{B_\rho(0)} f(y)\,dy < \infty,$$

from which $f \in \mathbb{D}$ by Theorem 2.2. Notice that since $\int_{\mathbb{R}^N} (f(y))^r\,dy = \sum_{n=1}^{\infty} a_n^r |F_n| < \infty$, we have $f \in L^r(\mathbb{R}^N)$ for all $0 < r < 1$.

Example 2.5 Let us put $a_n = (2^{(r+1)/2r})^n$ with some fixed $0 < r < 1$ in the previous example. Then $f \in L^1_{\mathrm{loc}}(\mathbb{R}^N) \cap L^r(\mathbb{R}^N)$, $f \notin \mathbb{D}$. Note that $f \notin L^1(\mathbb{R}^N)$.

Remark 2.6 Let Ω be an open set in $\mathbb{R}^N$. The space $\mathrm{BLO}(\Omega)$ is defined as the subspace of $\mathrm{BMO}(\Omega)$ of the functions f such that

$$\sup_{Q\subset\Omega}\left(\frac{1}{|Q|}\int_Q f(x)\,dx - \operatorname*{ess\,inf}_Q f\right) < +\infty.$$

It is possible to prove that (cf. [4], (8.38) p. 400; see also [29], p. 204)

$$M : \mathbb{D} \cap \mathrm{BMO}(\mathbb{R}^N) \to \mathrm{BLO}(\mathbb{R}^N) \subset \mathrm{BMO}(\mathbb{R}^N),$$

and that such operator is subjective, i.e.

$$\mathrm{BLO}(\mathbb{R}^N) = M(\mathbb{D} \cap \mathrm{BMO}(\mathbb{R}^N)) + L^\infty(\mathbb{R}^N) \tag{2.2}$$

Remark 2.7 The Sobolev sonces are in $\mathbb{D}$, but also functions in $L^1_{\mathrm{loc}}(\mathbb{R}^N)$ with weak derivatives in $L^p(\mathbb{R}^N)$, $1 \le p < N$ are in $\mathbb{D}$ since they, up to a constant, belong to some Lebesgue or Orlicz space (cf. [25], p. 40 or [21], p. 424).

Example 2.8 A natural question arises, namely, whether $\mathbb{D}$ is connected in some way with the measure of level sets of a function in question, and, consequently, whether the rearrangement invariant spaces can be used to characterizations of $\mathbb{D}$ (cf. also comments preceding Example 2.4). Functions in $\mathbb{D}$ can be very bad indeed, for instance, the measure of every level set can be infinite, hence these functions cannot be rearranged (as to general references about rearrangements see for instance [1, 18, 23]. Let us have a look at the following one-dimensional example. Put

$$f(t) = \sum_{n=1}^{\infty} n\chi_{(n^3, n^3+1)}(t), \quad t \in \mathbb{R}^1.$$

Clearly $f \in L^1_{\mathrm{loc}}(\mathbb{R}^1)$ and

$$0 \le \lim_{t\to\infty} \int_{-t}^{t} f(y)\,dy \le \lim_{n\to\infty} \frac{1}{2n^3} \int_{-n^3-1}^{n^3+1} f(y)\,dy = \lim_{n\to\infty} \frac{n(n+1)}{4n^3} = 0,$$

therefore, $f \in \mathbb{D}$ by Theorem 2.2. Notice that the measure of every level set of f is infinite, and, since $f \le Mf$, also the measure of every level set of Mf is infinite. Finally, it is $f \notin \mathrm{BMO}(\mathbb{R}^1)$.

3. Range of the Maximal Operator

Let us start with surveying the properties of Mf. If $f \in \mathbb{D}$ it may happen that $Mf \notin L^1_{\mathrm{loc}}(\mathbb{R}^N)$. The following example is well-known:

Example 3.1 Let us put $f(x) = \chi_{]0,1/2[}(x)/(x \log^2 x)$, $x \in \mathbb{R}^1$a.e. (cf. [4], p. 118 or [29], p. 77). Then $f \in \mathbb{D}$, $Mf \notin L^1_{\mathrm{loc}}(\mathbb{R}^N)$. We observe that $Mf \in L^1(\log L)^{-1}$ $(]0, 1/2[)$.

Example 3.1 shows that also starting from $f \in L^1(\mathbb{R}^N)$ generally we have not $Mf \in L^1_{\mathrm{loc}}(\mathbb{R}^N)$. As we are going to see, in this case it is possible to find optimal spaces to which Mf belongs.

First of all, by virtue of the Kolmogorov inequality (cf. e.g. [24], Theorem 6.1, p. 190 or [28], p. 43)

$$\| Mf \|^r_{L^r(A)} \le \frac{c(N) \mid A \mid^{1-r}}{1-r} \| f \|^r_{L^1(\mathbb{R}^N)} \quad \forall f \in L^1(\mathbb{R}^N), \quad r \in]\,0, 1[, \tag{3.1}$$

true for every set $A \subset \mathbb{R}^N$, $|A| < \infty$, the function Mf is in every $L^r(A)$ with $0 < r < 1$ provided $f \in L^1(\mathbb{R}^N)$, where A is any subset of $\mathbb{R}^N$ with finite measure. Furthermore, Mf belongs to any space on A resulting from an extrapolation procedure based on (3.1), which preserves a finite (quasi) norm on the right hand side. The extrapolation theory as developed up to now offers two reasonable candidates. The first is given by using an analogue of the approach considered for $p > 1$ in [7]

$$\int_0^{\varepsilon_0} \varepsilon^{\sigma-1} \| Mf \|_{L^{1-\varepsilon}(A)} \, d\varepsilon \le c(N, |A|, \varepsilon_0, \sigma) \, \|f\|_{L^1(\mathbb{R}^N)}, \tag{3.2}$$

where $\varepsilon_0 \in (0, 1)$ is chosen arbitrarily, $\sigma > 1$ is a parameter, or by using its discrete variant, the Σ-method due to Milman [22]. Going along the lines of the proof in [7], where spaces $L^p(\log L)^{-\sigma}(A)$ with $1 < p < \infty$ are considered, it is not difficult to prove that the quasinorm on the left hand side of (3.2) is equivalent to the quasinorm in the (generalized) Orlicz space $L^1(\log L)^{-\sigma}(A)$. As a result we have the following

Proposition 3.2 If $f \in L^1(\mathbb{R}^N)$, then $Mf \in \bigcap_{\sigma>1} L^1(\log L)^{-\sigma}(A)$ for every $A \subset \mathbb{R}^N$ of finite measure.

The above assertion is optimal in the framework of logarithmic Lebesgue spaces as one can see from the next example.

Example 3.3 Let us consider the function

$$f(x) = \frac{1}{x\,|\log x|\,\log^2|\log x|}\,\chi_{]0,a[}(x), \quad x \in \mathbb{R}^1 \text{ a.e.,}$$

where $a = \exp(-\exp(1))$. Then $f \in L^1(\mathbb{R}^N)$, but on the other hand we have

$$Mf\left(\frac{x}{2}\right) \ge \frac{1}{x}\int_0^x f(t)\,dt = \frac{1}{x\log|\log x|}, \; x \in \left]0, \frac{a}{2}\right[,$$

and therefore, since $\log|\log x| > 1 \;\forall\, x \in]0, a/2[$, we get

$$\frac{Mf\left(\frac{x}{2}\right)}{\log\left(e + Mf\left(\frac{x}{2}\right)\right)} \ge \frac{\dfrac{1}{x\log|\log x|}}{\log\left(e + \dfrac{1}{x\log|\log x|}\right)}$$

$$\ge \frac{1}{x\log|\log x|\log\left(e + \dfrac{1}{x}\right)} \ge \frac{1}{x\log|\log x|\log\dfrac{2}{x}}$$

$$\ge \frac{1}{2x|\log x|\log|\log x|},$$

from which $Mf \notin L^1(\log L)^{-1}(]0, a[)$.

Another estimate following from (3.1) results in establishing a bound for the quasinorm of Mf in $L^{1)}(A)$, the grand L^1 space—a particular case of spaces introduced by Iwaniec and Sbordone [17] and investigated in details e.g. in Greco [15]:

$$\| Mf \|_{L^{1)}(A)} \leq c(N, |A|) \, \| f \|_{L^1(\mathbb{R}^N)}.$$

Let us recall that the quasinorm in $L^{1)}(A)$ is given by

$$\| f \|_{L^{1)}(A)} = \sup_{0<\varepsilon<\varepsilon_0} \left(\varepsilon \frac{1}{|A|} \int_A |f(y)|^{1-\varepsilon} \, dy \right)^{1/(1-\varepsilon)}$$

where $\varepsilon_0 \in (0, 1)$; observe that only values of ε near zero are relevant. We have:

Proposition 3.4 If $f \in L^1(\mathbb{R}^N)$, then $Mf \in L^{1)}(A)$ for every $A \subset \mathbb{R}^N$ of finite measure.

Let us observe that this extrapolation gives a better result in terms of inclusions of functions spaces since

$$L^1(\log L)^{-1}(A) \subset L^{1)}(A) \subset \bigcap_{\sigma>1} L^1(\log L)^{-\sigma}(A)$$

for every A of finite measure.

The following Theorem 3.5 gives a number of equivalent conditions for the range of Mf. We refer to [10] for the proof.

Theorem 3.5 Let $f \in L^1_{\text{loc}}(\mathbb{R}^N)$ and $\varphi : [0, \infty[\to [0, \infty[$, φ strictly increasing, $\varphi(\infty) = \infty$, $\lim_{t\to\infty} \varphi(t)/t^s = 0$ for some $0 < s < 1$. Then the following statements are equivalent:

(i) $f \in L^1(\mathbb{R}^N) + L^\infty(\mathbb{R}^N)$;
(ii) there is $\alpha > 0$ such that $f \in L^1(\{f > \alpha\})$;
(iii) there is $\alpha > 0$ such that $|\{Mf > \alpha\}| < \infty$;
(iv) there is $\alpha > 0$ and $0 < r < 1$ such that $Mf \in L^r(\{Mf > \alpha\})$;
(v) there is $\alpha > 0$ such that $Mf \in L^r(\{Mf > \alpha\})$ for all $0 < r < 1$;
(vi) there is $0 < r < 1$ such that $Mf \in L^r(\mathbb{R}^N) + L^\infty(\mathbb{R}^N)$;
(vii) $Mf \in \bigcap_{0<r<1} L^r(\mathbb{R}^N) + L^\infty(\mathbb{R}^N)$;
(viii) $\varphi(Mf) \in L^1(\mathbb{R}^N) + L^\infty(\mathbb{R}^N)$;
(ix) there is $\alpha > 0$ such that $|\{Mf > \alpha\}| < \infty$ and

$$Mf \in L^{1)}(\{Mf > \alpha\}) + L^\infty(\mathbb{R}^N);$$

(x) there is $\alpha > 0$ such that $|\{Mf > \alpha\}| < \infty$ and

$$Mf \in \bigcap_{\sigma>1} L^1(\log L)^{-\sigma}(\{Mf > \alpha\}) + L^\infty(\mathbb{R}^N).$$

Remark 3.6 In Example 2.8 we saw that if $f \in \mathbb{D}$, then the measure of every level set of Mf can still be infinite—it was sufficient to consider a function f

having the same property. The next theorem shows among others that such phenomenon may occur even if the measure of every level set of, f is finite (namely, when $f \notin L^1(\mathbb{R}^N) + L^\infty(\mathbb{R}^N)$, as in Example 2.4.

The condition (ii) in Theorem 3.5 says that f is integrable over a special set of a finite measure. Examples 2.4 and 2.5 show that this cannot be replaced by integrability of f over any set of finite measure. Furthermore, the condition (iii) in Theorem 3.5 implies that all level sets of the maximal functions in the examples recalled are of infinite measure.

Remark 3.7 Let us observe that if $f \in L^{1-\varepsilon}(\mathbb{R}^N) \cap L^1_{\mathrm{loc}}(\mathbb{R}^N)$ for some $\varepsilon \in (0, 1)$ and $Mf \in L^r(\{Mf > \alpha\})$ for some $0 < r < 1$ and $\alpha > 0$, then by virtue of Theorem 3.5 we get $f \in L^1(\mathbb{R}^N)$.

For completeness we state here a well-known variant of Stein's $L \log L$ theorem [26, 27] for the maximal function.

Theorem 3.8 Let $f \in L^1_{\mathrm{loc}}(\mathbb{R}^N)$ and $\sigma \geq 1$. Then the following statements are equivalent:

(i) $f[\log(1 + f)]^\sigma \in L^1(\mathbb{R}^N) + L^\infty(\mathbb{R}^N)$;
(ii) there is $\alpha > 0$ such that $f[\log(1 + f)]^\sigma \in L^1(\{f > \alpha\})$;
(iii) there is $\alpha > 0$ such that $Mf \in L^1(\log L)^{\sigma-1}(\{f > \alpha\})$;
(iv) $Mf \in L^1(\log L)^{\sigma-1}(\mathbb{R}^N) + L^\infty(\mathbb{R}^N)$.

The preceding theorems and examples make it possible to present a scheme of some of the sets of functions we have considered; we shall order the conditions according to their strength:

$$
\begin{array}{ccc}
f \text{ measurable in } \mathbb{R}^N & \rightarrow & f \notin L^1_{\mathrm{loc}}(\mathbb{R}^N) \\
\downarrow & & \\
f \in L^1_{\mathrm{loc}}(\mathbb{R}^N) & \rightarrow & f \notin \mathbb{D} \\
\downarrow & (\text{Ex. } 2.5) & \\
f \in \mathbb{D} & \rightarrow & |\{f > \alpha\}| = \infty \ \forall \ \alpha \geq 0 \\
\downarrow & (\text{Ex. } 2.8) & \\
\exists \ \alpha > 0 : |\{f > \alpha\}| < \infty & \rightarrow & |\{Mf > \alpha\}| = \infty \ \forall \ \alpha \geq 0 \\
\downarrow & (\text{Ex. } 2.4) & \\
\exists \ \alpha > 0 : |\{Mf > \alpha\}| < \infty & \rightarrow & Mf \notin \dfrac{L^1}{\log L}(\{f > \alpha\}) \ \forall \ \alpha \geq 0 \\
\downarrow & (\text{Ex. } 3.3) & \\
\exists \ \alpha > 0 : Mf \in \dfrac{L^1}{\log L}(\{f > \alpha\}) & \rightarrow & Mf \notin L^1(\{f > \alpha\}) \ \forall \ \alpha \geq 0 \\
\downarrow & (\text{Ex. } 3.1) & \\
\exists \ \alpha > 0 : Mf \in L^1(\{f > \alpha\}) & &
\end{array}
$$

4. On the Local Maximal Function

For the rest of the paper let Ω be an open bounded subset of $\mathbb{R}^N$. Functions in Ω will be assumed to be measurable and non-negative.

The local maximal function of f is defined by

$$M_\Omega f(x) = \sup_{\substack{Q \ni x \\ Q \subset \Omega \\ Q \text{cube}}} \frac{1}{|Q|} \int_Q f(y)\, dy, \qquad x \in \Omega,$$

where edges of cubes Q are parallel with coordinate axes. If $\Omega \neq \mathbb{R}^N$, then M_Ω preserves only some of the properties of M.

We shall first consider the case $\Omega = Q_0$, cube in $\mathbb{R}^N$. Putting

$$\bar{f} = \begin{cases} f & \text{in } Q_0, \\ 0 & \text{in } \mathrm{R}^N \setminus Q_0, \end{cases}$$

it is easy to prove that

$$M_{Q_0} f = (M_{\mathbb{R}^N} \bar{f})_{|Q_0}.$$

As a consequence of Theorem 2.2, we have

Proposition 4.1 It is $M_{Q_0} f < \infty$ a.e. in Q_0 if and only if $f \in L^1(Q_0)$.

The considerations of Section 3 make it possible to draw some conclusion about the range of M_{Q_0}: if $f \in L^1(Q_0)$, then we have $Mf \in L^{(1)}(Q_0)$ and $Mf \in \bigcap_{\sigma > 1} L^1(\log L)^{-\sigma}(Q_0)$. In particular, $Mf \in \bigcap_{0<r<1} L^r(Q_0)$. Moreover, Example 3.1 shows that Mf need not be locally integrable in Q_0.

Let us recall some known results about the range of M_{Q_0} when f is "better" than $L^1(Q_0)$: by the Stein's $L \log L$ result [26], $f \in L \log L\, (Q_0)$ iff $M_{Q_0} f \in L^1(Q_0)$ and f belongs to the Orlicz space $L_A(Q_0)$, where $\inf_{t>0} \dfrac{tA'(t)}{A(t)} > 1$, iff M_{Q_0} belongs to the same space. For other imbedding properties of the maximal operator see [14, 20, 29].

We shall turn our attention to the local maximal function with respect to a general domain Ω. In this case M_Ω is different from $(M_{\mathbb{R}^N} \bar{f})_{|\Omega}$. In general $M_\Omega f \leq (M_{\mathbb{R}^N} \bar{f})_{|\Omega}$ and these functions need not be equivalent. We shall give an example:

Example 4.2 Let $N = 2$,

$$\Omega = \{z = (x, y);\, |\, z - 1\,| < 1\}$$

and let $f = 1$ in $\Omega \cap \{y > 2/\sqrt{5}\}$ and 0 otherwise in Ω. Then $(M_\Omega f)_{|\Omega \cap \{y \leq 0\}} = 0$ while $(M_{\mathbb{R}^N} \bar{f})_{|\Omega \cap \{y \leq 0\}} > 0$.

Hence $\bar{f} \in \mathbb{D}$ implies $f \in \mathbb{D}_\Omega$. The next example shows, however, that the converse is not necessarily true. A necessary and sufficient condition for a function f to be in $\mathbb{D}_\Omega$ will be given in Theorem 4.5 below.

Example 4.3 Let us consider upper part of Fig. 1, that is, let $N = 2$ and consider a sequence of open cubes $Q_1, Q_2, \ldots$, laying on the positive axe x, and such that the right vertical side of Q_{k+1} is subset of the left vertical side of Q_k and the left upper corners of all Q_k lay on a line $y = kx$ with some $k > 0$. Let Ω be the triangle domain whose boundary is contained in the positive axe x, the line $y = kx$, and the line containing the right vertical side of Q_1. Denote by $Q_1/2, Q_2/2, \ldots$, concentric cubes with sidelength equal to the half of the sides of $Q_1, Q_2, \ldots$ Let (a_i), be any sequence of positive real numbers satisfying

$$\sum_{i=1}^{\infty} a_i \mid Q_i \mid = \infty.$$

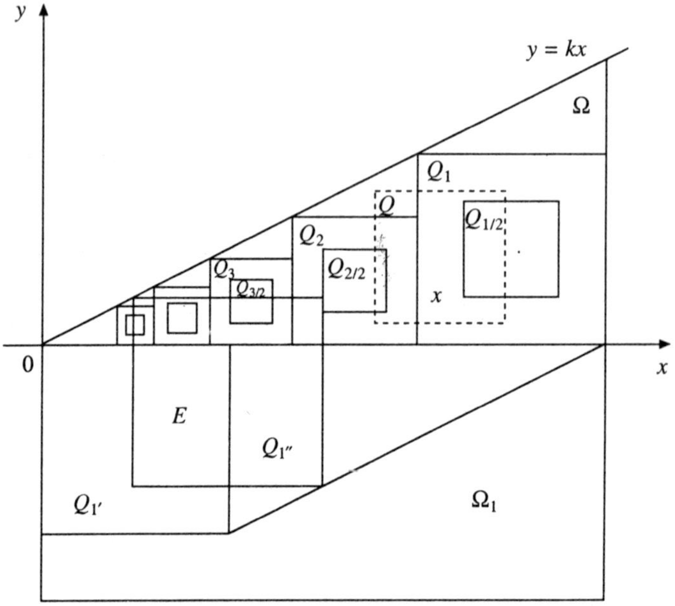

Fig. 1

Put
$$f = \sum_{i=1}^{\infty} a_i \chi_{Q_i/2}.$$

Then f is supported in a compact set and $f \notin L^1(\Omega)$. If we fix $x \in \Omega$, then every cube Q such that $Q \ni x$, $Q \subset \Omega$, intersect at most two of the cubes Q_i, thus M_Ω is finite a.e.

Hence f need not be integrable over every compact subset of its support in order to have $M_\Omega f < \infty$ a.e. in Ω. Of course f must be integrable over cubes contained in Ω, but this is, however, not sufficient for $M_\Omega f < \infty$ a.e. in Ω. Indeed, f can be integrable over cubes in Ω, and still M_Ω need not be a.e. finite. This is demonstrated by the following:

Example 4.4 Let Ω_1 be Ω from Example 4.3 united with a rectangle pasted from below to Ω, with the left vertical side on the axe y and the upper horizontal side on the axe x and with the appropriate part of the positive axe x. Put

$$f = \sum_{i=1}^{\infty} \frac{i}{|Q_i|} \chi_{Q_i/2}.$$

Let Q_1' be the translation of Q_1 having the left upper corner on the origin, and $Q_1'' \neq Q_1'$ be any fixed translation of Q_1, contained in Ω_1, such that the left upper corner of Q_1'' stays on the line $y = kx$ and such that the set $E = Q_1' \cap Q_1''$ has positive measure. Then for every Q_i, i sufficiently large, there exists a translation of Q_1 containing $E \cup Q_i$, therefore, if $x \in E$, since

$$M_\Omega f(x) \geq \frac{1}{|Q_1|} \int_{Q_i} f(y)\, dy = \frac{i}{4\,|Q_1|},$$

we have $M_\Omega f = \infty$ in E. Hence $M_\Omega f$ is not finite a.e. On the other hand f is integrable over every cube contained in Ω.

The foregoing two examples indicate a crucial point: in both cases there we have a sequence of cubes such that the averages of f blow up. In the former case the edges tend to zero and this is not necessarily the case in the latter example. The following theorem holds true (for the proof see [10]; the proof can be modified to work in the case of unbounded Ω, too.).

Theorem 4.5 Let $f \in L^1(Q)$ for all $Q \subset \Omega$. Then the following statements are equivalent:

(i) $M_\Omega f < \infty$ a.e. in Ω.

(ii) $$\sup_{\substack{|Q| > \varepsilon \\ Q \text{cube} \\ Q \subset \Omega}} \frac{1}{|Q|} \int_Q f(y)\, dy < \infty \qquad \text{for all } \varepsilon > 0.$$

Remark 4.6 If Ω is a cube Q_0, then (ii) holds if and only if $f \in L^1(Q_0)$. Furthermore, Theorem 4.5 remains to be true for $\Omega = \mathbb{R}^N$ by virtue of Theorem 2.2.

In the rest of this paper we turn our attention to the problem of the range of the local maximal function. If $f \in \mathbb{D}_\Omega$, then it is easy to see that M_Ω, enjoys the standard properties, namely, it is lower semicontinuous and $f \leq M_\Omega f$ a.e. in Ω. On the other hand, in contrast to the behaviour of $M_{\mathbb{R}^N}$ and M_{Q_0}, it is not generally true that $M_\Omega f \in L^r(\Omega)$. Let us consider, for instance, Example 4.3 with $a_i = i \exp(1/|Q_i|^2)$. Then

$$\int_\Omega f^r\, dx = \sum_{i=1}^{\infty} \frac{|Q_i|}{4} i^r \exp\left(\frac{r}{|Q_i|^2}\right) = \infty, \qquad 0 < r < 1,$$

therefore $M_\Omega f \notin L^r(\Omega)$. This example indicates that the integrability properties of M_Ω depend heavily on the geometric properties of Ω. In particular, if Ω is a cube Q_0, then $M_{Q_0} f \in L^r(Q_0)$ for all $0 < r < 1$.

Let us remark also that if Ω is a cube Q_0, in contrast to the behaviour of $M_{\mathbb{R}^N}$, $\mathrm{BMO}(Q_0) \subset \mathbb{D}_{Q_0}$ because $\mathrm{BMO}(Q_0) \subset L^1(Q_0)$, and, modulo bounded functions, $\mathrm{BLO}(Q_0)$ is exactly the range of M_{Q_0} on $\mathrm{BMO}(Q_0)$: in fact we have the following analogous of (2.2) (see [4] p. 390).

$$\mathrm{BLO}(Q_0) = M_{Q_0}(\mathrm{BMO}(Q_0)) + L^\infty(Q_0).$$

Despite of this, one can prove the following easy analogue of (x) of Theorem 3.5 (see [10]):

Theorem 4.7 Let $f \in L^1(\Omega)$. Then $M_\Omega f \in L^{(1)}(\Omega)$. Therefore, $M_\Omega f \in \bigcap_{\sigma>1} L^1(\log L)^{-\sigma}(\Omega)$.

Proof Let $Q_0 \supset \Omega$ be a cube and put $\bar{f}(x) = f(x)$ for $x \in \Omega$ and $\bar{f}(x) = 0$ in $\mathbb{R}^N \setminus \Omega$. The claims now follow since $M_\Omega f \leq M_{Q_0}\bar{f} = (M_{\mathbb{R}^N}\bar{f})\,|_{Q_0}$ and $f \in L^1(Q_0)$. $\square$

Remark 4.8 If Ω is a cube Q_0 and $M_{Q_0} f \in L^1(\log L)^{-1}(Q_0)$, then the L^1 norm of f can be estimated as follows [9]:

$$\int_{\Omega_0} f(x)\,dx \leq 2^{N+1} \int_{\Omega_0} \frac{M_{Q_0} f(x)}{\log\left(e + \dfrac{M_{Q_0} f(x)}{|M_{Q_0} f(x)|_{Q_0}}\right)}\,dx,$$

where $|M_{Q_0} f(x)|_{Q_0} = \dfrac{1}{|Q_0|}\displaystyle\int_{Q_0} M_{Q_0} f(x)\,dx$.

References

1. C. Bandle, *Isoperimetric Inequalities and Applications*. Monographs and Studies in Math. 7, Pitman, London 1980.
2. B. Bojarski and T. Iwaniec, *Analytical foundations of the theory of quasiconformal mappings in* $\mathbb{R}^N$. Ann. A. Sc. Fen. 8(1983), 257–324.
3. C. Bennett and K. Rudnick, *On Lorentz-Zygmund spaces*. Dissertations Math. (Rozprawy Mat.) 175 (1980), 1–72.
4. C. Bennett and R. Sharpley, *Interpolation of Operators*. Academic Press, Boston 1988.
5. L.A. Caffarelli and I. Peral, *On* $W^{1,p}$ *estimates of elliptic equations*. To appear.
6. R.A. DeVore and R.C. Sharpley, *Maximal functions measuring smoothness*. American Mathematical Society, Memoirs of the AMS vol. 47, 293, 1984.
7. D.E. Edmunds and H. Triebel, *Logarithmic Sobolev spaces and their applications to spectral theory*. Proc. London Math. Soc. (3) 71 (1995), 333–371.
8. A. Fiorenza, *A. note on the spherical maximal function*. Rend. Accad. Sci. Fis. Mat. Napoli 65 (1987), 77–83.

9. A. Fiorenza, *Some remarks on Stein's LlogL result.* Diff. and Int. Equations 5 (6) (1992), 1355–1362.
10. A. Fiorenza and M. Krbec, *On the domain and range of the maximal operator.* To appear in Nagoya Math. J.
11. N. Fusco and C. Sbordone, *Higher integrability from reverse Jensen inequalities with different supports.* "Partial Differential Equations and the Calculus of Variations: Essays in Honor of Ennio De Giorgi", Progress in Non-linear Differential Equations and their Applications. Birkhaüser Boston Inc. 1989, 541–561.
12. N. Fusco and C. Sbordone, *Higher integrability of the gradient of minimizers of functionals with nonstandard growth conditions.* Comm. Pure Appl. Math. 43 (1990), 673–683.
13. J. García-Cuerva and J.L. Rubio de Francia, *Weighted Norm Inequalities and Related Topics.* North Holland, Amsterdam 1985.
14. L. Greco, T. Iwaniec and G. Moscariello, *Limits of the improved integrability of the volume forms.* Indian Univ. Math. J. 44 (1995), 305–339.
15. L. Greco, *A Remark on the Equality* det Du = Det Du. Diff. and Int. Equations 6 (1993), 1989–1100.
16. M. de Guzmán, *Differentiation of Integrals in* R^n. Lecture Notes in Math. vol. 481, Springer-Verlag, Berlin-Heidelberg-New York 1975.
17. T. Iwaniec and C. Sbordone, *On the integrability of the Jacobian under minimal hypothesis.* Arch. Rat. Mech. Anal. 119 (1992), 129–143.
18. B. Kawohl, *Rearrangements and Convexity of Level Sets in PDE.* Lecture Notes in Math. vol. 1150, Springer-Verlag, Berlin-New York 1985.
19. V. Kokilashvili and M. Krbec, *Weighted Inequalities in Lorentz and Orlicz Spaces.* World Scientific, Singapore 1991.
20. C. Li and K. Zhang, *Higher integrability of certain bilinear forms on Orlicz Spaces.* Macquarie Math. Reports 94–157, Sydney 1994.
21. V.G. Maz'ja, *Sobolev Spaces.* Springer-Verlag, Berlin 1985.
22. M. Milman, *Extrapolation and Optimal Decompositions.* Springer-Verlag, Berlin 1994.
23. J. Mossino, *Inégalités Isopérimétriques et Applications en Physique.* Collection Travaux en Cours, Hermann, Paris 1984.
24. C. Sadosky, *Interpolation of Operators and Singular Integrals.* M. Dekker Inc., New York 1979.
25. L. Schwartz, *Théorie des Distributions. Tome II.* Publications de l'Institut de Mathématique de l'Université de Strasbourg, vol X, Hermann, Paris 1957.
26. E. Stein, *Note on the class L* log *L.* Studia Math. 31 (1969), 305–310.
27. E.M. Stein, *Singular Integrals and Differentiability Properties of Functions.* Princeton Univ. Press, Princeton, New Jersey 1970.
28. E.M. Stein, *Harmonic Analysis: Real-Variable Methods, Orthogonality, and Oscillatory Integrals.* Princeton Univ. Press, Princeton, New Jersey 1993.
29. A. Torchinsky, *Real Variable Methods in Harmonic Analysis.* Academic Press, San Diego 1986.
30. I. Wik, *A comparison of the integrability of* f *and* Mf *with that of* $f^{\#}$ Preprint No. 2, University of Umeå, 1983, pp. 1–30.
31. N. Wiener, *The ergodic theorem.* Duke Math. J. 5 (1939), 1–18.

Function Spaces and Applications
D.E. Edmunds et al (Eds)

7. Nontangential Approach Regions on Groups

José L. García[*1] **and Javier Soria**[*2]

[1]Departament de Matemàtica i Informàtica, Universitat de Vic, E-08500 Vic, E-08500 Vic, Spain

[2]Departament de Matemàtica Aplicada i Anàlisi, Universitat de Barcelona, E-08071 Barcelona, Spain

1. Introduction

In the history of analysis, many problems are concerned with the existence of boundary limits for certain classes of functions defined on a general domain. The beginning of the study of these questions goes back to 1872, when Schwarz [17] proved the existence of radial limits for all boundary points in the unit disc D for the Poisson integral of a continuous 2π-periodic function. Fatou proved in 1906 [6] the existence of boundary limits, for almost every point in the boundary of D, for bounded holomorphic functions if we approach the boundary point within certain regions called *nontangential* approach regions. The nontangential condition became strongly linked with boundary convergence when Littlewood [12] proved the failure of convergence for bounded holomorphic functions on D along tangential curves. In 1930, Hardy and Littlewood [9] introduced the idea of studying the convergence of a sequence of operators by means of estimates on a maximal function. The natural setting to study estimates on maximal functions is the spaces of homogeneous type.

In [11], Korànyi gave the boundary convergence for H^p functions on the generalized half-plane of $\mathbb{C}^{n+1}$ within the so-called admissible domains. These regions allow parabolic tangential approach to the boundary along certain directions. In 1984, Nagel and Stein [13] completely characterized the approach regions in the half space $\mathbb{R}^{n+1}_+$ where we have boundary convergence for the Poisson integral of L^p functions on the boundary. They studied estimates on a maximal operator M_Ω (see (2)) related to the approach region Ω, and the result is quite surprising: there exist nontangential regions (without tangential directions) not contained in any cone for which convergence is allowed.

*This work has been partially supported by DGICYT PB 97-0986 and by CIRIT 1997SGR00185.

These are the so-called *non nontangential* approach regions. Sueiro [20] generalizes this result by studying the problem in the general setting of the spaces of homogeneous type, and he applies it in the case of the generalized half-plane in $\mathbb{C}^{n+1}$ which can be seen as the product $\mathbf{H}^n \times (0, \infty)$, where $\mathbf{H}^n$ is the Heisenberg group. Later, Andersson and Carlsson [1] gave an easy proof of the Nagel-Stein result using the key concept of Carleson measures. A complete and original overview of this subject can be found in ([5], Chapter 2), where new interesting results are given.

Following the ideas of [20], Pan [14] studied the weak-type weighted norm estimates for M_Ω also in spaces of homogeneous type. Later, in [15], A. Sánchez-Colomer and J. Soria gave and answer to the strong-type weighted estimates for M_Ω in the Euclidean space, and they also studied the relationship between weighted inequalities for this operator and the geometry of the region Ω [16]. This research is followed by García and Soria [8] and solved in a more general context. The work presented here deals with this last paper, and it is devoted to the special case of the existence of a group structure (this was also considered in [8] with a more complicated argument).

We would like to thank the referee for the valuable comments which have improved the final version of this work.

2. Definitions

Let X be a topological space with a nonnegative Borel measure μ. Suppose we have a nonnegative real-valued function d defined in $X \times X$ that satisfies the following properties:

(i) $d(x, y) = 0$ if and only if $x = y$.

(ii) $d(x, y) = d(y, x)$ for all $x, y \in X$.

(iii) There is a constant $A \geq 1$ such that $d(x, y) \leq A\, (d(x, z) + d(y, z))$, for all x, y, z in X.

(iv) The balls $B(x, r) = \{y \in X : d(x, y) < r\}$ are measurable sets for all x in X and $r > 0$. Moreover, $\{B(x, r)\}_{r>0}$ is a basis of open neighborhoods for all x in X.

(v) There is a constant $K > 1$ such that $0 < \mu(B(x, 2r)) \leq K\mu(B(x, r))$ for all x in X and $r > 0$.

The function d is called a quasidistance, and a measure space (X, μ, d) satisfying conditions (i) to (v) is called a space of homogeneous type (see [4] for a more general definition). In the sequel, we will write every positive constant as C and it may change from one occurrence to the next. If f and g are two given functions defined in a space Y, we say that they are equivalent, and we write $f \approx g$, if there exist two positive constants C and C' such that $Cf(y) \leq g(y) \leq C'f(y)$ for all $y \in Y$.

It is proved in [8], using arguments of [19] (see Lemma I.23), that we can always find, under the hypothesis that there is a constant $M > 1$ such that $B(x, Mr) \backslash B(x, r) \neq \phi$ for all x in X and $r > 0$ (if X is bounded this should only be considered for r small enough), an equivalent quasimetric d' in such a way that (X, μ, d') is a space of homogeneous type, topologically equivalent to the given one, such that the boundedness of the corresponding maximal operators in either case is also equivalent, and for this quasimetric, one has that the measure of a ball is (essentially) independent of the center (in fact it is comparable to the radius). From now on, we take this condition for granted. Thus, without loss of generality, we can assume our measure μ to be invariant in the sense that there exist two postive constants D and D' such that

$$D \le \frac{\mu(B(x, r))}{\mu(B(y, r))} \le D', \tag{1}$$

for all $x, y \in X$ and $r > 0$.

Finally, assume that X is a group with a multiplicative law (not necessary commutative) such that d is left-invariant and μ is left-invariant and invariant under inversion, that is respectively,

$$x \cdot B(y, r) = B(x \cdot y, r) \ \forall\, x, y \in X, r > 0$$

$$\mu(x \cdot E) = \mu(E) \ \forall\, x \in X, E \subset X \text{ measurable}$$

$$\mu(E) = \mu(E^{-1}) \ \forall\, E \subset X \text{ measurable}$$

(e.g. this is the case if X is unimodular and μ is the Haar measure [7] Proposition 2.9). We denote by e the identity element of the group.

3. Approach Regions in X

We say that a family of measurable sets $\Omega = \{\Omega_x\}_{x \in X}$ in $X \times (0, \infty)$ is a family of approach regions if $(x, 0)$ is in the closure of Ω_x with respect to the product topology in $X \times (0, \infty)$. The natural example of approach region is the cone of width $\theta > 0$, $\Gamma_\theta(x) = \{(y, t) \in X \times (0, \infty) : d(y, x) < \theta t\}$. A simple way of constructing a family of approach regions is the following: for a fixed approach region Ω_e of e, consider for every $x \in X$ the set

$$\Omega_x = \{(x \cdot y, t) : (y, t) \in \Omega_e\}.$$

The maximal operator (related to Ω) for a measurable function f is:

$$M_\Omega f(x) = \sup_{(y,t) \in \Omega_x} \frac{1}{\mu(B(y, t))} \int_{B(y,t)} |f(z)|\, d\mu(z). \tag{2}$$

A weight u in (X, μ, d) is a positive and locally integrable function. We denote $u(D) = \int_D u(z)\, d\mu(z)$ for $D \subset X$ measurable. A weight u is in the A_p class, $1 \le p < \infty$, if $M : L^p(u) \to L^{p,\infty}(u)$ is bounded, where M is the Hardy-Littlewood maximal operator

$$Mf(x) = \sup_{B \ni x} \frac{1}{\mu(B)} \int_B |f(z)|\, d\mu(z).$$

Define A_p^Ω to be the class of weights u such that $M_\Omega : L^p(u) \to L^{p,\infty}(u)$ is bounded, $1 \le p < \infty$. Set

$$W(\Omega) = \{u \in L^1_{\text{loc}}(\mu), u \ge 0 : \exists C > 0, u(S(x,t)) \le C\, u(B(x,t)), \forall\ (x,t)\},$$

where $S(x,t) = \{y \in X : \Omega_y(t) \cap B(x,t) \ne \phi\}$ and $\Omega_x(t) = \{y \in X : (y,t) \in \Omega_x\}$ is the section of Ω_x at height $t > 0$.

It is proved in [20] that for the boundedness of M_Ω we can always assume (with no loss of generality) that Ω is full on vertical directions, that is $(y, s) \in \Omega_x$ implies $(y, t) \in \Omega_x$ for all $t \ge s$. From now on, we will always assume that this condition holds.

The next result is proved in [8]:

Theorem 3.1 For $1 \le p < \infty$, we have $A_p^\Omega = A_p \cap W(\Omega)$.

If $\Omega_x = \Gamma_\theta(x)$ for some $\theta > 0$ and for all $x \in X$, then it is known that M_{Γ_θ} and M are equivalent operators, that is, there exist two positive constants c and c' such that

$$cMf(x) \le M_{\Gamma_\theta} f(x) \le c'\, Mf(x),$$

for all measurable functions f and for all $x \in X$, and hence, we have that $A_p = A_p^\Omega$. We will see that the converse is also true.

It is essentially proved in [3] that $M\nu^\varepsilon$ is an A_1 weight for all Borel measures ν in X such that $M\nu \not\equiv \infty$, and for all $0 \le \varepsilon < 1$. Then, taking $\nu = \delta_x$, the Dirac delta at x, the weight $u(y) = d(x,y)^{-\varepsilon}$ is in A_1 for all $x \in X$, because $M\delta_x(y)$ is pointwise equivalent to $d(x,y)^{-1}$. If u_1 and u_2 are two A_1 weights, Hölder's inequality gives us that $u_1 u_2^{1-p}$ is an A_p weight for $1 < p < \infty$. Then, there exists $0 < \gamma = \gamma(p) \le 1$ such that $u(y) = d(x,y)^\gamma$ is an A_p weight for all x.

The main result we are going to prove is the next theorem.

Theorem 3.2 If $\Omega_x = \{(x \cdot y, t) : (y,t) \in \Omega_e\}$ for a given approach region Ω_e of e, then, the following conditions are equivalent:

(a) There exist $C > 0$ and $\theta > 0$ such that $M_\Omega f(x) \le CM_{\Gamma_\theta} f(x)$, for all $x \in X$ and all measurable functions f.

(b) $A_p^\Omega = A_p$ for all $1 \le p < \infty$.

(c) There is $p \ge 1$ such that $A_p = A_p^\Omega$.

(d) There exists $0 < \gamma \le 1$ such that $u(y) = d(e,y)^\gamma \in W(\Omega)$.

(e) There exists $\theta > 0$ such that $\Omega_e \subset \Gamma_\theta(e)$.

Proof It is obvious that (b) implies (c). Because of the left-invariance of d, if $\Omega_e \subset \Gamma_\theta(e)$ then $\Omega_x \subset \Gamma_\theta(x)$ for all x, and so, (e) implies (a). The implication (a) $\Rightarrow$ (b) is easy if we recall Theorem 3.1 (one can also show that since Ω is full on vertical directions, then $Mf(x) \le CM_\Omega f(x)$). Now, suppose

$A_p^\Omega = A_p$ for some $p \geq 1$.We can assume that $p > 1$ by the extrapolation theorem of Rubio de Francia, as proved in ([10], Theorem 6.7). We have seen that there is $0 < \gamma = \gamma(p) \leq 1$ such that $u(y) = d(e, y)^\gamma$ is in A_p, and so, by hypothesis and Theorem 3.1, $u \in W(\Omega)$.

Suppose that $u(y) = d(e, y)^\gamma \in W(\Omega)$. Take $(y, t) \in \Omega_e$. Using the triangle inequality, we have

$$\begin{aligned} d(e,y)^\gamma &= \frac{1}{\mu(B(y,t)^{-1})}\int_{B(y,t)^{-1}} d(e,y)^\gamma \, d\mu(z) \\ &\leq \frac{A^\gamma}{\mu(B(y,t)^{-1})}\int_{B(y,t)^{-1}} (d(e,z^{-1})^\gamma + d(z^{-1},y)^\gamma)\, d\mu(z) \\ &= \frac{A^\gamma}{\mu(B(y,t))}\int_{B(y,t)^{-1}} (d(e,z)^\gamma + d(z^{-1},y)^\gamma)\, d\mu(z) \\ &\leq A^\gamma\left(\frac{1}{\mu(B(y,t))} u(B(y,t)^{-1}) + t^\gamma\right). \end{aligned}$$

We now use that $S(e, t) = B(e, t) \quad [\Omega_e(t)]^{-1}$ (see [20]), that is

$$S(e,t) = \bigcup_{y\in\Omega_e(t)} B(y,t)^{-1},$$

and so, by hypothesis, we have

$$\begin{aligned} d(e,y)^\gamma &\leq A^\gamma\left(\frac{u(S(e,t))}{\mu(B(y,t))} + t^\gamma\right) \\ &\leq CA^\gamma\left(\frac{u(B(e,t))}{\mu(B(y,t))} + t^\gamma\right). \end{aligned}$$

Finally, observe that $u(B(e, t)) = \int_{B(e,t)} d(e, z)^\gamma \, d\mu(z) \leq t^\gamma \mu(B(e, t))$, and then using (1),

$$d(e, y)^\gamma \leq 2CA^\gamma t^\gamma.$$

Hence, $\Omega_e \subset \Gamma_\theta(e)$, with $\theta = A(2C)^{1/\gamma}$. □

Remark 3.3 The restriction $\gamma \leq 1$ in statement (d) is not really needed.

4. Examples

(1) Take $X = \mathbb{R}^n$ and μ the Lebesgue measure. Consider the nonisotropic quasidistance

$$d(x,y) = \sum_{k=1}^{n} |x_k - y_k|^{1/a_k},$$

where $a_1, \ldots, a_n$ are strictly positive constants, or the equivalent

$$d'(x, y) = \sup_k | x_k - y_k |^{1/a_k}.$$

It is easy to see that all the required conditions are satisfied. We write $|x| = d(x, 0) \approx d'(x, 0)$. Our theorem completes now the result proved in [16]:

Theorem 4.1 If $\Omega_x = \{(x + y, t): (y, t) \in \Omega_e\}$ for a given approach region Ω_0 of 0, then, the following conditions are equivalent:

(a) There exist $C > 0$ and $\theta > 0$ such that $M_\Omega f(x) \leq C M_{\Gamma_\theta} f(x)$, for all $x \in X$ and all measurable functions f.

(b) $A_p^\Omega = A_p$ for all $1 \leq p < \infty$.

(c) There is $p \geq 1$ such that $A_p = A_p^\Omega$.

(d) There exists $0 < \gamma \leq 1$ such that $u(x) = |x|^\gamma \in W(\Omega)$.

(e) There exists $\theta > 0$ such that $\Omega_e \subset \Gamma_\theta(e)$.

If $a_k = 2$ for all k, Nagel and Stein proved in [13] that if the weight $u(y) = |y|^0 = 1$ is in A_p^Ω, then Ω_0 need not be contained in a cone: in fact, it can contain a sequence of points approaching 0 tangentially. Our theorem states that this is the extremal situation, because if A_p^Ω contains a power weight with positive exponent, Ω_0 is necessarily a subset of a cone.

(2) Take $X = \mathbf{H}^n$ the Heisenberg group (see [18], XII, 1.4 or [20] for details). This is the set

$$\mathbb{C}^n \times \mathbb{R} = \{[\zeta, t] : \zeta \in \mathbb{C}^n, t \in \mathbb{R}\},$$

with the (noncommutative) multiplicative law

$$[\zeta, t] \cdot [\eta, s] = [\zeta + \eta, t + s + 2\mathrm{Im}(\zeta \overline{\eta})].$$

The identity element is $e = [0, 0]$, and we have $[\zeta, t]^{-1} = [-\zeta, -t]$.

Consider the generalized half-plane $D = \{z \in \mathbb{C}^{n+1} : h(z) > 0\}$, where

$$h(z) = \mathrm{Im}\, z_{n+1} - \sum_{k=1}^{n} | z_k |^2,$$

for $z = (z_1, \ldots, z_{n+1}) \in \mathbb{C}^{n+1}$. Then, we can see $\mathbf{H}^n$ as the boundary $\partial D = \{z \in \mathbb{C}^{n+1} : h(z) = 0\}$, by considering the map $\psi: \mathbb{C}^{n+1} \to \mathbf{H}^n$ defined as $\psi(z) = (z_1, \ldots, z_n, \mathrm{Re}\, z_{n+1})$, where the restriction to ∂D gives the desired bijection. Now, we are able to think of D as the product $\mathbf{H}^n \times (0, \infty)$ using the identification

$$\Phi : D \to \mathbf{H}^n \times (0, \infty),$$

defined by $\Phi(z) = (\Psi(z), h(z))$.

The group $\mathbf{H}^n$ acts on D (and ∂D) associating to each element $[\zeta, t] \in \mathbf{H}^n$ the following affine self-mapping:

$$[\zeta, t] : (z', z_{n+1}) \to (z' + \zeta, z_{n+1} + t + 2iz'\overline{\zeta} + i\,|\,\zeta\,|^2).$$

It is not difficult to see that $[\eta, s]\,([\zeta, t]z) = ([\eta, s] \cdot [\zeta, t])z$, and that this action is simply transitive on ∂D: for every two points in ∂D, there is exactly one element in $\mathbf{H}^n$ mapping the first to the second. So, we can also identify $\mathbf{H}^n$ as the *translations* on ∂D.

We consider the pseudo-norm function defined in $\mathbf{H}^n$

$$\|\,[\zeta, t]\,\| = \max\,(\,|\,\zeta\,|^2, |\,t\,|\,),$$

satisfying the quasi-triangle inequality

$$\|\,x \cdot y\,\| \leq c(\,\|\,x\,\| + \|\,y\,\|\,),$$

for all $x, y \in \mathbf{H}^n$. Observe the different homogeneity in ζ and t. We can define the quasidistance in $\mathbf{H}^n$ by $d(x, y) = \|\,y^{-1} \cdot x\,\|$, symmetric because $\|\,x\,\| = \|\,x^{-1}\,\|$, and left-invariant with respect to the group action.

We take as the underlying measure $d\mu$ to be the Euclidean Lebesgue measure on $\mathbb{C}^n \times \mathbb{R}$, which is left-invariant and invariant under inversion.

With these definitions, $(\mathbf{H}^n, \mu, d)$ becomes a space of homogeneous type, with the conditions required in Theorem 3.2. For the following result we will use the trivial fact that $d(e, [\zeta, t]) = \max\,(\,|\,\zeta\,|^2, |\,t\,|\,) \approx (\,|\,\zeta\,|^2 + |\,t\,|\,)$.

Theorem 4.2 If $\Omega_x = \{(x \cdot z, t) : (z, t) \in \Omega_e\}$ for *a* given approach region Ω_e of e, then, the following conditions are equivalent:

(a) There exist $C > 0$ and $\theta > 0$ such that $M_\Omega f(x) \leq CM_{\Gamma_\theta} f(x)$, for all $x \in \mathbf{H}^n$ and all measurable functions f.

(b) $A_p^\Omega = A_p$ for all $1 \leq p < \infty$.

(c) There is $p \geq 1$ such that $A_p = A_p^\Omega$.

(d) There exists $0 < \gamma \leq 1$ such that $u\,([\zeta, t]) = (\,|\,\zeta\,|^2 + |\,t\,|\,)^\gamma \in W(\Omega)$.

(e) There exists $\theta > 0$ such that $\Omega_e \subset \Gamma_\theta(e)$.

The Korányi admissible regions for D are defined by

$$\tilde{\Gamma}_\theta(0) = \{z \in D : \|\,\psi(z)\,\| < \theta h(z)\},$$

and $\tilde{\Gamma}_\theta(g \cdot 0) = g \cdot \tilde{\Gamma}_\theta(0)$ if $g \in \mathbf{H}^n$. We have

$$\Phi(\tilde{\Gamma}_\theta(\zeta)) = \Gamma_\theta(\psi(\zeta)),$$

where $\Gamma_\theta(x) = \{(y, s) \in \mathbf{H}^n \times (0, \infty) : d(y, x) < \theta s\}$ is a cone in $\mathbf{H}^n \times (0, \infty)$. If $\tilde{\Omega}$ is an approach region of $0 \in \partial D$, we can translate it by the action of the Heisenberg group: for $\zeta \in \partial D$, there exists a unique element $x \in \mathbf{H}^n$ such that $\zeta = x \cdot 0$, and then consider $\tilde{\Omega}_\zeta = x \cdot \tilde{\Omega}$. Then, taking $\Phi(\tilde{\Omega}) = \Omega$ we have

$$\Phi(\tilde{\Omega}_\zeta) = \Omega_x,$$

with $\Omega_x = \{(x \cdot y, t) : (y, t) \in \Omega\}$. We have that Ω is contained in a cone if and only if $\Phi^{-1}(\Omega)$ is contained in an admissible region (see [20]). With this notations, we can give the next result.

Corollary 4.3 For a given approach region $\tilde{\Omega}$ of $0 \in \partial D$, the following conditions are equivalent:

(a) There exist $C > 0$ and $\theta > 0$ such that $M_\Omega f(x) \leq CM_{\Gamma_\theta} f(x)$, for all $x \in \mathbf{H}^n$ and all measurable functions f.

(b) $A_p^\Omega = A_p$ for all $1 \leq p < \infty$.

(c) There is $p \geq 1$ such that $A_p = A_p^\Omega$.

(d) There exists $0 < \gamma \leq 1$ such that $u([\zeta, t]) = (|\zeta|^2 + |t|)^\gamma \in W(\Omega)$.

(e) There exists $\theta > 0$ such that $\tilde{\Omega} \subset \tilde{\Gamma}_\theta(0)$.

(3) Consider X as the set of all real 3×3 upper-triangular matrices having ones along the diagonal. The group multiplicative law is the usual matrix product (see [19, XIII.5.2.3). The norm function is defined by

$$\| x \| = \max \{ |a|, |b|^{1/2}, |c| \},$$

where

$$x = \begin{pmatrix} 1 & a & b \\ 0 & 1 & c \\ 0 & 0 & 1 \end{pmatrix}.$$

The function $d(x, y) = \| y^{-1} \cdot x \|$ is left-invariant but nonsymmetric. We can consider the equivalent quasidistance $d'(x, y) = (d(x, y) + d(y, x))/2$, since there exists a constant C such that $d(x, y) \leq C\, d(y, x)$ for all $x, y \in X$. Notice that we can realize X as $\mathbb{R}^3$ with the inner product

$$(a, b, c) \cdot (d, e, f) = (a + d, b + e + af, c + f).$$

We write $x = (a, b, c) \in X$ and observe that $\| x \| \approx (a^2 + |b| + c^2)^{1/2} \approx d'(x, 0)$. Then we take as the underlying measure $d\mu$ to be the Lebesgue measure, which is left-invariant and invariant under inversion. Also, $(X, \mu\, d')$ is a space of homogeneous type. We can state the theorem

Theorem 4.4 If $\Omega_x = \{(x \cdot y, t) : (y, t) \in \Omega_0\}$ for a given approach region Ω_0 of 0, then, the following conditions are equivalent:

(a) There exist $C > 0$ and $\theta > 0$ such that $M_\Omega f(x) \leq CM_{\Gamma_\theta} f(x)$, for all $x \in X$ and all measurable functions f.

(b) $A_p^\Omega = A_p$ for all $1 \leq p < \infty$.

(c) There is $p \geq 1$ such that $A_p = A_p^\Omega$.

(d) There exists $0 < \gamma \leq 1$ such that $u(x) = (a^2 + |b| + c^2)^{\gamma/2} \in W(\Omega)$.

(e) There exists $\theta > 0$ such that $\Omega_0 \subset \Gamma_\theta(0)$.

(4) Every connected nilpotent unimodular Lie group with a left-invariant

Riemannian metric and the induced measure is a space of homogeneous type satisfying our conditions (see [2], Example VI.7).

References

1. M. Andersson and H. Carlsson, *Boundary convergence in non-tangential and non-admissible approach regions,* Math. Scand. **70** (1992), 293–301.
2. M. Christ, *Lectures on singular integral operators,* CBMS vol. 77, Amer. Math. Soc. (1990).
3. R.R. Coifman and R. Rochberg, *Another characterization of BMO,* Proc. Amer. Math. Soc. **79** (1980), 249–254.
4. R.R. Coifman and G. Weiss, *Analyse harmonique non-commutative sur certain espaces homogenes,* Lecture Notes in Mathematics, vol. 242 (1971).
5. F. Di Biase, *Fatou Type Theorems,* Progress in Mathematics vol. 147, Birkhäuser (1998).
6. P. Fatou, *Series trigonometriques et series de Taylor,* Acta Math. **30** (1906), 335–400.
7. G. Folland, *A course in obstract harmonic analysis,* CRC press (1995).
8. J.L. García and J. Soria, *Weighted inequalities and the shape of approach regions* Studia Math. **133** (1999), 261–274.
9. G.H. Hardy and J. Littlewood, *A maximal inequality with function theoretic applications,* Acta. Math. **54** (1930), 81–116.
10. B. Jawerth, *Weighted inequalities for maximal operators: linerization, localization and factorization,* Amer. J. Math. **108** (1986), 361–414.
11. A. Korányi, *Harmonic functions on hermitian hyperbolic space,* Trans. Amer. Math. Soc. **135** (1969), 507–516.
12. J. Littlewood, *On a theorem of Fatou,* J. London Math. Soc. **2** (1927), 222–234.
13. A. Nagel and E.M. Stein, *On certain maximal functions and approach regions,* Adv. Math. **54** (1984), 83–106.
14. Pan Wenjie, *Weighted norm estimates for certain maximal operators with approach regions,* Lecture Notes in Mathematics, vol. 1949 (1992), 169–175.
15. A. Sánchez-Colomer and J. Soria, *Weighted norm inequalities for general maximal operators and approach regions,* Math. Nachr. **172** (1995), 249–260.
16. A. Sánchez-Colomer and J. Soria, *A_p and approach regions,* Fourier Analysis and Partial Differential Equations, CRC Press (1995), 311–315.
17. H.A. Schwarz, *Zur integration der partiellen differential gleichung* $\frac{\partial^2 u}{\partial x^2} + \frac{\partial^2 u}{\partial x^2} = 0$, J. Reine Angew. Math. **74** (1872), 218–253.
18. E.M. Stein, *Harmonic Analysis,* Princeton University Press (1993).
19. J.O. Strömberg and A. Torchinsky, *Weighted Hardy spaces,* Lecture Notes in Mathematics, vol. 1381 (1989).
20. J. Sueiro, *On maximal functions and Poisson-Szegö integrals,* Trans. Amer Math. Soc. **289** (1986), 653–669.

Function Spaces and Applications
D.E. Edmunds et al (Eds)

8. Stability of Sobolev Spaces with Zero Boundary Values

Lars Inge Hedberg

Department of Mathematics, Linköping University, S-58183 Linköping, Sweden

The new results in this talk are contained in a joint paper with Tero Kilpeläinen, University of Jyväskylä, Finland, see [4]. The stability problem alluded to in the title, which is solved in [4], is the following:

For which bounded open $\Omega \subset \mathbf{R}^n$ is it true that

$$W_0^{1,p}(\Omega) = W^{1,p}(\Omega) \cap \bigcap_{q<p} W_0^{1,q}(\Omega)? \tag{1}$$

Here $1 < p < \infty$, and $W_0^{1,p}(\Omega)$ is the closure of $C_0^\infty(\Omega)$, the compactly supported C^∞-functions on Ω, in the $W^{1,p}$-norm $\| u \|_{1,p}$, defined by $\| u \|_{1,p}^p = \int (| u |^p + | \nabla u |^p)\, dx$. Clearly $W_0^{1,p}(\Omega)$ can be considered as a subspace of $W^{1,p}(\mathbf{R}^n) = W^{1,p}$, but in general $W^{1,p}(\Omega)$ cannot be embedded in $W^{1,p}$, unless Ω satisfies some regularity condition, such as a Lipschitz condition.

The above problem was posed by P. Lindqvist in connection with his work on the p-Laplace equation, see [7], [8]. It is easy to see that for a bounded domain Ω the minimum of a Rayleigh quotient

$$\lambda_p = \lambda_p(\Omega) = \min_{\substack{0 \neq v \in \\ W_0^{1,p}(\Omega)}} \frac{\int_\Omega | \nabla v |^p\, dx}{\int_\Omega | v |^p\, dx},$$

is also the least real λ such that the equation

$$\operatorname{div}\,(| \nabla u |^{p-2} \nabla u) + \lambda | u |^{p-2} u = 0$$

has a nontrivial solution $u \in W_0^{1,p}(\Omega)$. The nonlinear operator $\Delta_p : u \to \operatorname{div}\,(| \nabla u |^{p-2} \nabla u)$ is known as p-Laplacian, and λ_p as its first eigenvalue. As in the classical case $p = 2$ this eigenvalue is simple, and the corresponding eigenfunction u_p is strictly positive, see [7]. The purpose of [8] was to study the continuity of u_p and λ_p as functions of p.

Lindqvist [8] proved the following theorem.

Theorem 1 For any bounded domain $\Omega \subset \mathbf{R}^n$ the limits $\lim_{q \to p-} \lambda_q(\Omega)$, and $\lim_{q \to p+} \lambda_q(\Omega)$ exist. Moreover, $\underline{\lambda}_p(\Omega) = \lim_{q \to p-} \lambda_q(\Omega) \le \lambda_p(\Omega)$, and $\lim_{q \to p+} \lambda_q(\Omega) = \lambda_p(\Omega)$.

He also observed that there are irregular domains such that $\underline{\lambda}_p(\Omega) < \lambda_p(\Omega)$, a fact which considerably complicates the situation. For example, if $K \subset \Omega$ is a compact set such that $C_p(K) > 0$, and $C_q(K) = 0$ for all $q < p$, then $\underline{\lambda}_p(\Omega \backslash K) < \lambda_p(\Omega \backslash K)$.

Here $C_p(K)$ is the p-capacity, defined for an arbitrary set E and $p > 1$ by $C_p(E) = \inf \{ \| u \|_{1,p}^p : u \in W^{1,p}, u \ge 1$ a.e. on a neighborhood of $E\}$.

It is easy to see that a set of positive Lebesgue measure always has positive p-capacity, and that for $p > n$ every non-empty set has positive p-capacity. Properties of p-capacity have been studied in great detail, and there are very precise criteria which allow the construction of sets of Cantor type satisfying the above. See e.g. Chapter 5 in [2].

The following theorem follows from the results of Lindqvist, see 3.6 and 3.12 in [8]:

Theorem 2 If the bounded domain Ω is such that the equality (1) holds, then

$$\underline{\lambda}_p(\Omega) = \lambda_p(\Omega),$$

and
$$\lim_{q \to p-} \int_\Omega | \nabla u_p - \nabla u_q |^q \, dx = \lim_{q \to p+} \int_\Omega | \nabla u_p - \nabla u_q |^p \, dx = 0.$$

Whether (1) is a necessary condition here, does not seem to be known. On the other hand, one can give a complete characterization of the bounded open sets Ω satisfying (1), and this is done in [4].

Before giving this description we will take the opportunity to review a few of the local properties of Sobolev functions that have been found during the last thirty years. In the Hilbert space case, $p = 2$, the results are older, and closely tied to classical potential theory. The extension to $p \ne 2$ (and more general spaces) required the development of a new, nonlinear potential theory. Many mathematicians contributed; B. Fuglede, V.P. Havin, V.G. Maz'ya, N.G. Meyers, Yu. G. Reshetnyak, and J. Serrin were the founders of the theory. Details and many references are found in the books [2], and, from a different point of view, [5].

The statement $f \in W_0^{1,p}(\Omega)$ should be understood as meaning that f belongs to $W^{1,p}(\Omega)$ and vanishes at the boundary of Ω (and on the exterior) in a generalized sense, and if $f \in W^{1,p}(\Omega) \cap \bigcap_{q<p} W_0^{1,q}(\Omega)$ this means that f is zero at the boundary in a slightly different generalized sense, but we need to make the meaning of this more concrete.

The way the Sobolev spaces $W^{1,p}$ are usually defined, their elements are functions defined almost everywhere (a.e.), or more precisely, equivalence classes of functions defined a.e. with the equivalence relation being equality a.e. For $p > n$ the Sobolev imbedding theorem says that each element can be represented by a continuous function. In the general case one can define a representative $\tilde{f}$ by setting

$$\tilde{f}(x) = \lim_{r\to 0} \frac{1}{|B(x,r)|} \int_{B(x,r)} f(y)\,dy,$$

whenever this limit exists. ($B(x, r)$ is the ball with radius r centered at x.) One can show that $\tilde{f}(x)$ exists everywhere outside an exceptional set of zero p-capacity (p-*quasi everywhere*, p-q.e.). It is even true that p-quasievery x is a Lebesgue point in the sense that

$$\lim_{r\to 0} \frac{1}{|B(x,r)|} \int_{B(x,r)} |f(y) - \tilde{f}(x)|\,dy = 0.$$

For every $\varepsilon > 0$ there is also an open G such that $C_p(G) < \varepsilon$ and such that the averages defining $\tilde{f}(x)$ converge uniformly on the complement G^c, which makes the restriction $\tilde{f}|_{G^c}$ continuous on G^c.

This property of $\tilde{f}$ is called *p-quasicontinuity*, so $\tilde{f}$ is a *p-quasicontinuous representative* of f. There is also a uniqueness theorem to the effect that if g_1 and g_2 are p-quasicontinuous and agree almost everywhere, then they agree p-q.e., so $\tilde{f}$ is essentially unique. From now on we drop the "tilde" symbol, and assume that any f in $W^{1,p}$ is chosen p-quasicontinuous.

The following characterization of $W_0^{1,p}(\Omega)$ is due to V.P. Havin for $p > n$ and $p = 2$, and to T. Bagby in the general case. The proof is a consequence of the well known fact that $W^{1,p}$ is closed under truncation, i.e., if $f \in W^{1,p}$, then the same is true, e.g., for its positive part $f^+ = \max\{f, 0\}$.

Theorem 3 For any open $\Omega \subset \mathbf{R}^n$ and $1 < p < \infty$, $f \in W_0^{1,p}(\Omega)$ if and only if $f \in W^{1,p}(\mathbf{R}^n)$, and $f(x) = 0$ p-q.e. on Ω^c.

For $p > n$ the condition just means that $f(x) = 0$ everywhere on Ω^c. D. Swanson and W.P. Ziemer [10] recently proved an interesting sharpening of this theorem.

Theorem 4 For any open $\Omega \subset \mathbf{R}^n$ and $1 < p < \infty$, $f \in W_0^{1,p}(\Omega)$ if and only if $f \in W^{1,p}(\Omega)$, and

$$\lim_{r\to 0} \frac{1}{|B(x,r)|} \int_{B(x,r)\cap\Omega} f(y)\,dy = 0$$

for p-q.e. $x \in \partial\Omega$.

Thus, the difference between Theorems 3 and 4 is that in the latter the function is not assumed to have an extension to $W^{1,p}(\mathbf{R}^n)$.

A different way for f to be zero at $\partial\Omega$ is:

$$f \in W^{1,p}(\mathbf{R}^n), \text{ and } f(x) = 0 \ p\text{-q.e. on } \overline{\Omega}^c.$$

If $p \le n$ this does not always imply that $f \in W_0^{1,p}(\Omega)$. Provided that Ω equals the interior of its closure $\overline{\Omega}$, we say that Ω is *p-stable* if the implication holds. (This differs slightly from [2], Definition 11.1.7. There it is the set $\overline{\Omega}$ that is called stable.)

The p-stable sets can be characterized in terms of p-capacity.

Theorem 5 An open set Ω that equals the interior of its closure is p-stable if and only if

$$C_p(G \cap \Omega^c) = C_p(G \cap \overline{\Omega}^c)$$

for all open G.

This description can be used for manufacturing non-stable sets. The following example comes from [3], p. 244.

Example Let B_0 be an open ball in $\mathbf{R}^n$, and let $K \subset B_0$ be a compact set without interior such that $C_p(K) > 0$, for example a suitable set of Cantor type. Let $\{B_i\}_1^\infty$ be disjoint open balls in $B_0 \backslash K$, chosen so that each point of K is an accumulation point for $\cup_1^\infty B_i$, and so that for any neighborhood U of K, all but a finite number of the B_i are contained in U. Set $\Omega = B_0 \backslash \overline{\cup_1^\infty B_i}$. Then $K \subset \partial\Omega$, and $\overline{\Omega} = \overline{B}_0 \backslash \cup_1^\infty B_i$. Choosing G in the theorem to be B_0, we find $C_p(G \cap \Omega^c) \ge C_p(K) > 0$. On the other hand $C_p(G \cap \overline{\Omega}^c) = C_p(\cup_1^\infty B_i) \le \sum_1^\infty C_p(B_i)$, which can be made arbitrarily small by choosing the radii small enough, so that we can make $C_p(G \cap \overline{\Omega}^c) < C_p(G \cap \Omega^c)$.

In the converse direction the theorem is difficult to apply. Other characterizations can be obtained by means of the concept of *p-thinness*. Its definition is a generalization of the classical Wiener condition, which characterizes the boundary points that are regular for the Dirichlet problem.

Definition 1 A set $E \subset \mathbf{R}^n$ is *p-thin* at a point x if

$$\int_0^1 \left(\frac{C_p(E \cap B(x, r))}{r^{n-p}} \right)^{1/(p-1)} \frac{dr}{r} < \infty .$$

The set of such points is denoted $e_p(E)$. If E is not p-thin at x, it is *p-thick*, and the set of such points is $b_p(E)$.

For $p < n$ the quantity r^{n-p} can be replaced by $C_p(B(x, r))$. Clearly, $e_p(E)$ contains the exterior of E, and its complement $b_p(E)$ contains the interior.

We will need the fact that $b_q(E) \subset b_p(E)$ if $1 < q \le p$; see [1] and 6.5.8 in [2].

The following important result shows that the definition is the right one.

Theorem 6 A point $x \in \partial\Omega$, where Ω is a region in $\mathbf{R}^n$, is regular for the Dirichlet problem for the p-Laplace equation $\Delta_p u = 0$, $1 < p \leq n$, if and only if $\partial\Omega$ is p-thick at x.

The sufficiency part of the theorem was proved in 1970 by V.G. Maz'ya [9], and the necessity almost a quarter of a century later by T. Kilpeläinen and J. Malý [6]. We refer to the original papers for the precise definition of boundary regularity, and for more general results. See also [2], Sections 6.3 and 6.5.5.

Another important fact is that the set of irregular boundary points has p-capacity zero. This results from combining the sufficiency part of the above theorem with the following *Kellogg property*, first proved in this generality by T.H. Wolff (see [2], Section 6.3).

Theorem 7 For an arbitrary set E, $C_p(E \cap e_p(E)) = 0$.

There is also another continuity property, called *p-fine continuity*, see [2], Section 6.4.

Theorem 8 If f is a p-quasicontinuous function, then f is p-finely continuous at p-q.e. x, i.e., for all $\varepsilon > 0$ the set $\{y : |f(y) - f(x)| \geq \varepsilon\}$ is p-thin at x. Conversely, f is p-quasicontinuous if it is p-finely continuous p-q.e.

We can now formulate other equivalent conditions for Ω to be p-stable. Proofs are found in [2], Section 11.4.

Theorem 9 Let $\Omega \subset \mathbf{R}^n$ be open and equal to the interior of its closure. Then the following conditions are equivalent:

- Ω is p-stable;
- $C_p(G \cap \overline{\Omega}^c) = C_p(G \cap \Omega^c)$ for all open G;
- there is $\eta > 0$ such that $C_p(G \cap \overline{\Omega}^c) \geq \eta C_p(G \cap \Omega^c)$ for all open G;
- $\liminf_{r \to 0} C_p(B(x, r) \cap \overline{\Omega}^c)/C_p(B(x, r) \cap \Omega^c) > 0$ for p-q.e. $x \in \partial\Omega$;
- $C_p(\partial\Omega \cap e_p(\overline{\Omega}^c)) = 0$, i.e., $\overline{\Omega}^c$ is p-thick p-q.e. on $\partial\Omega$;
- $e_p(\overline{\Omega}^c) = e_p(\Omega^c)$;
- $C_p(e_p(\overline{\Omega}^c) \backslash e_p(\Omega^c)) = 0$.

In order to give a brief indication of how the proof of this theorem is related to the concepts defined above, we observe that if $f \in W^{1,p}$ and $f = 0$ on $\overline{\Omega}^c$, then by fine continuity $f(x) = 0$ p-q.e. on $\partial\Omega \cap b_p(\overline{\Omega}^c)$. If one of the last three conditions is satisfied, it follows (by means of the Kellogg property in the case of the last two) that $f(x) = 0$ p-q.e. on $\partial\Omega$, and thus $f \in W_0^{1,p}(\Omega)$ by the Havin–Bagby theorem, Theorem 3.

We have stated this result, because the solution of the problem stated initially is somewhat analogous. We have, in fact the following theorem.

Theorem 10 Let $\Omega \subset \mathbf{R}^n$ be open and bounded, and let $1 < q < p < \infty$. Then the following conditions are equivalent:

- $W_0^{1,p}(\Omega) = W^{1,p}(\Omega) \cap W_0^{1,q}(\Omega)$;
- $C_p(G \cap b_q(\Omega^c)) = C_p(G \cap \Omega^c)$ for all open G;
- there is $\eta > 0$ such that $C_p(G \cap b_q(\Omega^c)) \geq \eta C_p(G \cap \Omega^c)$ for all open G;
- $\liminf_{r \to 0} C_p(B(x, r) \cap b_q(\Omega^c))/C_p(B(x, r) \cap \Omega^c) > 0$ for p-q.e. $x \in \partial\Omega$;
- $C_p(\partial\Omega \cap e_q(\Omega^c)) = 0$, i.e., $b_q(\Omega^c)$ is p-thick p-q.e. on $\partial\Omega$;
- $e_p(b_q(\Omega^c)) = e_p(\Omega^c)$;
- $C_p(e_p(b_q(\Omega^c))\backslash e_p(\Omega^c)) = 0$.

The similarity is that if $f \in W_0^{1,q}(\Omega)$, then $f(x) = 0$ q-q.e. on Ω^c, and we want to conclude that $f(x) = 0$ p-q.e. on Ω^c. But by the definition of thickness the set $\{x : f(x) = 0\}$ is q-thick, and hence also p-thick, everywhere on $b_q(\Omega^c)$. It follows that $f(x) = 0$ at all points of $b_q(\Omega^c)$ where f is p-finely continuous, i.e., p-q.e. If one of the last three conditions is satisfied, the desired conclusion follows as before.

Finally, we obtain in the same way the following solution to the problem posed initially.

Theorem 11 Let $\Omega \subset \mathbf{R}^n$ be open and bounded, and let $1 < p < \infty$. Then the following conditions are equivalent:

- $W_0^{1,p}(\Omega) = W^{1,p}(\Omega) \cap \bigcap_{q<p} W_0^{1,q}(\Omega)$;
- $C_p(G \cap \bigcup_{q<p} b_q(\Omega^c)) = C_p(G \cap \Omega^c)$ for all open G;
- there is $\eta > 0$ such that $C_p(G \cap \bigcup_{q<p} b_q(\Omega^c)) \geq \eta C_p(G \cap \Omega^c)$ for all open G;
- $\liminf_{r \to 0} C_p(B(x, r) \cap \bigcup_{q<p} b_q(\Omega^c))/C_p(B(x, r) \cap \Omega^c) > 0$ for p-q.e. $x \in \partial\Omega$;
- $C_p(\partial\Omega \cap \bigcap_{q<p} e_q(\Omega^c)) = 0$, i.e., $\bigcup_{q<p} b_q(\Omega^c)$ is p-thick p-q.e. on $\partial\Omega$;
- $e_p(\bigcup_{q<p} b_q(\Omega^c)) = e_p(\Omega^c)$;
- $C_p(e_p(\bigcup_{q<p} b_q(\Omega^c))\backslash e_p(\Omega^c)) = 0$.

In particular, it is a sufficient condition for (1) that Ω is regular for the q-Laplace equation for some $q < p$, since then $b_q(\Omega^c) = \Omega^c$, and Ω^c is p-thick p-q.e. on $\partial\Omega$ by the Kellogg property. Proofs, and other, easier to apply, sufficient conditions are given in [4].

References

1. Adams, D.R. and Hedberg, L.I., Inclusion relations among fine topologies in non-linear potential theory, *Indiana Univ. Math. J.* **33** (1984), 117–126.
2. Adams, D.R. and Hedberg, L.I., *Function Spaces and Potential Theory,* Springer, Berlin Heidelberg, 1996.
3. Hedberg, L.I., Approximation by harmonic functions, and stability of the Dirichlet problem, *Expo. Math.* **11** (1993), 193–259.
4. Hedberg, L.I. and Kilpeläinen, T., On the stability of Sobolev spaces with zero boundary values, *Math. Scand.,* to appear.
5. Heinnonen, J., Kilpeläinen, T. and Martio, O., *Nonlinear Potential Theory of Degenerate Elliptic Equations,* Oxford University Press, Oxford, 1993.
6. Kilpeläinen, T. and Malý, J., The Wiener test and potential estimates for quasilinear elliptic equations, *Acta Math.* **172** (1994), 137–161.
7. Lindqvist, P., On the equation div $(|\nabla u|^{p-2}\nabla u) + \lambda |u|^{p-2}u = 0$, *Proc. Amer. Math. Soc.* **109** (1990), 157–164; Addendum, *Ibid.* **116** (1992), 583–584.
8. Lindqvist, P., On non-linear Rayleigh quotients, *Potential Analysis* **2** (1993), 199–218.
9. Maz'ya, V.G., O nepreryvnosti v granichnoĭ tochke resheniĭ kvazilineĭnykh èllipticheskikh uravneniĭ, *Vestnik Leningrad. Univ. Mat. Mekh. Astronom.* **25** 13 (1970), 42–55. Correction, *ibid.* **27**: 1 (1972), 160. English translation: On the continuity at a boundary point of solutions of quasilinear equations, *Vestnik Leningrad Univ. Math.* **3** (1976), 225–242.
10. Swanson, D. and Ziemer, W.P., Sobolev functions whose inner trace at the boundary is zero, *Ark. Mat.,* **37** (1999), 373–380.

Function Spaces and Applications
D.E. Edmunds et al (Eds)

9. Imbeddings of Weighted Sobolev Spaces

Pawan K. Jain* and Pankaj Jain†

Department of Mathematics, University of Delhi, Delhi, India

1. Introduction

The study of 'Function Spaces' has become an important tool in modern mathematics; specially, in the areas where partial differential equations and boundary value problems are involved. In fact, these spaces offer us the facility to provide a weak solution of certain PDE or BVP where, in the classical sense, there might not be any solution.

Function spaces have been studied and are being studied for many aspects. Some of them to mention are basic properties of different function spaces (equivalent norms, separability, reflexibility etc.); density of smooth functions; imbedding properties; inequalities; trace theory; extension theorems; duality theory; interpolation theory; entropy and approximation numbers; measure of non-compactness etc. Tremendous amount of work has been done in these areas and it is not possible to mention the references of all of them. However, one may refer to the monographs [1, 2, 3, 8, 18–21, 24–27] and the references therein.

Weights play an important role in the theory of function spaces, e.g., while dealing with degenerate boundary value problems. On considering unbounded domains, weighted spaces do a remarkable job.

Here, the imbedding aspect of weighted Sobolev spaces is considered. In fact, the purpose of this article is to collect certain results characterizing weights for which the corresponding weighted Sobolev spaces are continuously (or, compactly) imbedded into a suitable weighted Lebesgue space or another weighted Sobolev space. This includes a very recent work done by a group consisting of P.K. Jain, Pankaj Jain and Bindu Bansal at the University of Delhi, Delhi. Since all the results have either been published or under publication, we omit the details of them. Our precise aim is the following:

We deal with the continuous imbedding

$$W^{1,p}(\Omega; v) \to L^q(\Omega; w)$$

$$(\text{or } W_0^{1,p}(\Omega; v) \to L^q(\Omega; w) \tag{1.1}$$

*This work has been supported by the U.G.C. (India) vide F.8–2/97 (SR–I).
†This work was partially supported by the U.G.C. (India) vide 13–1/97.

and the compact imbedding

$$W^{1,p}(\Omega; v) \to \to L^q(\Omega; w)$$

$$(\text{or } W_0^{1,p}(\Omega; v) \to \to L^q(\Omega; w).$$

Many authors have considered these imbeddings with various combinations of different parameters involved. Let us consider the weights which are some form of the distance from the boundary $\partial\Omega$ of Ω. Three situations arise:

(a) the distance, to be denoted by d, is taken from the whole boundary $\partial\Omega$;
(b) the distance, to be denoted by d_0, is taken from a point $x_0 \in \partial\Omega$;
(c) the distance, to be denoted by d_M, is taken, in general, from a proper subset $M \subset \partial\Omega$.

Situation (a) has been dealt by Kufner [18] for the case $p = q$ and with the weights which are either power type, d^α with $\alpha \in \mathbb{R}$ or some function of the distance d. In fact, he has obtained sufficient conditions for the continuous imbeddings (1.1) to hold. For $p \neq q$, both continuous and compact imbeddings were obtained with power type weights in [6]. These results have also been collected in the monograph [20] and so we avoid repetition.

As regards situation (b), again, Kufner [18] considered weights of the type d_0^α and $s(d_0)$ and obtained sufficient conditions for the continuous imbeddings (1.1) to hold for $p = q$. In [15] and [16], these results are extended so as to cover the case $1 \leq q < p < \infty$ as well as it is shown that the conditions are both necessary and sufficient. Moreover, the compactness of the imbeddings are also studied. These results are collected in Sections 2 and 3. In the case of power type weights d_0^α, the imbeddings are further extended to higher order weighted Sobolev spaces.

For the last situation (c), Kufner [18] conjuctured a sufficient condition for the imbedding (1.1) to hold when $p = q$. This conjucture was proved by Rákosnik [23]. The weights considered either by Kufner or by Rákosnik were power type d_M^α. The results were extended in [17] in many ways. The case $1 \leq q < p < \infty$ has been considered, weights are either power type or of the type $s(d_M)$. The compact imbeddings have also been considered and above all, the conditions obtained are both necessary and sufficient. Furthermore, it is shown that some of imbeddings could be extended to higher order weighted spaces. Section 4 has been devoted for these results.

So far, all the weights considered have been some form of the distance from the boundary $\partial\Omega$ of the bounded domain Ω. Let us, now, consider weights which are fairly general in nature. Let S denotes the set of all weight functions on Ω, i.e., the set of all functions which are positive, finite and measurable a.e. on Ω. In [11], Opic has given necessary and sufficient conditions so that the imbeddings

$$W^{1,p}(\Omega; S) \rightarrow L^q(\Omega; w)$$

and

$$W^{1,p}(\Omega; S) \rightarrow\rightarrow L^q(\Omega; w)$$

hold, where $1 \leq p, q < \infty$ and $w \in S$. About the domain Ω, it has been assumed that $\Omega = \bigcup_{n=1}^{\infty} \Omega_n$, where Ω_n are domains in $\mathbb{R}^N$ such that $\Omega_n \subset \Omega_{n+1} \subset \Omega$, $\Omega_{n+1} \neq \Omega$.

Using the characterizations for the above imbeddings, Gurka and Opic [5] obtained necessary and sufficient conditions for the weighted Sobolev space $W^{1,p}(\Omega; v_0, v_1)$ to imbedd continuously or compactly into a weighted Lebesgue space $L^q(\Omega; w)$, the domain Ω being bounded in $\mathscr{C}^{0,k}$, $0 < \kappa \leq 1$ and $1 \leq p \leq q < \infty$. This case has also been investigated by Lizorkin and Otelbaev in [9, 10]. The case for unbounded domains has been studied, again, by Gurka and Opic [7].

The second author has studied necessary and sufficient conditions, using the techniques of Gurka and Opic, so that the imbeddings

$$W^{1,p}(\Omega; v_0, v_1) \rightarrow W^{1,p}(\Omega; w)$$

and

$$W^{1,p}(\Omega; v_0, v_1) \rightarrow\rightarrow W^{1,p}(\Omega; w)$$

take place, for bounded domains in [13] and for unbounded domains in [12]. The precise results are presented in Section 5.

For different notations and terminology used in this article one may refer to [1, 12, 13, 14, 18, 23]. However, some of the concepts will be given at different places just to maintain the continuty of the subject matter.

2. Imbeddings for the Domains in $\mathscr{K}(x_0)$

Let $x_0 \in \partial\Omega$, the boundary of Ω. The domain Ω is said to be of class $\mathscr{K}(x_0)$, denoted by $\Omega \in \mathscr{K}(x_0)$, if $\Omega \in \mathscr{C}^0$ and there is a bounded (open) cone K with its vertex at the point x_0 such that $K \cap \overline{\Omega} = \phi$. It can easily be seen that $\mathscr{C}^{0,1} \subset \mathscr{K}(x_0)$, and that the inclusion holds for every $x_0 \in \partial\Omega$.

Recall that d_0 is the distance given by

$$d_0(x) = d_{\{x_0\}}(x) = |x - x_0|, x \in \Omega.$$

Kufner [18] gave the following imbedding results:

Theorem 1 Let $\Omega \in \mathscr{K}(x_0)$, $x_0 \in \partial\Omega$, $1 < p < \infty$ and $\varepsilon > p - 1$. Then, the imbedding

$$W^{1,p}(\Omega; d_0^{\varepsilon}) \rightarrow L^p(\Omega; d_0^{\varepsilon - p})$$

holds.

Theorem 2 Let $\Omega \in \mathscr{K}(x_0)$, $x_0 \in \partial\Omega$, $1 < p < \infty$ and $\varepsilon \neq p - 1$. Then, the imbedding

$$W_0^{1,p}(\Omega; d_0^{\varepsilon}) \to L^p(\Omega; d_0^{\varepsilon-p})$$

holds.

Remark Imbedding in Theorem 1 has also been obtained by Kufner [18] for $\varepsilon \leq p - 1$. More precisely, in that case, the imbedding

$$W^{1,p}(\Omega, d_0^{\varepsilon}) \to L^p(\Omega; d_0^{-1+w})$$

holds, where $w > 0$ is arbitrary.

It can be noted that Theorems 1 and 2 give only the sufficient conditions for the corresponding imbeddings to hold and also nothing has been said about the compactness of these imbeddings. An attempt has been made to fill this gap in [15]. The results are the following:

Theorem 3 Let $1 \leq q < p < \infty$, $\Omega \in \mathscr{K}(x_0)$, $x_0 \in \partial\Omega$ and $\alpha, \beta \in \mathbb{R}$. Then, the imbedding

$$W^{1,p}(\Omega; d_0^{\beta}) \to L^q(\Omega; d_0^{-1+\alpha})$$

holds if and only if

$$\left.\begin{array}{ll} \beta > p - 1, & \alpha > \beta \dfrac{q}{p} - \dfrac{q}{p'} - 1, \\ \text{or} & \\ \beta \leq p - 1, & \alpha > -1, \end{array}\right\} \tag{2.1}$$

where p' satisfies $\dfrac{1}{p} + \dfrac{1}{p'} = 1$.

Theorem 4 Let $1 \leq p, q < \infty$, $\Omega \in \mathscr{K}(x_0)$, $x_0 \in \partial\Omega$ and $\alpha, \beta \in \mathbb{R}$. Then, the compact imbedding

$$W^{1,p}(\Omega; d_0^{\beta}) \to \to L^q(\Omega; d_0^{\alpha})$$

holds if and only if (2.1) is satisfied.

Remark It is important to note, from Theorems 3 and 4, that when $1 \leq q < p < \infty$, the necessary and sufficient conditions are the same for continuous imbedding as well as for compact imbedding. In fact, it turns out that, under the condition, either the space $W^{1,p}(\Omega; d_0^{\beta})$ is imbedded compactly into the space $L^q(\Omega; d_0^{\alpha})$ or it is not even imbedded continuously. This fact will be realized more closely later in Section 4 when we shall deal with imbeddings in different domains.

Also, necessary and sufficient conditions for the space $W_0^{1,p}(\Omega; d_0^{\beta})$ to be imbedded continuously and compactly into the space $L^q(\Omega; d_0^{\alpha})$ have been obtained, again, in [15]. The results are the following:

Theorem 5 Let $1 \le q < p < \infty$, $\Omega \in \mathscr{K}(x_0)$, $x_0 \in \partial\Omega$ and $\alpha, \beta \in \mathbb{R}$. Then, the imbedding

$$W_0^{1,p}(\Omega; d_0^{\beta}) \to L^q(\Omega; d_0^{\alpha})$$

holds if and only if

$$\beta \in \mathbb{R}, \alpha > \beta \frac{q}{p} - \frac{q}{p'} - 1. \tag{2.2}$$

In view of the last remark, the compact imbedding

$$W_0^{1,p}(\Omega; d_0^{\beta}) \to\to L^q(M; d^{\alpha})$$

holds if and only if, (2.2) holds.

Theorem 5 can be extended in the following way:

Theorem 6 Let $1 \le q < p < \infty$, $\Omega \in \mathscr{K}(x_0)$, $x_0 \in \partial\Omega$ and $\alpha, \beta \in \mathbb{R}$. Then, the imbedding

$$W_0^{1,p}(\Omega; d_0^{\beta}) \to L^q(\Omega; d_0^{\alpha})$$

holds if and only if

$$\beta \in \mathbb{R}, \alpha > (\beta + N - 1) \frac{q}{p} - \frac{q}{p'} - N.$$

So far, the discussion has been for the case $1 \le q < p < \infty$. In case $1 \le p \le q < \infty$, the imbeddings $W^{1,p}(\Omega; d_0^{\beta}) \to L^q(\Omega; d_0^{\alpha})$ and $W_0^{1,p}(\Omega; d_0^{\beta}) \to L^q(\Omega; d_0^{\alpha})$ can be obtained by extending and using the results of Kufner ([18], Theorems 8.10 and 8.11) as done by Gurka and Opic ([6], Theorems 8.4 and 8.5) alongwith a special case of the imbedding $W^{1,p}(\Omega, v_0, v_1) \to L^q(\Omega; w)$ given by Gurka and Opic [5]. More precisely, we have

Theorem 7 Let $1 \le p \le q < \infty$, $\Omega \in \mathscr{K}(x_0)$, $x_0 \in \partial\Omega$ and $\alpha, \beta \in \mathbb{R}$. Then, the imbedding

$$W^{1,p}(\Omega; d_0^{\beta}) \to L^q(\Omega; d_0^{\alpha})$$

hods if and only if

$$\beta > p - 1, \alpha \ge \beta \frac{q}{p} - \frac{q}{p'} - 1,$$

or

$$\beta \le p - 1, \quad \alpha > -1.$$

Theorem 8 Let $1 \le p \le q < \infty$, $\Omega \in \mathscr{K}(x_0)$, $x_0 \in \partial\Omega$ and $\alpha, \beta \in \mathbb{R}$. Then, the imbedding

$$W_0^{1,p}(\Omega; d_0^{\beta}) \to L^q(\Omega; d_0^{\alpha})$$

holds if and only if

$$\beta \neq p - 1, \alpha \geq \beta \frac{q}{p} - \frac{q}{p'} - 1,$$

or

$$\beta = p - 1, \quad \alpha > - 1.$$

As Theorem 5 has been extended to yield Theorem 6, the same treatment can be given to Theorems 7 and 8 but we have to assume, in addition, that $\frac{1}{N} \geq \frac{1}{p} - \frac{1}{q}$. Let us formulate the extension of Theorem 7 only.

Theorem 9 Let $1 \leq p \leq q < \infty$, $\Omega \in \mathscr{K}(x_0)$, $x_0 \in \partial\,\Omega$, $\frac{1}{N} \geq \frac{1}{p} - \frac{1}{q}$ and $\alpha, \beta \in \mathbb{R}$. Then, the imbedding

$$W^{1,p}(\Omega; d_0^{\beta}) \to L^q(\Omega; d_0^{\alpha})$$

holds if and only if

$$\beta > p - 1, \alpha \geq (\beta + N - 1)\frac{q}{p} - \frac{q}{p'} - N,$$

or

$$\beta \leq p - 1, \quad \alpha > (N - 1)\frac{q}{p} - N.$$

On the lines of Theorem 4, one can easily formulate and prove the corresponding compact imbeddings for the imbeddings in Theorems 6, 7, 8 and 9.

Is it possible to extend Theorem 3 also as has been done for Theorems 5, 7 and 8? The answer to this question is affermative but a different kind of domain is required i.e. $\mathscr{K}^{0,1}(x_0)$.

Let $x_0 \in \partial\Omega$. The domain Ω is said to be of class $\mathscr{K}^{0,1}(x_0)$, denoted by $\Omega \in \mathscr{K}^{0,1}(x_0)$, if $\Omega \in \mathscr{K}(x_0)$ and the functions $a_i = a_i(y_i')$ in the definition of $\mathscr{C}^0$, which describes the boundary $\partial\Omega$ in a neighbourhood of the point x_0, satisfies the Lipschitz condition for $i = 1$.

We can, now, state the extension of Theorem 3.

Theorem 10 Let $1 \leq q < p < \infty$, $\Omega \in \mathscr{K}^{0,1}(x_0)$, $x_0 \in \partial\Omega$ and $\alpha, \beta \in \mathbb{R}$. Then, the imbedding

$$W^{1,p}(\Omega; d_0^{\beta}) \to L^q(\Omega; d_0^{\alpha})$$

holds if and only if

$$\beta > p - 1, \alpha > (\beta + N - 1)\frac{q}{p} - \frac{q}{p'} - N,$$

or

$$\beta \leq p - 1, \quad \alpha > - N.$$

It is also possible to obtain the imbeddings discussed in this section, in certain situations, for the Sobolev spaces of higher order. For example, we state the following two results:

Theorem 11 Let $1 \le q < p < \infty$, $\Omega \in \mathscr{K}(x_0)$, $x_0 \in \partial\Omega$ and $\alpha, \beta \in \mathbb{R}$, $\alpha \ge \beta$. Then, for $\beta > p - 1$, the imbedding

$$W^{k,p}(\Omega; d_0^{\beta}) \to W^{s,q}(\Omega; d_0^{\alpha})$$

holds if and anly if

$$\alpha > \beta\frac{q}{p} - \frac{q}{p'} - (k - s - 1)\, q - 1,$$

where k and s are non-negative integers.

Theorem 12 Let $1 \le p \le q < \infty$, $\Omega \in \mathscr{K}(x_0)$, $x_0 \in \partial\Omega$ and $\alpha, \beta \in \mathbb{R}$, $\alpha \le \beta$. Then, for $\beta \le p - 1$, the imbedding

$$W^{k,p}(\Omega; d_0^{\beta}) \to W^{s,q}(\Omega; d_0^{\alpha})$$

holds if and only if $\alpha > -1$, where k and s are non-negative integers.

3. Imbeddings with Some General Weights

In the previous section, we discussed continuous and compact imbeddings of weighted Sobolev spaces with power type weights, where the powers were of the distance measured from a point $x_0 \in \partial\Omega$. Instead of taking powers, let us consider a general functions of this distance i.e. consider the function $s_0 \equiv s(d_0) = s(d_0(x))$.

In this section, we wish to study continuous imbeddings of Sobolev spaces with weights of the type $s(d_0)$ and with the domain Ω in $\mathscr{C}^{0,1}$ or $\mathscr{K}(x_0)$. Kufner [18] has dealt with these type of imbeddings for the case $p = q$ and has given only the sufficient conditions for the imbeddings to hold. We need the following concepts.

We say that a positive continuous function $f \equiv f(t)$ defined on $(0, \infty)$ has the Property (H) if for every pair of positive constants c_1, c_2, there exist a pair of positive constants C_1, C_2 such that

$$c_1 \le \frac{t_1}{t_2} \le c_2 \Rightarrow C_1 \le \frac{f(t_1)}{f(t_2)} \le C_2.$$

Clearly, all power type weights d^{α} and d_0^{α} satisfy Property (H).

Let $f \equiv f(t)$ be a positive continuous function defined on $(0, \infty)$. Let f be non-increasing in the interval $(0, c)$, $c > 0$ and $\lim_{t\to 0} f(t) = \infty$. Then, f is said to be of type I (or, of type II) if $\int_0^c f(t)\,dt < \infty$ $\left(\text{or, } \int_0^c f(t)\,dt = \infty\right)$. Note that power type weights with negative powers are of type I, or of type II.

We, now, discuss the results of this section.

Theorem 13 Let $1 \le q < p < \infty$ and $\Omega \in \mathscr{C}^{0,1}$.

(a) If either s_1 is non-decreasing in a neighbourhood of zero with $\lim_{t\to 0} s_1(t) = 0$, or s_1 is non-increasing in a neighbourhood of zero with $\lim_{t\to 0} s_1(t) = \infty$ and s_1 is of type I, then the imbedding

$$W^{1,p}(\Omega; s_1(d)) \to L^q(\Omega; s_0(d)) \tag{3.1}$$

holds if and only if

$$\left\{\int_0^{\lambda}\left[\left(\int_0^z s_0(x)\,dx\right)^{1/q}\left(\int_z^{\lambda}(s_1(x))^{1-p'}\,dx\right)^{1/q'}\right]^r (s_1(z))^{1-p'}\,dz\right\}^{1/r} < \infty. \tag{3.2}$$

(b) If s_1 is non-increasing in a neighbourhood of zero with $\lim_{t\to 0} s_1(t) = \infty$ and s_1 is of type II, then the imbedding (3.1) holds if and only if (3.2) and

$$\left\{\int_0^{\lambda}\left[\left(\int_z^{\lambda} s_0(x)\,dx\right)^{1/q}\left(\int_0^z \left(s_1(x)\right)^{1-p'}dx\right)^{1/q'}\right]^r \left(s_1(z)\right)^{1-p'}dz\right\}^{1/r} < \infty \tag{3.3}$$

hold.

Theorem 14 Let $1 \le q < p < \infty$ and $\Omega \in \mathscr{C}^{0,1}$. Then, the imbedding

$$W_0^{1,p}(\Omega; s_1(d)) \to L^q(\Omega; s_0(d))$$

holds if and only if (3.2) or (3.3) hold.

Note. The functions s_0 and s_1 in Theorems 13 and 14 satisfy Property (H). Further, the number λ in Theorems 13 and 14 has been taken from the definition of $\mathscr{C}^0$.

Example 1 Let $1 \le q < p < \infty$, $s_0(d) = x^{\alpha}$, $s_1(d) = x^{\beta}$ where $\alpha, \beta \in \mathbb{R}$. Clearly, the functions s_0 and s_1 satisfy conditions of Theorem 13(a) and consequently the imbedding

$$W^{1,p}(\Omega, d^{\beta}) \to L^q(\Omega; d^{\alpha})$$

holds if and only if (3.2) is satisfied which reduces to

$$\beta > p - 1, \qquad \alpha > \beta\frac{q}{p} - \frac{q}{p'} - 1$$

or

$$\beta \le p - 1, \qquad \alpha > -1.$$

This result matches one of the results proved by Gurka and Opic for $\kappa = 1$ ([6], Theorem 9.3).

In what follows, we give imbeddings for the domains $\Omega \in \mathscr{K}(x_0)$ which, in fact, is a larger class than $\mathscr{C}^{0,1}$.

Theorem 15 Let $1 \le q < p < \infty$, $\Omega \in \mathscr{K}(x_0)$, $x_0 \in \partial\Omega$, $\lim_{t\to 0} s_1(t) = 0$ and the functions s_0 and s_1 satisfy Property (H). Then, the imbedding

$$W^{1,p}(\Omega;\ s_1(d_0)) \to L^q(\Omega;\ s_0(d_0)) \tag{3.4}$$

holds if and only if

$$\left\{\int_0^{\lambda+|y^{**}|}\left[\left(\int_z^z s_0(x_0)\,dx\right)^{1/q}\left(\int_z^{\lambda+|y^{**}|}\left(s_1(x_0)\right)^{1-p'}dx\right)^{1/q'}\right]^r \times \left(s_1(z_0)\right)^{1-p'}dz\right\}^{1/r} < \infty \tag{3.5}$$

Theorem 16 Let $1 \le q < p < \infty$, $\Omega \in \mathscr{K}(x_0)$, $x_0 \in \partial\Omega$ and the functions s_0 and s_1 satisfy Property (H). Then, the imbedding

$$W_0^{1,p}(\Omega; s_1(d_0)) \to L^q(\Omega; s_0(d_0)) \tag{3.6}$$

holds if and only if (3.5) or

$$\left\{\int_0^{\lambda+|y^{**}|}\left[\left(\int_z^{\lambda+|y^{**}|} s_0(x_0)\,dx\right)^{1/q} \times \left(\int_0^z \left(s_1(x_0)\right)^{1-p'}dz\right)^{1/q'}\right]^r \left(s_1(z_0)\right)^{1-p'}dz\right\}^{1/r} < \infty$$

hold.

The expression y^{**} in Theorems 15 and 16 is given by $y^{**} = (y_1', a_1(y_1'))$, where the numbers λ, y_1', $a_1(y_1')$ are taken from the definition of $\mathscr{C}^0$.

Towards an example for Theorem 15, we can consider Example 1 but with weights taken in terms of the distance from $x_0 \in \partial\Omega$. Then, the corresponding conditions are those which were obtained in Theorem 3. Similar examples for Theorems 14 and 16 can be worked out.

Theorem 16 can be extended to give a different set of conditions for the imbedding (3.6) to hold. In fact, the resulting conditions are stronger but the property (H) is not required here which was necessary for Theorem 16. More precisely, we have

Theorem 17 Let $1 \le q < p < \infty$ $\Omega \in \mathscr{K}(x_0)$, $x_0 \in \partial\Omega$ and $0 < b < \infty$. Then, the imbedding (3.6) holds if and only if

$$\left\{\int_0^b \left[\left(\int_0^z d_0^{N-1} s_0(x_0)\, dx\right)^{1/q} \left(\int_0^b \left(d_0^{N-1} s_1(x_0)\right)^{1-p'} dx\right)^{1/q'}\right]^r \left(d_0^{N-1} s_1(z_0)\right)^{1-p'} dz\right\}^{1/r} < \infty \quad (3.7)$$

or

$$\left\{\int_0^b \left[\left(\int_z^b d_0^{N-1} s_0(x_0)\, dx\right)^{1/q} \left(\int_0^z \left(d_0^{N-1} s_1(x_0)\right)^{1-p'} dx\right)^{1/q'}\right]^r \left(d_0^{N-1} s_1(z_0)\right)^{1-p'} dz\right\}^{1/r} < \infty$$

In a similar fashion, Theorem 15 can also be extended out in that case again, we assume Property (H) on functions s_0 and s_1 and restrict our domain $\mathscr{K}(x_0)$ to $\mathscr{K}^{0,1}(x_0)$. The result is the following:

Theorem 18 Let $1 \le q < p < \infty$, $\Omega \in \mathscr{K}^{0,1}(x_0)$, $x_0 \in \partial\Omega$ and $0 < b < \infty$. Then, the imbedding (3.4) holds if and only if (3.7) holds.

Examples for Theorems 17 and 18 can also be obtained. For details regarding the proofs of the theorems and examples presented in this section, one may refer to [16].

4. Imbeddings for the Domains in $Q^{0,1}(M)$

Let $M \subset \partial\Omega$ be a closed set. Put

$$C_M^\infty(\Omega) = \{u \in C^\infty(\Omega);\ \text{supp}\ u \cap M = \phi\}.$$

We write $W_M^{k,p}(\Omega; w)$ for the closure of $C_M^\infty(\Omega)$ with respect to the norm

$$\| u \|_{k,p,w;\Omega} = \Big(\sum_{|\alpha| \le k} \| D^\alpha u \|_{p,w;\Omega}^p\Big)^{1/p}$$

which, in fact, is the norm of $W^{k,p}(\Omega; w)$. It can be observed that $C_0^\infty(\Omega) = C_{\partial\Omega}^\infty(\Omega)$ and that $W_0^{k,p}(\Omega; w) = W_{\partial\Omega}^{k,p}(\Omega; w)$.

Set $d_M(x) = \text{dist}\,(x, M)$, $x \in \Omega$. Clearly, any power d_M^α and the function $s(d_M)$ are weight functions on Ω.

Put $Q = (0, 1)^N$. For $m = 0, 1, \ldots, N-1$, define

$$Q(m) = \{x \in \overline{Q} : x_{m+1} = x_{m+2} = \ldots = x_N = 0\}.$$

We say that a closed subset $M \subset \partial\Omega$ is a manifold of dimension m and that the bounded domain Ω is of class $Q^{0,1}(M)$, if there exists an open covering $\{U_i\}_{i=0}^{\bar{k}}$ ($\bar{k}$ may be finite or infinite), of $\overline{\Omega}$ with the follwing properties:

(i) $M \subset \bigcup_{i=1}^{\bar{k}} U_i$ and these exists $n_0 \in \{3, 4, \dots\}$ such that every system of $n_0 + 1$ sets U_i is disjoint;

(ii) there exists a $\delta > 0$ such that $d_M(x) \geq \delta$, $x \in U_0$;

(iii) there exist numbers $C_2 \geq C_1 > 0$ and a system of one-to-one mappings $T_i : Q \to \Omega \cap U_i$ such that $T_i(Q(m)) = \overline{M \cap U_i}$ and

$$C_1 \mid x - y \mid \leq \mid T_i(x) - T_i(y) \mid \leq C_2 \mid x - y \mid,$$

where $x, y \in Q$ and $i = 1, 2, \dots, \bar{k}$.

In what follows, we give results in which the weights are of the type d_M^{α}.

Theorem 19 Let $1 \leq q < p < \infty$, $\alpha, \beta \in \mathbb{R}$, $0 < m \leq N - 1$, $\Omega \in Q^{0,1}(M)$ and $M \subset \partial\Omega$ be a manifold of dimension m. Then, the imbedding

$$W^{1,p}(\Omega; d_M^{\beta}) \to L^q(\Omega; d_M^{\alpha}) \tag{4.1}$$

holds if and only if

$$\left.\begin{array}{ll} \beta > p - N + m, & \alpha > (\beta + N - m - 1)\dfrac{q}{p} - \dfrac{q}{p'} - N + m \\ \text{or} & \\ \beta \leq p - N + m, & \alpha > -N + m \end{array}\right\} \tag{4.2}$$

Remark In Theorems 3 and 4, we observed that the necessary and sufficient conditions for the space $W^{1,p}(\Omega;\ d_0^{\beta})$ to imbedd continuously and compactly into the space $L^q(\Omega; d_0^{\alpha})$ are the same. Also in Theorem 19 the conditions for the continuous imbedding are same as for the compact imbedding $W^{1,p}(\Omega;\ d_0^{\beta}) \to\to L^q(\Omega; d_M^{\alpha})$. In fact, the condition $1 \leq q < p < \infty$ plays an important role. Indeed, if we take $\alpha = \beta = 0$ and also $m = 0$ in Theorem 19, we get that the imbedding $W^{1,p}(\Omega) \to L^q(\Omega)$ holds if and only if

$$p < N, \quad q < \frac{Np}{N-p},$$

or

$$p \geq N, \quad N > 0.$$

On the other hand, the classical imbedding theorem says that the space $W^{1,p}(\Omega)$ is continuously imbedded into the space $L^{\frac{Np}{N-p}}(\Omega)$ and for any

$q \in \left[p, \frac{Np}{N-p}\right)$, the imbedding is essentially compact (Ω needs to be bounded for compactness). Moreover, these imbeddings holds even when $1 \leq q < p < \infty$ provided the domain Ω is of finite volume. Thus, in our case because of the strictness of the inequality $1 \leq q < p < \infty$ and the fact that $p < \frac{Np}{N-p}$, we claim that the said imbeddings are compact as well, the domain Ω being bounded.

We also have the following imbedding results:

Theorem 20 Let $1 \leq q < p < \infty$, $\alpha, \beta \in \mathbb{R}$, $0 \leq m \leq N-1$, $\Omega \in Q^{0,1}(M)$ and $M \subset \partial\Omega$ be a manifold of dimension m. Then, the imbedding

$$W_M^{1,p}(\Omega; d_M^\beta) \to L^q(\Omega; d_M^\alpha)$$

holds if and only if

$$\beta \in \mathbb{R}, \qquad \alpha > (\beta + N - m - 1)\,\frac{q}{p} - \frac{q}{p'} - N + m.$$

Theorem 21 Under the assumptions of Theorem 20, the imbedding

$$W_0^{1,p}(\Omega; d_M^\beta) \to L^q(\Omega; d_M^\alpha)$$

holds if and only if

$$\beta \in \mathbb{R}, \qquad \alpha > \beta\frac{q}{p} - \frac{q}{p'} - 1.$$

It is also possible, in certain situations, to extend the imbedding $W^{1,p}(\Omega; d_M^\beta) \to L^q(\Omega; d_M^\alpha)$ to higher order weighted Sobolev spaces as we did in Theorems 11 and 12.

Theorem 22 Let $1 \leq q < p < \infty$, $\alpha, \beta \in \mathbb{R}$, $\alpha \geq \beta$, $0 \leq m \leq N-1$, $\Omega \in Q^{0,1}(M)$ and $M \subset \partial\Omega$ be a manifold of dimension m. Then, for $\beta > p - N + m$, the imbedding

$$W^{k,p}(\Omega; d_M^\beta) \to W^{s,q}(\Omega; d_M^\alpha)$$

holds if and only if

$$\alpha > (\beta + N - m - 1)\,\frac{q}{p} - \frac{q}{p'} - (k - s - 1) - N + m.$$

In view of the remark after Theorem 19, the imbeddings in Theorems 20, 21 and 22 are also compact under the corresponding conditions.

Now, we consider the weights of the type $s(d_M)$, i.e. function of the distance d_M. We assume that such functions satisfy Property (H) and that $\frac{1}{r} = \frac{1}{q} - \frac{1}{p}$. We have the following:

Theorem 23 Assume that the assumptions of Theorem 20 are satisfied. Further assume that the functions s_0 and s_1 satisfy Property (H) and that $\frac{1}{r} = \frac{1}{q} - \frac{1}{p}$. Then, the imbedding

$$W^{1,p}(\Omega; s_1(d_M)) \to L^q(\Omega; s_0(d_M))$$

holds if and only if

$$\left\{\int_0^b \left[\left(\int_0^z x^{N-m-1} s_0(x)dx\right)^{1/q} \left(\int_z^0 \left(x^{N-m-1} s_1(x)\right)^{1-p'} dx\right)^{1/q'}\right]^r \left(z^{N-m-1} s_1(z)\right)^{1-p'} dz\right\}^{1/r} < \infty \tag{4.3}$$

holds and the imbedding

$$W_M^{1,p}(\Omega; d_M^{\beta}) \to L^q(\Omega; d_M^{\alpha})$$

holds if and only if (4.3) or

$$\left\{\int_0^b \left[\left(\int_z^b x^{N-m-1} s_0(x)dx\right)^{1/q} \left(\int_0^z \left(x^{N-m-1} s_1(x)\right)^{1-p'} dx\right)^{1/q'}\right]^r \left(z^{N-m-1} s_1(z)\right)^{1-p'} dz\right\}^{1/r} < \infty$$

hold.

Example If we take either

$$s_0(t) = \alpha\, t^{q-N+m},\ s_1(t) = \beta^{(1-p)} t^{(1-p)(1-q')-N+m+1}$$

with $2 < q < p < \infty$, $\alpha, \beta > 0$, or

$$s_0(t) = \alpha t^{q-N+m},\ s_1(t) = \beta^{(1-p)} t^{q'-pq'+p-N+m}$$

with $1 < q < p < \infty$, $\alpha, \beta > 0$, then the corresponding imbedding can be formulated to hold.

5. Imbeddings with More General Weights

Gurka and Opic [5] obtained necessary and sufficient conditions for the space $W^{1,p}(\Omega;\, v_0, v_1)$ to be continuously and compactly imbedded into the space $L^q(\Omega;\, w)$, where v_0, v_1, w are weight functions, $1 \le p \le q < \infty$ and Ω is a bounded domain. For unbounded domains, the treatment has been given in [7]. Similar study with different conditions has been investigated by Lizorkin and Otelbaev [9, 10].

We give below certain results concerning the imbeddings of the type $W^{1,p}(\Omega;\, v_0, v_1) \to W^{1,p}(\Omega;\, w)$. For the proofs of these results one may refer to [12], [13] and [14]. Throughout this section, Ω is assumed to be broken up into a sequence of subdomains $\{\Omega_n\}$ in $\mathbb{R}^N$ such that $\Omega = \bigcup_{n=1}^{\infty} \Omega_n$ and $\Omega_n \subset \Omega_{n+1} \subset \Omega$, $\Omega_{n+1} \neq \Omega$. We first deal with the case when Ω is bounded. In this case, we make the following assumptions, unless specified otherwise:

A1. Ω is bounded domain in $\mathbb{R}^N$.

A2. $\{\Omega_n\} \subset \zeta^{0,1}$ is a sequence of domains such that

$$\left\{x \in \Omega, \frac{1}{n} < d(x)\right\} \subset \Omega_n \subset \left\{x \in \Omega, \frac{1}{n+1} < d(x)\right\},$$

where $d(x) = \text{dist}\,(x, \partial\Omega)$, $x \in \Omega$.

A3. there exist $n_0 \ge 3$ and a positive measurable function r defined on Ω^{n0} such that

$$r(x) \le d(x)/3, \quad x \in \Omega^{n_0}$$

where $\Omega^{n_0} = \Omega \setminus \Omega_{n_0}$.

Theorem 24 Let the following conditions be satisfied:

S1. $W^{1,p}(\Omega_n, v_0, v_1) \to W^{1,p}(\Omega_n;\, w), \quad n \ge n_0$.

S2. there exist positive measurable functions a_0, a_1 defined on Ω^{n_0} such that

$$w(y) \le a_0(x), \text{ and} \tag{5.1}$$

$$(1 + r^{-p}(y))\, a_1(x) \le v_1(y), \tag{5.2}$$

for all $x \in \Omega^{n_0}$ and for a.e. $y \in B(x, r(x))$.

S3. there exists a constant $K_0 > 0$ such that

$$v_1(x)\, r^{-p}(x) \le K_0 v_0(x), \text{ for a.e. } x \in \Omega^{n_0}.$$

S4. there exists a constant C_r (depending upon r) such that

$$x \in \Omega^{n_0}, y \in B(x, r(x)) \Rightarrow C_r^{-1} \le \frac{r(y)}{r(x)} \le C$$

S5. $\lim_{n\to\infty} A_n = A < \infty$, where

$$A_n = \sup_{x\in\Omega^n} \frac{a_0(x)}{a_1(x)} r^p(x).$$

Then, the imbedding

$$W^{1,p}(\Omega; v_0, v_1) \to W^{1,p}(\Omega; w) \tag{5.3}$$

holods.

The conditions given in Theorem 24 are only sufficient for the imbedding (5.3) to hold. Towards the necessary conditions, we have the following:

Theorem 25 Let S4 and the following conditions be satisfied:

*N*2. there exist positive measurable functions $\hat{a}_0, \hat{a}_1$ defined on Ω^{n_0} such that

$$w(y) \geq \hat{a}_0(x), \quad \hat{a}_1(x) \geq v_1(y)$$

for all $x \in \Omega^{n_0}$ and for a.e. $y \in B\,(x, r(x))$.

N3. there exists a constant $K_0 > 0$ such that

$$K_0 v_0(x) \leq v_1(x) r^{-p}(x), \text{ for a.e. } x \in \Omega^{n_0}.$$

*N*5. $\lim_{n\to\infty} \hat{A}_n = \infty$, where

$$\hat{A}_n = \sup_{x\in\Omega^n} \frac{\hat{a}_0(x)}{\hat{a}_1(x)} r^p(x).$$

Then, $W^{1,p}(\Omega, v_0, v_1)$ is not continuously imbedded in $\dot{W}^{1,p}(\Omega; w)$.

Under the special case $v_0 = v_1 = v$, the conditions for the imbedding in Theorems 24 and 25 can be weekend. More precisely, we have the following:

Theorem 26 Let *S*1, *S*3 with $v_0 = v_1 = v$ along with *S*5 and the following condition be satisfied:

$\overline{\text{S2}}$. there exist positive measurable functions a_0, a_1 defined on Ω^{n_0} such that (5.1) and

$$a_1(x) \leq v(y)$$

for all $x \in \Omega^{n_0}$ and for a.e. $y \in B(x, r(x))$.

Then, the imbedding

$$W^{1,p}(\Omega; v, v) \to W^{1,p}(\Omega; w)$$

holds.

Theorem 27 Let N2 with $v_0 = v_1 = v$ along with N5 be satisfied. Then, the space $W^{1,p}(\Omega; v\ , v)$ is not continuously imbedded in the space $W^{1,p}(\Omega; w)$.

The sufficient condition for the imbedding $W^{1,p}(\Omega; v, v) \to W^{1,p}(\Omega; w)$ given in Theorem 26 can further be weekend but in that case the Condition S5 changes. In fact, we have

Theorem 28 Let S1 with $v_0 = v_1 = v$ along with S2 and the following condition be satisfied:

$\overline{\text{S5}}$ $\lim_{n\to\infty} \overline{A}_n = \overline{A} < \infty$, where

$$\overline{A}_n = \sup_{x\in\Omega^n} \frac{a_0(x)}{a_1(x)}.$$

Then, the imbedding (5.3) holds.

The results for compact imbeddings corresponding to the continuous imbedding (5.3) can be formulated. We give below results corresponding to Theorems 24 and 25.

Theorem 29 Let S2, S3, S4 and the following conditions be satisfied:

S1*. $W^{1,p}(\Omega_n, v_0, v_1) \to\to W^{1,p}(\Omega_n; w)$, $n \geq n_0$

S5*. $\lim_{n\to\infty} A_n = 0$, where A_n is given by the expression in condition S5.
Then, we have the imbedding

$$W^{1,p}(\Omega; v_0, v_1) \to\to W^{1,p}(\Omega; w). \tag{5.4}$$

Theorem 30 Let S4, N2, N3 and the following condition be satisfied.

N5* $\lim_{n\to\infty} \hat{A}_n > 0$, where $\hat{A}_n$ is given in N5.

Then, the space $W^{1,p}(\Omega; v_0, v_1)$ is not compactly imbedded in the space $W^{1,p}(\Omega; w)$.

The special cases of the compact imbeddings in Theorems 29 and 30 can be obtained as we did for continuous imbeddings in Theorems 24 and 25 (see Theorems 26, 27 and 28).

Let us now consider the imbeddings (5.3) and (5.4) when the domain Ω is unbounded. In this case also, we shall be taking the domain Ω such that $\Omega = \bigcup_{n=1}^{\infty} \Omega_n$, where Ω_n are domains in $\mathbb{R}^N$ such that $\Omega_n \subset \Omega_{n+1} \subset \Omega$, $\Omega_{n+1} \neq \Omega$.

Suppose $n \in \mathbb{N}$, $I = (n, \infty)$ and $r : I \to (0, \infty)$. For $x \in \mathbb{R}$, write $I(x) = [x - r(x), x + r(x)]$. We say that the function r has the property $V(n)$ (written $r \in V(n)$) if

(i) r is continuous and non decreasing on I
(ii) $x - r(x)$ is non-decreasing on I
(iii) $\lim_{x\to\infty} [x - r(x)] = \infty$, $\lim_{x\to\bar{n}} r(x) > 0$

(iv) $r(x) \le \frac{x}{2}$ for $x \in I$

(v) there is a constant $C_r \ge 1$ such that

$$C_r^{-1} \le \frac{r(y)}{r(x)} \le C_r,$$

for all $x \in I$ and $y \in I(x) \cap I$.

We have the following imbedding result:

Theorem 31 Let Ω be a domain in $\mathbb{R}^N$, $1 \le p \le \infty$ and let the following conditions be satisfied:

$\tilde{S1}$ there exists $n_0 \in \mathbb{N}$ such that $\Omega^{n_0} = \{x \in \mathbb{R}^N ; |x| > n_0\}$;

$\tilde{S2}$ $W^{1,p}(\Omega_n; v_0, v_1) \to W^{1,p}(\Omega_n; w)$, $n \ge n_0$;

$\tilde{S3}$ there exists positive measurable functions a_0, a_1 defined on Ω^{n_0} and a function $r \in V(n_0)$ such that

$$w(y) \le a_0(x)$$

and

$$(1 + r^{-p}(|y|))\, a_1(x) \le v_1(y),$$

for all $x \in \Omega^{n_0}$ and for a.e. $y \in B(x, r(|x|))$.

$\tilde{S4}$ there exists a constant $K_0 > 0$ such that

$$v_1(x)\, r^{-p}(|x|) \le K_0 v_0(x), \text{ for a.e. } x \in \Omega^{n_0}.$$

$\tilde{S5}$ $\lim_{n\to\infty} A_n = A < \infty$, where

$$A_n = \sup_{x \in \Omega^n} \frac{a_0(x)}{a_1(x)} r^p(|x|).$$

Then, the imbedding (5.3) holds.

Towards the converse of Theorem 31, we have the following:

Theorem 32 Let Ω be a domain in $\mathbb{R}^N$. Let S1 and the following conditions be satisfied:

$\tilde{N3}$ there exist positive measurable functions $\hat{a}_0$, $\hat{a}_1$ defined on Ω^{n_0} and a function $r \in V(n_0)$ such that

$$w(y) \ge \hat{a}_0(x) \text{ and } \hat{a}_1(x) \ge v_1(y)$$

for all $x \in \Omega^{n_0}$ and for a.e. $y \in B(x, r(|x|))$.

$\tilde{N4}$ there exists a constant $K_0 > 0$ such that

$$K_0 v_0(x) \le v_1(x) r^{-p}(|x|), \text{ for a.e. } x \in \Omega^{n_0}.$$

$\tilde{\text{N}}5$ $\lim_{n\to\infty} \hat{A}_n = \infty$, where

$$\hat{A}_n = \sup_{x\in\Omega^n} \frac{\hat{a}_0(x)}{\hat{a}_1(x)} r^p(|x|).$$

Then, the space $W^{1,p}(\Omega;\, v_0,\, v_1)$ is not continuously imbedded in the space $W^{1,p}(\Omega;\, w)$.

Remarks (i) Special cases of Theorems 31 and 32, under weaker assumptions, can be formulated on the lines of Theorems 26, 27 and 28.

(ii) Compact imbeddings for the case of unbounded domains can be obtained as was done for bounded domains.

References

1. R.A. Adams, *Sobolev Spaces,* Academic Press, New York-San Francisco-London, 1975.
2. D.E. Edmunds and W.D. Evans, *Spectral Theory and Differential Operators,* Oxford, Oxford University Press, 1987.
3. D.E. Edmunds and H. Triebel, *Function Spaces, Entropy Numbers, Differential Operators,* Cambridge University Press, 1996.
4. P. Gurka and B. Opic, *A_r-condition for two weight functions and compact imbeddings of weighted Sobolev spaces,* Czechoslovak Math. J. 38 (133) (1988), 611–617.
5. P. Gurka and B. Opic, *Continuous and compact imbeddings of weighted Sobolev spaces I,* Czechoslovak Math. J. 38 (133) (1988), no. 4, 730–744.
6. P. Gurka and B. Opic, *Continuous and compact imbeddings of weighted Sobolev spaces II,* Czechoslovak Math. J. 39 (134) (1989), 78–94.
7. P. Gurka and B. Opic, *Continuous and compact imbeddings of Weighted Sobolev spaces III,* Czechoslovak Math. J. 41 (116) (1991), 317–341.
8. G.H. Hardy, J.E. Littlewood and G. Poyla, *Inequalities,* Cambridge University Press, 1952.
9. P.I. Lizorkin and M. Otelbaev, *Imbeddings and compactness theorems for Sobolev type spaces with weights I (Russian),* Mat. Sb. (*N.S.*), 108 (150), (1979), no. 3, 358–377.
10. P.I. Lizorkin and M. Otelbaev, *Imbeddings and compactness theorems for Sobolev type spaces with Weights II (Russian),* Mat. Sb. (N.S.), 112 (154), (1980), no. 1, 56–85.
11. B. Opic, *Necessary and sufficient conditions for imbeddings in weighted Sobolev spaces,* Časopis Pro Pěst. Mat. 114, no. 4, (1989) 343–355.
12. P. Jain, *On imbeddings of Weighted Sobolev spaces on an unbounded domain,* Publications De L'inst. Math. 56 (70) (1994), 79–89.
13. P. Jain, *On imbeddings of weighted Sobolev spaces,* Note di Matematica, Vol. 15, no. 1, (1995), 31–43.
14. P. Jain, *Certain Imbeddings of Sobolev Like Spaces,* Thesis, University of Delhi, 1996.
15. P. Jain, B. Bansal and P.K. Jain, *Certain imbeddings of Sobolev spaces* with power type weights, Preprint.

16. P. Jain, B. Bansal and P.K. Jain, *Continuous and compact imbeddings of weighted Sobolev spaces, Acta Sci. Math.* (sezed) (to appear).
17. P. Jain, B. Bansal and P.K. Jain, *Certain imbeddings of weighted Sobolev spaces,* Preprint.
18. A. Kufner, *Weighted Sobolev spaces,* Teubner-Texte zur Math., 31, Teubner, Leipzig, 1980 (first edition); John Wiley and Sons, Chichester-New York-Brisbane-Toronto-Singapore, 1985.
19. A. Kufner, O. John and S. Fučik, *Function Spaces,* Noordhoff International Publishing, Leyden, 1977.
20. B. Opic and A. Kufner, *Hardy-Type Inequalities,* Pitman Research Notes in Mathematics, Vol. 219, Longman, Harlow, 1990.
21. V.G. Maz′ya, *Sobolev Spaces,* Springer, Berlin, 1985.
22. J. Rákosník, *On imbeddings of Sobolev spaces with power type weights,* Theory of Approximation of Functions, 505–507, Nauka, Moscow, 1987.
23. J. Rákosník, *Weighted Sobolev spaces,* Preprint.
24. H. Triebel, *Interpolation Theory, Function Spaces, Differential Operators,* Berlin: VEB Deutsch. Vert. Wissenschaften, 1978 : Amsterdam : North-Holland, 1978. 2nd (revised) edition Leipzig : Barth 1995.
25. H. Triebel, *Theory of Function Spaces,* Leipzig: Geest & Porting 1983; Basel : Birkhäuser, 1983.
26. H. Triebel, *Theory of Function Spaces II,* Basel : Birkhäuser, 1992.
27. W.P. Ziemer, *Weakly Differebtiable Functions,* Springer-Verlag, 1989.

Function Spaces and Applications
D.E. Edmunds et al (Eds)

10. From Hardy to Carleman and General Mean-Type Inequalities

Pankaj Jain[1], Lars-Erik Persson[2] and Anna Wedestig[3]

[1]Department of Mathematics, University of Delhi-110 007, India

[2]Department of Mathematics, Luleå University, S-971 87 Luleå, Sweden

[3]Department of Mathematics, Luleå University, S-971 87 Luleå, Sweden

1. Introduction

The classical Hardy inequality is the following

$$\int_0^\infty F^p(t)t^{\varepsilon-p}dt \le C\int_0^\infty f^p(t)t^\varepsilon dt,$$

where $1 < p < \infty$, f is a non-negative measurable function defined on $(0, \infty)$ and

$$F(t) = \begin{cases} \displaystyle\int_0^t f(x)\,dx, & \varepsilon < p-1 \\ \displaystyle\int_t^\infty f(x)\,dx, & \varepsilon > p-1. \end{cases}$$

This inequality was first established by Hardy [5] in 1920, but he did not give the estimate for the best constant C. Later in 1926 it was Landau [12] who proved that the value of the best constant in the Hardy inequality is $\left(\frac{p}{|\varepsilon-p+1|}\right)^p$. Since then, the Hardy inequality has been considered by several authors and many applications have been pointed out. Here we only refer to the book [15] and mention the following facts:

Many alternative proofs of the Hardy inequality have been given, e.g. by Broadbent [1], Elliot [4], Hardy [6], Kaluza and Szegö [10], and Knopp [11].

One modern form of the Hardy inequality is

$$\left(\int_0^\infty \left(A(f,x)\right)^q w(x)\,dx\right)^{1/q} \le C\left(\int_0^\infty f^p(x)v(x)\,dx\right)^{1/p}, \tag{1.1}$$

1991 Mathematics Subject Classification. 26D15 26D07.

where $1 \le p,\ q \le \infty$, w, v are weight functions and $A(f, x)$ is some of the "averaging operators"

$$A(f,x) = \frac{1}{x}\int_0^x f(t)\,dt \quad \text{or} \quad A(f,x) = \frac{1}{x}\int_x^\infty f(t)\,dt.$$

Concerning necessary and sufficient conditions on the weights w and v so that (1.1) holds see [15] (c.f. also Theorem A for $p \le q$). Also, other types of averaging operators have been considered e.g.

$$A(f,x,u) = \exp\left(\frac{1}{\int_0^x u(t)\,dt}\int_0^x u(t)\ln f(t)\,dt\right).$$

Such operators have been discussed in [14], [17] for $u(t) \equiv 1$ and in [8], [9] for $u(t) \equiv t^k$. It is worth mentioning that for the operator $A(f, x, u)$, the inequality (1.1) reads

$$\left(\int_0^\infty \left(\exp\left(\frac{1}{\int_0^x u(t)\,dt}\int_0^x u(t)\ln f(t)\,dt\right)\right)^q w(x)\,dx\right)^{1/q}$$

$$\le C\left(\int_0^\infty f^p(x)v(x)\,dx\right)^{1/p} \tag{1.2}$$

which for $u(t) \equiv 1$, $w \equiv v \equiv 1$, $p = q = 1$ becomes

$$\int_0^\infty \exp\left(\frac{1}{x}\int_0^x \ln f(t)\,dt\right)dx \le C\int_0^\infty f(x)\,dx.$$

This is the well known Carleman inequality [2] (see also [7], p. 250) which holds for the constant $C = e$ and the constant is the best possible.

In this paper, we will study the inequality (1.1) when $A(f, x)$ and $A(f, x, u)$ are replaced by more general averaging operators e.g. the following :

$$\mathscr{A}_\alpha(f,x,u) = \begin{cases} \left(\dfrac{1}{\int_0^x u(t)\,dt}\displaystyle\int_0^x u(t) f^\alpha(t)\,dt\right)^{1/\alpha}, & \alpha \ne 0 \\ \exp\left(\dfrac{1}{\int_0^x u(t)\,dt}\displaystyle\int_0^x u(t)\ln f(t)\,dt\right), & \alpha = 0. \end{cases} \tag{1.3}$$

One guiding observation in our study is that $\mathscr{A}_\alpha(f, x, u)$ is a nondecreasing function of α. An elementary proof of this fact, in a more general form, is given in Appendix.

Section 2 discusses the possibility to obtain general weighted Carleman type inequalities by making limiting procedures of some modern forms of Hardy's inequality. Section 3 derives a new weighted Carleman type inequality. Section 4 is reserved for some concluding remarks and results. In particular, we derive "good" bounds of the best constants in the power weight case of the inequality obtained in Section 3. Moreover, we introduce the idea to study inequalities of "the Hardy type" when the arithmetic or geometric mean operators are replaced by general Gini-mean operators. One important observation is that the two parameter scale of Gini means is non-decreasing in both the parameters. Appendix includes an elementary proof of this fact.

2. From Hardy to Carleman Type Inequalities

First, we recall the following modern version of the Hardy inequality [15]:

Theorem A Let $1 < p \leq q < \infty$ and w, v be weight functions defined on $(0, \infty)$. Assume that, for every $x > 0$,

$$\int_0^x v^{\frac{1}{1-p}}(t)\,dt < \infty.$$

Then, the inequality

$$\left(\int_0^\infty \left(\int_0^x f(t)\,dt\right)^q w(x)\,dx\right)^{1/q} \leq C\left(\int_0^\infty f^p(x)\,v(x)\,dx\right)^{1/p} \tag{2.1}$$

holds for all positive functions f defined on $(0, \infty)$ if and only if

$$D \equiv \sup_x \left(\int_x^\infty w(t)dt\right)^{\frac{1}{q}} \left(\int_0^x v^{1/1-p}(t)dt\right)^{\frac{p-1}{p}} < \infty$$

Moreover, if C is the best possible constant in (2.1), then

$$D \leq C \leq k(p, q)\, D,$$

where

$$k(p, q) = \left(\frac{p + q(p-1)}{p}\right)^{\frac{1}{q}} \left(\frac{p + q(p-1)}{(p-1)q}\right)^{\frac{p-1}{p}}$$

Remark The inequalities (1.1) (with $1 < p \leq q < \infty$) and (2.1), are equivalent with the same best possible constant.

In accordance with the introduction and for later purposes we also present the following (formal) generalization of Theorem A:

Theorem 2.1 Let $0 < \alpha < p \leq q < \infty$ and w, v be weight functions defined on $(0, \infty)$. Assume that, for every $x > 0$,

$$\int_0^x v^{\frac{\alpha}{\alpha-p}}(t)dt < \infty.$$

Then, the inequality

$$\left(\int_0^\infty \left(\frac{1}{x}\int_0^x f^\alpha(t)dt\right)^{q/\alpha} w(x)dx\right)^{1/q} \leq C\left(\int_0^\infty f^p(x)\, v(x)dx\right)^{1/p} \qquad (2.2)$$

holds for all positive functions f on $(0, \infty)$ if and only if

$$D_\alpha \equiv \sup_x \left(\int_x^\infty w(t)\, t^{-q/\alpha}\, dt\right)^{1/q} \left(\int_0^x v^{\frac{\alpha}{\alpha-p}}\, (t)\, dt\right)^{\frac{p-\alpha}{\alpha p}} < \infty. \qquad (2.3)$$

Moreover, the best possible constant C in (2.2) satisfies the estimate

$$D_\alpha \leq C \leq k(p, q, \alpha)\, D_\alpha,$$

where $k(p, q, \alpha) = \left(\dfrac{\alpha p + qp - q\alpha}{\alpha p}\right)^{1/q} \left(\dfrac{\alpha p + qp - q\alpha}{(p-\alpha)q}\right)^{\frac{p-\alpha}{\alpha p}}$

The proof of Theorem 2.1 only consists of making some obvious substitutions in Theorem A.

Example 2.2 Apply Theorem 2.1 with $w(x) = x^a$, $v(x) = x^b$, $a, b \in \mathbb{R}$, $0 < \alpha < p \leq q < \infty$, and we see that

$$\left(\int_0^\infty \left(\frac{1}{x}\int_0^x f^\alpha(t)\, dt\right)^{q/\alpha} x^a\, dx\right)^{1/q} \leq C\left(\int_0^\infty f^p(x)\, x^b\, dx\right)^{1/p}$$

holds if and only if

$$\alpha b < p - \alpha, \quad a = (b+1)\,\frac{q}{p} - 1.$$

Remark The LHS in (2.2) is nondecreasing in α (see Appendix) and in fact converges to

$$\left(\int_0^\infty \left[\exp\left(\frac{1}{x}\int_0^x \ln f(t)\, dt\right)\right]^q w(x)\, dx\right)^{1/q}$$

as $\alpha \to 0$. However, when $\alpha = 0$, the first integral in (2.3) is equal to zero and the natural replacement of the other term is equal to $\exp\left(\dfrac{1}{x}\displaystyle\int_0^x \ln \frac{1}{v(t)}\, dt\right)^{1/p}$

We also see that the constant $k(p, q, \alpha)$ in Theorem 2.1 $\to \infty$ as $\alpha \to 0$.

Therefore, we cannot directly obtain an inequality of the type (1.2) (with $u(t) = 1$) by making a limiting procedure in Theorem 2.1. However, by using other arguments even the following theorem can be proved (see [9]):

Theorem 2.3 Let $0 < p \le q < \infty$ and $k \in [0, \infty)$. Then, the inequality

$$\left(\int_0^\infty \left[\exp\left(\frac{k+1}{x^{k+1}} \int_0^x t^k \ln f(t)\, dt \right) \right]^q w(x)\, dx \right)^{1/q} \le C \left(\int_0^\infty f^p(x)\, v(x)\, dx \right)^{1/p} \tag{2.4}$$

holds for positive functions f if and only if there exists $\theta > \frac{1}{k+1}$ such that

$$\mathscr{B}_0(p, q, \theta) = \sup_x x^{((k+1)\theta - 1)/p} \left(\int_x^\infty \frac{s(t)}{t^{(k+1)(\theta-1)\frac{q}{p}+1}}\, dt \right)^{1/q} < \infty, \tag{2.5}$$

where $\quad s(x) = \dfrac{w(x)}{x^{(k+1)q/p-1}} \left[\exp\left(\dfrac{k+1}{x^{k+1}} \displaystyle\int_0^x y^k \ln\left(\frac{1}{v(y)} \right) dy \right) \right]^{\frac{q}{p}}$

The best constant C in (2.4) can be estimated as follows:

$$\sup_\theta \left(\frac{(k+1)\,\theta - 1}{(k+1)\theta} \right)^{1/p} \mathscr{B}_0(p, q, \theta) \le C \le \inf_\theta \left((k+1)\, e^{(\theta-1)} \right)^{1/p} \mathscr{B}_0(p, q, \theta) \tag{2.6}$$

Remark Consider $\mathscr{B}_0$ defined by (2.5) with $k = 0$ and $\theta = p$, i.e.,

$$\mathscr{B}_0^q = \sup_x x^{(p-1)\frac{q}{p}} \int_x^\infty \frac{w(t)}{t^q} \left[\exp\left(\frac{1}{t} \int_0^t \ln \frac{1}{v(s)}\, ds \right) \right]^{q/p} dt.$$

We also note that (*c.f.* the Appendix)

$$\left(\frac{1}{t} \int_0^t v^{\alpha/\alpha - p}(s)\, ds \right)^{p-\alpha/\alpha} \to \exp\left(\frac{1}{t} \int_0^t \ln \frac{1}{v(s)}\, ds \right) \; as\ \alpha \to 0.$$

Thus, it is natural to consider the modified constant $\mathscr{B}_\alpha$ defined by

$$\mathscr{B}_\alpha^q = \sup_x x^{(p-1)\frac{q}{p}} \int_x^\infty \frac{u(t)}{t^q} \left(\frac{1}{t} \int_0^t v^{\alpha/\alpha - p}(s)\, ds \right)^{\frac{p-\alpha}{\alpha}\frac{q}{p}} dt.$$

In fact, we can prove the following:

Claim $D_\alpha \le C_0 \mathcal{B}_\alpha$, for some finite constant $C_0 \ge 1$, where D_α is the constant defined in (2.3).

Proof of the claim. Consider

$$f(x) = x^{(p-1)\frac{q}{p}} \int_x^\infty \frac{w(t)}{t^q} \left(\frac{1}{t} \int_0^t v^{\frac{\alpha}{\alpha-p}}(s)\, ds \right)^{\frac{p-\alpha}{\alpha}\cdot\frac{q}{p}} dt$$

and

$$g(x) = \left(\int_x^\infty w(t)\, t^{-q/\alpha}\, dt \right)^{1/q} \left(\int_0^x v^{\frac{\alpha}{\alpha-p}}(t)\, dt \right)^{\frac{p-\alpha}{\alpha}\cdot\frac{q}{p}}$$

By using the mean-value theorem for integrals and an elementary estimate, we find that for a fixed x, we can find $y = y(x)$ such that

$$f(x) = \left(\frac{x}{y(x)} \right)^{q(p-1)/p} \int_x^\infty \frac{w(t)}{t^{q/\alpha}} \left(\int_0^t v^{\frac{\alpha}{\alpha-p}}(s)\, ds \right)^{\frac{p-\alpha}{\alpha}\cdot\frac{q}{p}} dt$$

$$\ge \left(\frac{x}{y(x)} \right)^{q(p-1)/p} \left(\int_x^\infty w(t)\, t^{-q/\alpha} \right) \left(\int_0^x v^{\frac{\alpha}{\alpha-p}}(s)\, ds \right)^{\frac{p-\alpha}{\alpha}\cdot\frac{q}{p}} dt$$

$$= \left(\frac{x}{y(x)} \right)^{q(p-1)/p} g(x). \tag{2.7}$$

Moreover, according to the (supremum) definition of D_α, we have that there exists x_0, $0 < x_0 < \infty$, such that $D_\alpha \le 2g(x_0)$. Also by applying (2.7) with just this x_0, we find that

$$D_\alpha^q \le 2g(x_0) \le 2 \left(\frac{y(x_0)}{x_0} \right)^{q(p-1)/p} f(x_0) \le C_0^q \mathcal{B}_\alpha^q$$

and the claim is proved.

Summing up our investigations in this section, we see that

(a) a necessary and sufficient condition for the inequality (2.4) (or (1.3) with $\alpha = 0$ and $u(t) = t^k$) is that (2.5) holds (the condition $\mathcal{B}_0 < \infty$).

(b) this result cannot be obtained by making a limiting procedure from the Hardy inequality (c.f. the not useful condition $D_0 < \infty$ and our claim above).

In the next section, we shall prove that the definition of the constant D_α in (2.3) can be modified in such a way that it has sense also in the limit case $\alpha = 0$ and, in fact, then guarantees that (2.4) holds with $k = 0$.

3. A Generalized Carleman Type Inequality

Theorem 3.1 Let $0 < p \le q < \infty$ and w, v be weight functions on $(0, \infty)$. If

$$\mathscr{B}_* = \lim_{\alpha \to 0} \sup_x \left(\int_x^\infty w(t) \frac{t^{-q/\alpha}}{\alpha} \, dt \right)^{1/q} \left(\int_0^x v^{\frac{\alpha}{\alpha - p}}(t) \, dt \right)^{\frac{p-\alpha}{\alpha p}} < \infty, \tag{3.1}$$

then the inequality

$$\left(\int_0^\infty \left[\exp\left(\frac{1}{x} \int_0^x \ln f(t) \, dt \right) \right]^q w(x) \, dx \right)^{1/q} \leq C \left(\int_0^\infty f^p(x) v(x) \, dx \right)^{1/p} \tag{3.2}$$

holds for all positive functions f defined on $(0, \infty)$. Moreover, if C is the best possible constant in (3.1), then

$$C \leq q^{1/q} e^{1/q} \mathscr{B}_*.$$

Proof In view of (3.1), the integral $\int_0^t v^{\frac{\alpha}{\alpha - p}}(y) \, dy$ is finite for every $t \in (0, \infty)$. For a fixed $s \in (1, \infty)$, define a function A by

$$A^s(t) = \left(\int_0^t v^{\frac{\alpha}{\alpha - p}}(y) \, dy \right)^{\frac{p-\alpha}{p}}, \; t \in (0, \infty). \tag{3.3}$$

We note that $0 < A(t) < \infty$ and by Hölder's inequality for the conjugate exponents $\frac{p}{\alpha}, \frac{p}{p - \alpha}$ and (3.3), we have

$$\begin{aligned}
\int_0^x f^\alpha(t) \, dt &= \int_0^x f^\alpha(t) \, v^{\alpha/p}(t) A(t) \, A^{-1}(t) \, v^{-\alpha/p}(t) \, dt \\
&\leq \left(\int_0^x f^p(t) v(t) \, A^{p/\alpha}(t) \, dt \right)^{\alpha/p} \\
&\quad \times \left(\int_0^x A^{p/(p-\alpha)}(t) \, v^{\frac{\alpha}{\alpha - p}}(t) \, dt \right)^{\frac{p-\alpha}{p}} \\
&= \left(\int_0^x f^p(t) v(t) \, A^{p/\alpha}(t) \, dt \right)^{\alpha/p} \\
&\quad \times \left(\int_0^x \left(\int_0^t v^{\frac{\alpha}{\alpha - p}}(y) \, dy \right)^{-1/s} v^{\frac{\alpha}{\alpha - p}}(t) \, dt \right)^{\frac{p-\alpha}{\alpha}} \\
&= \left(\frac{s}{s-1} \right)^{\frac{p-\alpha}{p}} \left(\int_0^x f^p(t) \, v(t) \, A^{p/\alpha}(t) \, dt \right)^{\alpha/p} A^{s-1}(x).
\end{aligned}$$

Thus, according to the Minkowski integral inequality, we have

$$\left(\int_0^\infty \left(\frac{1}{x}\int_0^x f^\alpha(t)\,dt\right)^{q/\alpha} w(x)\,dx\right)^{p/q}$$

$$\le \left(\frac{s}{s-1}\right)^{\frac{p-\alpha}{\alpha}} \left(\int_0^\infty \left(\int_0^x f^p(t)v(t)\,A^{p/\alpha}(t)\,dt\right)^{q/p}\right.$$

$$\left. A^{(s-1)\frac{q}{\alpha}}(x)\,x^{-\frac{q}{\alpha}}\,w(x)\,dx\right)^{p/q}$$

$$\le \left(\frac{s}{s-1}\right)^{\frac{p-\alpha}{\alpha}} \int_0^\infty f^p(x)v(x)\,A^{p/\alpha}(x)$$

$$\left(\int_x^\infty A^{(s-1)\frac{q}{\alpha}}(t)\,t^{-\frac{q}{\alpha}}\,w(x)\,dt\right)^{q/p} dx. \tag{3.4}$$

Now let

$$\tilde{\mathscr{B}}_* = \sup_x \left(\int_x^\infty w(t)\,\frac{t^{-q/\alpha}}{\alpha}\,dt\right)^{1/q} \left(\int_0^x v^{\frac{\alpha}{\alpha-p}}(t)\,dt\right)^{\frac{p-\alpha}{\alpha p}} \tag{3.5}$$

From (3.3) and (3.5), we obtain

$$A^{(s-1)\frac{q}{\alpha}}(t) \le \tilde{\mathscr{B}}_*^{\frac{(s-1)}{s}q} \left(\int_t^\infty \frac{y^{-\frac{q}{\alpha}}}{\alpha}\,w(y)\,dy\right)^{-\frac{(s-1)}{s}}$$

and, thus

$$\int_x^\infty A^{(s-1)\frac{q}{\alpha}}(t)\,t^{-\frac{q}{\alpha}}\,w(t)\,dt \le \tilde{\mathscr{B}}_*^{\frac{(s-1)}{s}q}\,\alpha s \left(\int_x^\infty \frac{t^{-\frac{q}{\alpha}}}{\alpha}\,w(t)\,dt\right)^{1/s}$$

$$\le \tilde{\mathscr{B}}_*^q\,\alpha s A^{-\frac{q}{\alpha}}(x).$$

Hence, (3.4) and (3.5) give

$$\left(\int_0^\infty \left(\frac{1}{x}\int_0^x f^\alpha(t)\,dt\right)^{q/\alpha} w(x)\,dx\right)^{1/q} \le C\left(\int_0^\infty f^p(x)v(x)\,dx\right)^{1/p} \tag{3.6}$$

where $C = \left(\frac{s}{s-1}\right)^{\frac{p-\alpha}{\alpha p}} s^{1/q}\alpha^{1/q}\,\tilde{\mathscr{B}}_*$ is a constant. Now, taking the limit $\alpha \to 0$ in the last inequality, we can obtian (3.2). Note that, since s is arbitrary, the best constant C in (3.6) will always satisfy

$$C \le \liminf_{\alpha\to 0}\, \inf_{s>0} \left(\frac{s}{s-1}\right)^{\frac{p-\alpha}{\alpha p}} s^{1/q}\alpha^{1/q}\, \tilde{\mathcal{B}}_*$$

$$= \lim_{\alpha\to 0} \left(\frac{\alpha p + qp - q\alpha}{qp - q\alpha}\right)^{\frac{p-\alpha}{\alpha p}} \left(\frac{\alpha p + qp - q\alpha}{\alpha p}\right)^{1/q} \alpha^{1/q}$$

$$= q^{1/q} e^{1/q}\, \mathcal{B}_*.$$

4. Concluding Remarks and Results

In the power weight case the following more precise version of Theorems 2.3 and 3.1 holds:

Proposition 4.1 Let $0 < p \le q < \infty$, $a, b \in \mathbb{R}$. Then

$$\left(\int_0^\infty \left[\exp\left(\frac{1}{x}\int_0^x \ln f(t)\, dt\right)\right]^q x^a dx\right)^{1/q} \le C \left(\int_0^\infty f^p(x)\, x^b\, dx\right)^{1/p} \quad (4.1)$$

holds for some finite constant $C > 0$ if and only if

$$\frac{a+1}{q} = \frac{b+1}{p}. \quad (4.2)$$

Moreover, the best constant in (4.1) can be estimated by

$$C_0 \le C \le e^{\frac{1}{q}+\frac{b}{p}}, \quad (4.3)$$

where

$$C_0 = e^m \frac{p^{1/p}}{q^{1/q}} \sup_{\varepsilon>0} \left\{ e^{\varepsilon} \left(\frac{m+\varepsilon}{m\varepsilon}\right)^{1/q} \left(\frac{\varepsilon e^{mp} e^{\varepsilon p} + m}{m\varepsilon}\right)^{-1/p} \right\}, \quad m = \frac{b+1}{p}.$$

Proof First we assume that (4.2) holds and apply Theorem 3.1 with $w(x) = x^a$ and $v(x) = x^b$. We note that (3.1) is satisfied so that (4.1) holds and, moreover, according to a straightforward calculation

$$C \le e^{\frac{1}{q}+\frac{b}{p}} \sup_{x>0} x^{\frac{a+1}{q}-\frac{b+1}{p}} = e^{\frac{1}{q}+\frac{b}{p}}.$$

On the other hand, assume that (4.1) holds with some finite constant $C > 0$. We now choose $\alpha > \max(m_0, m_1)$, where $m_0 = \frac{a+1}{q}$ and $m_1 = \frac{b+1}{p}$, and a functin $f = f_{A,\alpha}$ defined as follows:

$$f(x) = \begin{cases} x^{-\alpha}, & x > A \\ 1, & 0 < x \le A. \end{cases} \quad (4.4)$$

By applying (4.1) with this function we find that

$$\left(\frac{A^{qm_0}}{a+1}+\frac{e^{\alpha q}A^{(-\alpha+m_0)q}}{q(\alpha-m_0)}\right)^{1/q}\left(\frac{A^{pm_1}}{b+1}+\frac{A^{(-\alpha+m_1)p}}{p(\alpha-m_1)}\right)^{-1/p}\leq C.$$

The only possibility that this can be true for all A, $0 < A < \infty$, is that $m_0 = m_1 = m$. Thus (4.2) holds and, moreover, we consider the function defined in (4.4) with $A = e$ and $\alpha = m + \varepsilon$, $\varepsilon > 0$. Then, we find that

$$C \geq e^m \frac{p^{1/p}}{q^{1/q}} e^{\varepsilon}\left(\frac{m+\varepsilon}{m\varepsilon}\right)^{1/q}\left(\frac{\varepsilon e^{mp}e^{\varepsilon p}+m}{m\varepsilon}\right)^{-1/p}$$

By taking supremum over $\varepsilon > 0$ we obtain that $C_0 \leq C$ and are done.

Remark We obtain the same upper bound $e^{\frac{b}{p}+\frac{1}{q}}$ in (4.3), in fact, by instead using Theorem 2.3. By applying the same Theorem we obtain the following lower bound:

$$C_1 = \begin{cases} \left(\frac{q-p}{q}\right)^{1/q}\left(\frac{p}{q-p}\right)^{1/p}\left(\frac{p}{q}\right)^{1/q} e^{b/p}, & q > p \\ e^{b/p}, & p = q. \end{cases}$$

Some straightforward (but fairly tedious) calculations show that C_0 is equal to the upper bound in (4.3) in all cases $p = q$ so that the inequality (4.1) is sharp for $p = q$. This is not true for the constant C_1.

Example 4.2 Let $p = q = 1$, $a = b = 0$. Then, $C_0 = e$ and $C_1 = 1$, which means that the sharpness of classical Carleman's inequality can be calculated from Proposition 4.1 (but not from Theorem 2.3 or 3.1).

Example 4.3 Let $p = 0.2$, $q = 1$, $b = 1$ and $a = 9$. Then, $C_0 \approx 6.16$ and $C_1 \approx 2.43$. However, none of these constants are close to the upper bound ≈ 403.

Remark The results of this paper reflect the possibility to study inequalities involving the more general so called "Gini-means" defined by

$$G_{\alpha,\beta}(f,x,u) = \begin{cases} \left(\dfrac{\int_0^x |f(t)|^{\alpha} u(t)\,dt}{\int_0^x |f(t)|^{\beta} u(t)\,dt}\right)^{\frac{1}{\alpha-\beta}} & \alpha \neq \beta, \\ \exp\left(\dfrac{\int_0^x |f(t)|^{\alpha} \ln |f(t)|\, u(t)\,dt}{\int_0^x |f(t)|^{\alpha} u(t)\,dt}\right), & \alpha = \beta. \end{cases} \tag{4.5}$$

It can be noted that $G_{\alpha,0}(f, x, u) = \mathscr{A}_\alpha(f, x, u)$, where $\mathscr{A}_\alpha(f, x, u)$ are the averages studied in Section 2. Moreover, $G_{\alpha,\beta}(f, x, u) = G_{\beta,\alpha}(f, x, u)$ and $G_{\alpha,\beta}(f, x, u)$ is non-decreasing in both α and β. Also, the following representation formula holds:

$$G_{\alpha,\beta}(f, x, u) = \exp\left(\frac{1}{\alpha - \beta}\int_\alpha^\beta \ln G_{a,a}(f, x, u)\, da\right).$$

For the reader's convenience, we include elementary proofs of these facts in an Appendix.

Obviously, the information given in the last remark give us a possibility to study Hardy type inequalities when the Hardy operator is replaced by more general operators like those defined in (4.5). We shall not continue in this direction in this paper but only give the following example of consequence of the remark and Theorem 3.1:

Corollary 4.4 Let $0 < p \leq q < \infty$ and w, v be weight functions on $(0, \infty)$. If (3.1) holds, then, for every $f > 0$ and all α, $\beta \leq 0$

$$\left(\int_0^\infty (G_{\alpha,\beta}(f, x, u))^q\, w(x)\, dx\right)^{1/q} \leq C\left(\int_0^\infty f^p(x)\, v(x)\, dx\right)^{1/p} \tag{4.6}$$

Open Question Find necessary and sufficient conditions on the weights w and v so that the (generalized Hardy-type) inequality (4.6) holds (in other cases than the cases $\beta = 0$, $0 \leq \alpha \leq p(\alpha = 0, 0 \leq \beta \leq p)$ in this paper).

Final Remark There exists also some other generalizations of Carleman's inequality in the literature. Here, we only refer to the recent paper [3] and the book [13] and the references therein.

APPENDIX

Let $G_{\alpha,\beta}(f, x, u)$ be the function defined in (4.5). We shall prove that $G_{\alpha,\beta}$ is non-decreasing in both α and β but first we prove a representation formula of independent interest.

Lemma Let $G_{\alpha,\beta} = G_{\alpha,\beta}(f, x, u)$ be defined by (4.5). Then

$$G_{\alpha,\beta} = \exp\left(\frac{1}{\alpha - \beta}\int_\alpha^\beta \ln (G_{a,a})\, da\right), \alpha \neq \beta.$$

Proof Consider

$$H(a) = \ln\left(\int_0^x |f(t)|^a\, u(t)\, dt\right)$$

and note that

$$\ln G_{\alpha,\beta} = \frac{H(\alpha) - H(\beta)}{\alpha - \beta}, \ \alpha \neq \beta. \tag{1}$$

We may assume that H is continuously differentiable. Then

$$\ln G_{\alpha,\alpha} = \lim_{\beta \to \alpha} \frac{H(\alpha) - H(\beta)}{\alpha - \beta} = H'(\alpha). \tag{2}$$

According to (1) and (2), we have

$$G_{\alpha,\beta} = \exp\left(\frac{1}{\alpha - \beta} \int_{\beta}^{\alpha} H'(a)\, da \right)$$

$$= \exp\left(\frac{1}{\alpha - \beta} \int_{\beta}^{\alpha} \ln\left(G_{a,a}\right) da \right)$$

and the proof is complete.

Theorem The function $G_{\alpha,\beta}$ defined by (4.5) is non-decreasing in both α and β.

Proof Consider the function

$$g(a) = \ln G_{a,a} = \frac{\int_0^x |f(t)|^a \ln|f(t)|\, u(t)\, dt}{\int_0^x |f(t)|^a\, u(t)\, dt}.$$

We differentiate and find that

$$g'(a) = \frac{\left(\int_0^x |f(t)|^a\, u(t)\, dt\right)\left(\int_0^x |f(t)|^a \ln^2|f(t)|\, u(t)\, dt\right) - \left(\int_0^x |f(t)|^a \ln|f(t)|\, u(t)\, dt\right)^2}{\left(\int_0^x |f(t)|^a\, u(t)\, dt\right)^2}.$$

Moreover, by Schwarz's inequality,

$$\int_0^x |f(t)|^a \ln|f(t)|\, u(t)\, dt$$

$$\leq \left(\int_0^x |f(t)|^a (\ln|f(t)|)^2\, u(t)\, dt\right)^{1/2} \left(\int_0^x |f(t)|^a\, u(t)\, dt\right)^{1/2}$$

which means that $g'(a) \geq 0$ so that g is non-decreasing. Thus, accordning to the lemma, we have the representation

$$G_{\alpha,\beta} = \exp\left(\frac{1}{\beta - \alpha} \int_{\alpha}^{\beta} g(a)\, da \right)$$

for some non-decreasing function g. Since for such function g, $\frac{1}{\beta - \alpha} \int_{\alpha}^{\beta} g(a)\, da$ is non-decreasing in α and β, we conclude that $G_{\alpha,\beta}$ is increasing both in α and β.

Remark The idea of the proofs in this Appendix are taken from [16]. It is obvious from these proofs that the theorem and representation formula above still hold if the integrals $\int_0^x \ldots dt$ are replaced by $\int_\Omega \ldots d\mu$, where (Ω, μ) is any measure space with positive measure μ.

Acknowledgment

The research of the first author is partially supported by the Department of Science and Technology (Ministry of HRD), India. The author also acknowledges with thanks the facilities provided to him by the Department of Mathematics, Luleå University of Technology, Sweden.

References

1. T.A.A. Broadbent, *A proof of Hardy's convergence theorem,* J. Lond. Math. Soc. **3** (1928), 242–243.
2. T. Carleman, *Sur les fonctions quasi-analytiques,* Conferences faites au cinquième congres des mathématiciens scandinaves (Helsingfors 1923), 181–196.
3. Čižmešija, J.B. Pečarié and D. Žubrinic, *On Levin's generalization of Carleman's inequalities,* Acta Sci. Math. (Szeged) 64 (1998), 473–481.
4. E.B. Elliot, *A simple exposition of some recently proved facts as to convergence,* J. Lond. Math. Soc. **1** (1926), 93–96.
5. G.H. Hardy, *Note on a theorem of Hilbert,* Math. Z. **6** (1920), 314–317.
6. G.H. Hardy, *Notes on some points in the integral Calculus (LX),* Messenger of Math. **54** (1925), 150–156.
7. G.H. Hardy, J.E. Littlewood and G. Pólya, *Inequalities,* University Press, Cambridge, 1964.
8. P. Jain and A.P. Singh, *A characterization for the boundedness of geometric mean operator,* Applied Math. Letters (Washington) (to appear).
9. A.M. Jarrah and A.P. Singh, *A limiting case of Hardy's inequality,* Indian J. Math. (In print).
10. T. Kaluza and G. Szegö, *Über Reihen mit lauter positive Gliedan,* J. Lond. Math. Soc. **2** (1927), 266–272.

11. K. Knopp, *Über Reihen mit positiven Gliedern,* Math. Z. **30** (1929), 387–413.
12. E. Landau, *A note on a theorem concerning series of positive terms,* J. Lond. Math. Soc. **1** (1926), 38–39.
13. D.S. Mitrinović, J.B. Pečarić and A.M. Fink, *Inequalities Involving Functions and Their Integrals and Derivatives,* Kluwer Acad. Publ., Dordrecht-Boston-London, 1991.
14. B. Opic and P. Gurka, *Weighted inequalities for geometric means,* Proc. Amer. Math. Soc. **3** (1994), 771–779.
15. B. Opic and A. Kufner, *Hardy-Type Inequalities,* Pitman Research Notes in Mathematics Series, Vol 211, Longman Scientific and Technical Harlow, 1990.
16. J. Peetre and L.E. Persson, *A general Beckenbach's inequality with applications,* In : Function Spaces, Differential Operators and Nonlinear Analysis, Pitman Research Notes in Math. ser 211 (1989), 125–139.
17. L. Pick and B. Opic, *On the geometric mean operator,* J. Math. Anal. Appl **183(3)** (1994), 652–662.

Function Spaces and Applications
D.E. Edmunds et al (Eds)

11. One-Dimensional Approximation of Eigenvalue Problems in Thin Rods

S. Kesavan and N. Sabu

Institute of Mathematical Sciences, C.I.T. Campus, Taramani,
Chennai-600113, India

1. Introduction

This article derives a one-dimensional eigenvalue problem that describes the limit behaviour of the three-dimensional eigenvalue problem of a thin linearly elastic rod when the thickness of the rod goes to zero.

The study of lower dimensional approximations of three dimensional eigenvalue problems from the mathematical viewpoint has been done in the works of Ciarlet and Kesavan [3] for plates, Bourquin and Ciarlet [1] for 3d-2d junctions, Le Dret [11] for folded plates and Kesavan and Sabu [10] for shallow shells.

In each instance, one or several portions of the whole three-dimensional structure have a small thickness which we denote by ε. Then if the various data behave as specific powers of ε as $\varepsilon \to 0$, one can establish the convergence of the (appropriately scaled) components of the displacement vector field towards the solution of a lower dimensional problem. In the present paper, we likewise establish the convergence of the eigenvalues and the associated eigenfunctions towards the solution of a one-dimensional eigenvalue problem in the case of thin rods. We now briefly outline the problem studied in this paper and describe the results obtained.

Let $\omega \subset \mathbb{R}^2$ be a bounded domain and let

$$\left.\begin{aligned} \Omega_\varepsilon &= \varepsilon\omega \times (0,1),\ \gamma_{\varepsilon_0} = \partial\Omega_\varepsilon \cap \{x_3 = 0\}, \\ \gamma_{\varepsilon_1} &= \partial\Omega_\varepsilon \cap \{x_3 = 1\} \text{ and } \gamma_\varepsilon = \gamma_{\varepsilon_0} \cup \gamma_{\varepsilon_1} \end{aligned}\right\} \quad (1.1)$$

For all 3×3 tensors $\underset{\sim}{\zeta}$, we define

$$(A\zeta)_{ij} = \lambda\zeta_{ll}\delta_{ij} + 2\mu\zeta_{ij} \quad (1.2)$$

where λ and μ are the Lamé constants of the material. We then define the space of admissible displacements as

$$V^{\varepsilon} = \{v^{\varepsilon} \in (H_0^1(\Omega_{\varepsilon}))^3 ; v^{\varepsilon} = 0 \text{ on } \gamma_{\varepsilon}\} \tag{1.3}$$

For each admissible displacement v^{ε}, we define the linearized stress tensor $e(v^{\varepsilon}) = (e_{ij}(v^{\varepsilon}))$ by

$$e_{ij}(v^{\varepsilon}) = \frac{1}{2}(\partial_i v_j^{\varepsilon} + \partial_j v_i^{\varepsilon}) \tag{1.4}$$

for $1 \leq i, j \leq 3$. Then, the eigenvalue problem consists of finding pairs $(u^{\varepsilon}, \xi^{\varepsilon}) \in V^{\varepsilon}\backslash\{0\} \times \mathbb{R}$ such that

$$\int_{\Omega^{\varepsilon}} (Ae(u^{\varepsilon}))_{ij}\, e_{ij}(v^{\varepsilon})\, dx = \xi^{\varepsilon} \int_{\Omega^{\varepsilon}} u_i^{\varepsilon} \cdot v_i^{\varepsilon}\, dx \tag{1.5}$$

for every $v^{\varepsilon} \in V^{\varepsilon}$ with the convention of summation over repeated indices. It can be shown that there exists a sequence of eigenpairs $\{(u^{\varepsilon,l}, \xi^{\varepsilon,l})\}_{l=1}^{\infty}$ such that

$$0 < \xi^{\varepsilon,1} \leq \xi^{\varepsilon,2} \leq \ldots \leq \xi^{\varepsilon,l} \leq \ldots \to \infty \tag{1.6}$$

and $\{u^{\varepsilon,l}\}$ forms a complete orthonormal basis for $(L^2(\Omega_{\varepsilon}))^3$.

Following Le Dret [12], we transform (1.5) into an equivalent problem over $\Omega = \omega \times (0, 1)$ after suitable scalings of the variables ξ^{ε} and u^{ε}.

In this fashion, we obtain scaled eigenpairs $(u^l(\varepsilon), \xi^l(\varepsilon)) \in V\backslash\{0\} \times \mathbb{R}$ where

$$V = \{v \in (H^1(\Omega))^3;\ v = 0 \text{ on } \gamma_1\} \tag{1.7}$$

which satisfy variational equations in which ε occurs as a parameter. We show that as $\varepsilon \to 0$, $u^l(\varepsilon) \to u^l$ in $(H^1(\Omega))^3$ and $\xi^l(\varepsilon) \to \xi^l$ for each fixed l for a suitable subsequence. We also show that

$$u_{\alpha}^l = \zeta_{\alpha}^l(x_3), \quad u_3^l = -\, x_{\alpha}(\zeta_{\alpha}^{l'}(x_3)) \tag{1.8}$$

for some $\underset{\sim}{\zeta}^l(\zeta_1^l, \zeta_2^l) \in (H_0^2(0, 1))^2$. The pair $(\underset{\sim}{\zeta}^l, \zeta^l)$ is an eigenpair for a fourth order elliptic problem posed over (0, 1). We can also prove that every eigensolution of the limit problem is a limit of a subsequence of $(u^l(\varepsilon), \xi^l(\varepsilon))$ for some integer $l \geq 1$.

There is an important difference between the one-dimensional model obtained by Le Dret [12] and the eigenvalue problem obtained in this paper. The former is a system of coupled fourth order equations involving all the components of the limit of $u(\varepsilon)$. The latter involves only the horizontal components. We will comment about this in a greater detail later.

This paper is organized as follows. In Section 2 below, we describe the three-dimensional problem. In Section 3, we transform the problem into one posed over a fixed domain and in Section 4, we study the limit problem. Section 5 is devoted to concluding remarks.

2. The three-dimensional problem

Throughout this article, the Latin indices will vary over the set {1, 2, 3} and

Greek indices will vary over the set $\{1, 2\}$ for the components of vectors and tensors. The convention of summation over repeated indices will be used in conjunction with the above rules.

We consider a family of three-dimensional, isotropic, homogeneous, linearly elastic bodies whose reference configurations are the sets Ω_ε defined for all $\varepsilon > 0$ by,

$$\Omega_\varepsilon = \{x = (x_1, x_2, x_3) \in \mathbb{R}^3, (x_1, x_2) \in \omega_\varepsilon, 0 < x_3 < 1\} \tag{2.1}$$

where $\omega_\varepsilon = \varepsilon\omega$ and ω is a bounded subset of $\mathbb{R}^2$, ie, straight cylinders in $\mathbb{R}^3$ with axes in the x_3 direction of length 1 and cross section ω_ε in the (x_1, x_2) plane. We refer to ε as the thickness of the rod under consideration.

Without loss of generality, we may assume that

$$\int_\omega x_1 dx_1 dx_2 = \int_\omega x_2 dx_1 dx_2 = \int_\omega x_1 x_2 dx_1 dx_2 = 0 \tag{2.2}$$

which means that we choose the origin of coordinates at the centre of gravity of ω, and the coordinate axes to be the principal axes of inertia of ω. Let I be the 2×2 inertia tensor of ω whose components are

$$I_{\alpha,\beta} = \int_\omega x_\alpha x_\beta dx_1 dx_2 \tag{2.3}$$

We let $\gamma_{\varepsilon_0} = \partial\Omega_\varepsilon \cap \{x_3 = 0\}$ and $\gamma_{\varepsilon_1} = \partial\Omega_\varepsilon \cap \{x_3 = 1\}$ denote the ends of the rod and $S_\varepsilon = \partial\Omega_\varepsilon \cap \{0 < x_3 < 1\}$ denote its lateral surface. We assume that the rod is clamped on both ends; if $\boldsymbol{v}^\varepsilon$ is an admissible displacement vector, then the stress tensor corresponding to this displacement is given by $\underset{\sim}{\tau} = A(e(\boldsymbol{v}^\varepsilon))$.

The usual scalar product of two vectors $\boldsymbol{u}$ and $\boldsymbol{v}$ will be denoted by $\boldsymbol{u} \cdot \boldsymbol{v} = u_i \cdot v_i$ and the usual scalar product of two tensors τ and σ will be denoted by $\underset{\sim}{\tau} : \underset{\sim}{\sigma} = \tau_{ij}\sigma_{ij}$. With this notation, the eigenvalue problem for the rods under consideration admits the following variational formulation.

Find $(\boldsymbol{u}^\varepsilon, \xi^\varepsilon) \in \boldsymbol{V}^\varepsilon \backslash \{0\} \times \mathbb{R}$ such that

$$\int_{\Omega_\varepsilon} A e(\boldsymbol{u}^\varepsilon) : e(\boldsymbol{v}^\varepsilon)\, dx = \xi^\varepsilon \int_{\Omega_\varepsilon} \boldsymbol{u}^\varepsilon \cdot \boldsymbol{v}^\varepsilon\, dx \tag{2.4}$$

for all $\boldsymbol{v}^\varepsilon \in \boldsymbol{V}^\varepsilon$.

The $\boldsymbol{V}^\varepsilon$ ellipticity of the bilinear form appearing in the left-hand side of (2.4) follows from Körn's inequality. Hence for each $\boldsymbol{f}^\varepsilon \in (L^2(\Omega_\varepsilon))^3$, there exists a unique $\boldsymbol{w}^\varepsilon \in \boldsymbol{V}^\varepsilon$ such that

$$\int_{\Omega_\varepsilon} A e(\boldsymbol{w}^\varepsilon) : e(\boldsymbol{v}^\varepsilon)\, dx = \int_{\Omega_\varepsilon} \boldsymbol{f}^\varepsilon \cdot \boldsymbol{v}^\varepsilon\, dx \tag{2.5}$$

for all $\boldsymbol{v}^\varepsilon \in \boldsymbol{V}^\varepsilon$.

We denote $w^\varepsilon = G^\varepsilon(f^\varepsilon)$ and thus $G^\varepsilon : (L^2(\Omega_\varepsilon))^3 \to V^\varepsilon$ defines a bounded linear operator. Since the inclusion $V^\varepsilon \to (L^2(\Omega_\varepsilon))^3$ is compact, we can consider G^ε as a compact linear operator of $(L^2(\Omega_\varepsilon))^3$ into itself. It is also clear that G^ε is selfadjoint. Thus problem (2.4) reduces to finding $u^\varepsilon \in V^\varepsilon$ and $\xi^\varepsilon \in \mathbb{R}$ such that $u^\varepsilon = \xi^\varepsilon G^\varepsilon(u^\varepsilon)$.

From the spectral theory of compact, selfadjoint, linear operators it follows that there exists a sequence of eigenpairs $\{(u^{\varepsilon,l}, \xi^{\varepsilon,l})_{l=1}^{\infty}\}$ such that

$$0 < \xi^{\varepsilon,1} \leq \xi^{\varepsilon,2} \leq \ldots \leq \xi^{\varepsilon,l} \leq \ldots \to \infty \tag{2.6}$$

$$\int_{\Omega_\varepsilon} u_i^{\varepsilon,l} \cdot u_i^{\varepsilon,k} \, dx = \delta_{lk} \tag{2.7}$$

The eigenvalues $(\xi^{\varepsilon,l})$ can be characterized via the min-max principle for the corresponding Rayleigh quotient by

$$\xi^{\varepsilon,l} = \min_{W \in \mathscr{V}_l^\varepsilon} \max_{v^\varepsilon \in W} \frac{\int_{\Omega_\varepsilon} Ae(v^\varepsilon) : e(v^\varepsilon) \, dx}{\int_{\Omega_\varepsilon} v^\varepsilon \cdot v^\varepsilon \, dx} \tag{2.8}$$

where $\mathscr{V}_l^\varepsilon$ denotes the collection of all l-dimensional subspaces of V^ε.

3. The Rescaled Problem

We set $\Omega = \omega \times (0, 1)$ and $\gamma = \partial\Omega \cap \{x_3 = 0, 1\}$ and with each point $x \in \overline{\Omega}$, we associate the point $x^\varepsilon \in \overline{\Omega}_\varepsilon$ through the bijection

$$\pi^\varepsilon : x = (x_i) \in \overline{\Omega} \to x^\varepsilon = (\varepsilon x_1, \varepsilon x_2, x_3) \in \overline{\Omega}_\varepsilon . \tag{3.1}$$

Given $v^\varepsilon \in V^\varepsilon$, we associate the scaled function $v(\varepsilon) \in V$, where

$$V = \{v \in (H^1(\Omega))^3 : v = 0 \text{ on } \gamma\}, \tag{3.2}$$

via the relations

$$v_\alpha^\varepsilon(x^\varepsilon) = v_\alpha(\varepsilon)(x), v_3^\varepsilon(x^\varepsilon) = \varepsilon v_3(\varepsilon)(x) \tag{3.3}$$

for all $x^\varepsilon = \pi^\varepsilon(x)$. We also scale the eigenvalues as follows:

$$\xi^\varepsilon = \varepsilon^2 \xi(\varepsilon). \tag{3.4}$$

Based on the above scalings and the routine calculations of change of variable in the integrals, we deduce that for a given ε, the rescaled eigensolutions of (2.4) satisfy the following variational problem.

Find $\{(u(\varepsilon), \xi(\varepsilon))\}_{l=1}^{\infty} \in V \backslash \{0\} \times \mathbb{R}$ such that

$$\int_\Omega b_\varepsilon(u(\varepsilon), v) \, dx = \xi(\varepsilon) \int_\Omega (u_\alpha(\varepsilon) v_\alpha + \varepsilon^2 u_3(\varepsilon) \, v_3) \, dx \tag{3.5}$$

for all $\boldsymbol{v} \in V$, where the bilinear form b_ε is given by

$$b_\varepsilon(\boldsymbol{u}, \boldsymbol{v}) = \varepsilon^{-4}[2\mu e_{\alpha\beta}(\boldsymbol{u})\, e_{\alpha\beta}(\boldsymbol{v}) + e_{\alpha\alpha}(\boldsymbol{u})\, e_{\beta\beta}(\boldsymbol{v})] + \varepsilon^{-2}[4\mu e_{\alpha 3}(\boldsymbol{u})\, e_{\alpha 3}(\boldsymbol{v}) + \lambda(e_{\alpha\alpha}(\boldsymbol{u})\, e_{33}(\boldsymbol{v}) + e_{33}(\boldsymbol{u})\, e_{\beta\beta}(\boldsymbol{v}))] + (2\mu + \lambda)\, e_{33}(\boldsymbol{u})\, e_{33}(\boldsymbol{v}) \tag{3.6}$$

and the rescaled eigenvectors $\{\boldsymbol{u}^l(\varepsilon)\}_{l=1}^{\infty}$ satisfy the normalization condition

$$\int_\Omega (u_\alpha^l(\varepsilon)\, u_\alpha^k(\varepsilon) + \varepsilon^2 u_3^l(\varepsilon)\, u_3^k(\varepsilon))\, dx = \delta_{lk} \tag{3.7}$$

The variational problem (3.5) may also be rewritten as

$$\int_\Omega \underset{\sim}{\sigma}(\varepsilon) : \boldsymbol{e}(\boldsymbol{v})\, dx = \xi(\varepsilon) \int_\Omega (u_\alpha(\varepsilon)\, v_\alpha + \varepsilon^2 u\,(\varepsilon)\, v_3)\, dx \tag{3.8}$$

for all $\boldsymbol{v} \in V$ where the rescaled stress tensor $\underset{\sim}{\sigma}(\varepsilon)$ is defined as

$$\left.\begin{aligned}
\sigma_{\alpha\beta}(\varepsilon) &= 2\mu\varepsilon^{-4} e_{\alpha\beta}(\boldsymbol{u}(\varepsilon)) + \lambda[\varepsilon^{-4} e_{\gamma\gamma}(\boldsymbol{u}(\varepsilon)) + \varepsilon^{-2} e_{33}(\boldsymbol{u}(\varepsilon))]\, \delta_{\alpha\beta} \\
\sigma_{\alpha 3}(\varepsilon) &= \sigma_{3\alpha}(\varepsilon) = 2\mu\varepsilon^{-2} e_{\alpha 3}(\boldsymbol{u}(\varepsilon)) \\
\sigma_{33}(\varepsilon) &= (2\mu + \lambda)\, e_{33}(\boldsymbol{u}(\varepsilon)) + \lambda\varepsilon^{-2} e_{\gamma\gamma}(\boldsymbol{u}(\varepsilon)).
\end{aligned}\right\} \tag{3.9}$$

If we introduce the auxiliary tensor $\underset{\sim}{\chi}^l(\varepsilon)$:

$$\left.\begin{aligned}
\chi_{\alpha\beta}(\varepsilon) &= \varepsilon^{-2}(e_{\alpha\beta}(\boldsymbol{u}(\varepsilon)) \\
\chi_{\alpha 3}(\varepsilon) &= \varepsilon^{-1} e_{\alpha 3}(\boldsymbol{u}(\varepsilon)) \\
\chi_{33}(\varepsilon) &= e_{33}(\boldsymbol{u}(\varepsilon))
\end{aligned}\right\} \tag{3.10}$$

the rescaled stresses assume the more homogeneous form

$$\left.\begin{aligned}
\sigma_{\alpha\beta}(\varepsilon) &= \varepsilon^{-2}(2\mu\chi_{\alpha\beta}(\varepsilon) + \lambda\chi_{ii}(\varepsilon)\,\delta_{\alpha\beta}) \\
\sigma_{\alpha 3}(\varepsilon) &= \sigma_{3\alpha}(\varepsilon) = 2\mu\varepsilon^{-1}\chi_{\alpha 3}(\varepsilon) \\
\sigma_{33}(\varepsilon) &= 2\mu\chi_{33}(\varepsilon) + \lambda\chi_{ii}(\varepsilon)
\end{aligned}\right\} \tag{3.11}$$

i.e.

$$\sigma_{\alpha\beta}(\varepsilon) = \varepsilon^{-2}(A\,\underset{\sim}{\chi}(\varepsilon))_{\alpha\beta};\ \sigma_{\alpha 3}(\varepsilon) = \varepsilon^{-1}(A\,\underset{\sim}{\chi}(\varepsilon))_{\alpha 3};\quad \sigma_{33}(\varepsilon) = (A\chi(\varepsilon))_{33}. \tag{3.12}$$

We denote the tensors $\underset{\sim}{\sigma}(\varepsilon)$ and $\underset{\sim}{\chi}(\varepsilon)$ associated to the eigenvector $\boldsymbol{u}^l(\varepsilon)$ using the above formulae by $\underset{\sim}{\sigma}^l(\varepsilon)$ and $\underset{\sim}{\chi}^l(\varepsilon)$ respectively.

4. Limit Problem

In this section we prove that the various unknowns involved ($\boldsymbol{u}^l(\varepsilon)$, $\xi^l(\varepsilon)$) satisfy appropriate bounds, which upon extraction of a subsequence will allow us to consider limits for these unknowns as $\varepsilon \to 0$. Then, we will identify the one-dimensional problem satisfied by the limits by using the techniques of Le Dret [12]. As these limit problems will turn out to be well-posed eigenvalue problems, we will thus be able to determine precisely the limit unknowns as being the eigenvalues and eigenvectors of the limit problem.

To begin with, let us consider the eigenvalues $\xi^l(\varepsilon)$.

Lemma 4.1 For each integer $l \geq 1$, there exists a constant $k(l)$ (independent of ε) such that $\xi^l(\varepsilon) \leq k(l)$.

Proof Since the Problem (3.5) was derived from (2.4) after a change of variable and change of scale, the eigenvalues can be characterized as follows.

$$\xi^l(\varepsilon) = \min_{W \in \mathscr{V}_l} \max_{v \in W} \frac{\int_\Omega b_\varepsilon(\boldsymbol{v}, \boldsymbol{v})\, dx}{\int_\Omega (v_\alpha^2 + \varepsilon^2 v_3^2)\, dx} \tag{4.1}$$

where $\mathscr{V}_l$ denotes the collection of al l-dimensional subspaces of $\boldsymbol{V}$.

Let $\mathscr{W}_l$ denote the collection of all l-dimensional subspaces of $H_0^2(0, 1)$. For $\varphi \in W \in \mathscr{W}_l$, we define

$$\boldsymbol{v}_\varphi = \{\varphi(x_3), \varphi(x_3), -(x_1 + x_2)\, \varphi'(x_3)\},\ \mathbf{W} = \{v_\varphi : \varphi \in W\} \tag{4.2}$$

Then it is easy to verify that $\mathbf{W} \in \mathscr{V}_l$ and $e_{\alpha\beta}(\boldsymbol{v}_\varphi) = e_{\alpha 3}(\boldsymbol{v}_\varphi) = 0$. Hence it follows from (4.1) that

$$\xi^l(\varepsilon) \leq \min_{W \in \mathscr{W}_l} \max_{\varphi \in W} \frac{\int_\Omega b_\varepsilon(\boldsymbol{v}_\varphi, \boldsymbol{v}_\varphi)\, dx}{\int_\Omega ((v_\varphi)_\alpha^2 + \varepsilon^2 (v_\varphi)_3^2)\, dx}$$

$$= \min_{W \in \mathscr{W}_l} \max_{\varphi \in W} \frac{\int_\Omega (\lambda + 2\mu)\, e_{33}(\boldsymbol{v}_\varphi)\, e_{33}(\boldsymbol{v}_\varphi)\, dx}{\int_\Omega ((v_\varphi)_\alpha^2 + \varepsilon^2 (v_\varphi)_3^2)\, dx}$$

$$\leq C \min_{W \in \mathscr{W}_l} \max_{\varphi \in W} \frac{\int_0^1 (\varphi'')^2\, dx_3}{\int_0^1 \varphi^2\, dx_3}$$

where C is a constant independent of ε. The expression on the right-hand side of the above relation gives exactly the l-th eigenvalue of the one-dimensional elliptic eigenvalue problem

$$\left.\begin{aligned} &\frac{d^4w}{dx^4} = \lambda w \\ &w(0) = w'(0) = w(1) = w'(1) = 0. \end{aligned}\right\} \tag{4.3}$$

This completes the proof of the lemma by setting $k(l) = C\lambda(l)$ where $\lambda(l)$ is the l-th eigenvalue of the problem (4.3). □

Let us now consider the eigenfunctions.

Theorem 4.2 (a) For each positive integer l, there exists a subsequence (still indexed by ε for convenience) such that $(\boldsymbol{u}^l(\varepsilon), \xi^l(\varepsilon))$ converges in $\boldsymbol{V} \times \mathbb{R}$ to $(\boldsymbol{u}^l, \xi^l)$; further there exists $(\zeta_\alpha^l) \in (H_0^2(0, 1))^2$ such that

$$u_\alpha^l = \zeta_\alpha^l(x_3),\ u_3^l = -\, x_\alpha \zeta_\alpha^{l'}(x_3) \tag{4.4}$$

(**b**) The pair (ζ_α^l, ξ^l) satisfies the following variational equation

$$\int_0^1 \frac{E}{a} I_{\alpha\beta} \zeta_\alpha^{l''} \eta_\beta'' \, dx = \xi^l \int_0^1 \zeta_\alpha^l \eta_\alpha \, dx \tag{4.5}$$

for all $(\eta_\alpha) \in (H_0^2(0, 1))^2$, where $E = \dfrac{\mu(3\lambda + 2\mu)}{\mu + \lambda}$ is the Young's modulus of the material, $I_{\alpha\beta}$ is the inertia tensor and $a = \displaystyle\int_\omega dx_1\, dx_2$ is the area of ω.

Proof For sake of clarity, the proof is divided into several steps.

Step 1 Boundedness of $\boldsymbol{u}^l(\varepsilon)$ and $\underset{\sim}{\chi}^l(\varepsilon)$:

Taking $\boldsymbol{v} = \boldsymbol{u}^l(\varepsilon)$ in (3.5), we get

$$\int_\Omega A\underset{\sim}{\chi}^l(\varepsilon) : \underset{\sim}{\chi}^l(\varepsilon)\, dx = \xi^l(\varepsilon) \int_\Omega ((u_\alpha^l(\varepsilon))^2 + \varepsilon^2 (u_3^l(\varepsilon))^2)\, dx \tag{4.6}$$

Due to positivity of the elasticity tensor $\boldsymbol{A}$, the left-hand side of (4.6) is bounded from below by $\mu \| \underset{\sim}{\chi}^l(\varepsilon) \|_{L^2(\Omega, M)}$. Therefore using Korn's inequality, we have

$$\begin{aligned} \mu \| \boldsymbol{u}^l(\varepsilon) \|^2_{H^1(\Omega, \mathbf{R}^3)} &\le \mu \| \boldsymbol{e}(\boldsymbol{u}^l(\varepsilon)) \|^2_{L^2(\Omega, M^3)} \\ &\le \mu \| \underset{\sim}{\chi}^l(\varepsilon) \|^2_{L^2(\Omega, M)} \le \xi^l(\varepsilon) \le k(l) \end{aligned} \tag{4.7}$$

for $\varepsilon \le 1$. This completes the proof of step 1.

Step 2 It follows from step 1 that (for a subsequence) $\xi^l(\varepsilon) \to \xi^l$ in $\mathbb{R}$, $\boldsymbol{u}^l(\varepsilon) \rightharpoonup \boldsymbol{u}^l$ weakly in $\boldsymbol{V}$. The boundedness of $\underset{\sim}{\chi}^l(\varepsilon)$ in $L^2(\Omega, M)$ implies that there exists a constant C (independent of ε) such that

$$\| e_{\alpha\beta}(\boldsymbol{u}^l(\varepsilon)) \|_{L^2(\Omega)} \le C\varepsilon^2,\ \| e_{\alpha 3}(\boldsymbol{u}^l(\varepsilon)) \|_{L^2(\Omega)} \le C\varepsilon \tag{4.8}$$

Therefore $e_{\alpha i}(\boldsymbol{u}^l(\varepsilon)) \to 0$ strongly in $L^2(\Omega)$ for all α and i and as $\boldsymbol{u}^l(\varepsilon) \rightharpoonup \boldsymbol{u}^l$ weakly in $L^2(\Omega)$, we have $e_{\alpha i}(\boldsymbol{u}^l(\varepsilon)) \rightharpoonup e_{\alpha i}(\boldsymbol{u}^l)$. Therefore $e_{\alpha i}(\boldsymbol{u}^l) = 0$ for all α and i. Then a standard argument (cf. Le Dret [12]) shows that there exists $(\zeta_\alpha^l) \in (H_0^2(0,1))^2$, $\zeta_3^l \in H_0^1(0,1)$ such that

$$u_\alpha^l = \zeta_\alpha^l(x_3), u_3^l = \zeta_3^l(x_3) - x_\alpha \zeta_\alpha^{l'}(x_3). \tag{4.9}$$

Step 3 Further (by going to a subsequence if necessary), $\underset{\sim}{\chi}^l(\varepsilon) \rightharpoonup \underset{\sim}{\chi}^l$ weakly in $L^2(\Omega, M)$. Then using lemma 5 in [12] we can show that

$$\chi_{11}^l = c_{22}^l = \frac{-\lambda(\zeta_3^{l'}(x_3) - x_\alpha \zeta_\alpha^{l''}(x_3))}{2(\lambda+\mu)}, \chi_{12}^l = 0$$

$$\chi_{33}^l = \zeta_3^{l'}(x_3) - x_\alpha \zeta_\alpha^{l''}(x_3) \tag{4.10}$$

(We do not identify the $\chi_{\alpha 3}^l$ components as they do not play any role in identification of the limit functions ζ_i^l.)

Step 4 We define

$$V_{BN}(\Omega) = \{\boldsymbol{v} \in V; \exists\, \eta_\alpha \in H_0^2(0,1), \exists\, \eta_3 \in H_0^1(0,1),$$

$$v_\alpha(x_1, x_2, x_3) = \eta_\alpha(x_3), v_3(x_1, x_2, x_3), = \eta_3(x_3) - x_\alpha \eta'(x_3)\}. \tag{4.11}$$

For test functions $\boldsymbol{v} \in V_{BN}$, the equation (3.8) becomes

$$\int_\Omega \sigma_{33}^l(\varepsilon)\, e_{33}(\boldsymbol{v})\, dx = \xi^l \int_\Omega (u_\alpha^l(\varepsilon)\, v_\alpha + \varepsilon^2 u_3^l(\varepsilon)\, v_3)\, dx \tag{4.12}$$

We know from (3.11) that $\sigma^l(\varepsilon)_{33} \rightharpoonup \sigma_{33}^l$ weakly in $L^2(\Omega)$ with

$$\sigma_{33}^l = 2\mu\chi_{33}^l + \lambda\chi_{ii}^l. \tag{4.13}$$

We can thus pass to the limit in (4.12) and obtain

$$\int_\Omega (2\mu\chi_{33}^l + \lambda\chi_{ii}^l)\, e_{33}(\boldsymbol{v})\, dx = \xi^l \int_\Omega u_\alpha^l v_\alpha\, dx \tag{4.14}$$

for all $\boldsymbol{v} \in V_{BN}(\Omega)$. Taking $\boldsymbol{v} = (0, 0, \eta(x_3))$ and substituting the values of χ_{ii}^l we get

$$\int_0^1 \zeta_3^{l'} \eta_3'\, dx_3 = 0 \text{ for all } \eta_3 \in H_0^1(0,1). \tag{4.15}$$

Taking $\eta_3 = \zeta_3^l$ in equation (4.15), we get

$$\int_0^1 (\zeta_3^{l'})^2\, dx_3 = 0. \tag{4.16}$$

This implies that $(\zeta_3^{l'}) = 0$ and since $\zeta_3^l(0) = \zeta_3^l(1) = 0$, we have

$$\zeta_3^l = 0 \text{ on } (0, 1). \tag{4.17}$$

Hence

$$u_\alpha^l = \zeta_\alpha^l(x_3),\ u_3^l = -\, x_\alpha \zeta_\alpha^{l'}(x_3) \tag{4.18}$$

$$\chi_{11}^l = \chi_{22}^l = \frac{\lambda x_\alpha(\zeta_\alpha^{l''}(x_3))}{2(\lambda+\mu)},\ \chi_{12}^l = 0,\ \chi_{33}^l = -\, x_\alpha \zeta^{l''}(x_3) \tag{4.19}$$

Substituting the above values of χ_{ii}^l and $\boldsymbol{u}^l$ and taking $\boldsymbol{v} = (\eta_1(x_3), \eta_2(x_3), -\, x_\alpha \eta_\alpha'(x_3))$ in equation (4.14), we get

$$\int_0^1 \frac{E}{a} I_{\alpha\beta} \zeta_\alpha^{l''} \eta_\alpha''\, dx_3 = \xi^l \int_0^1 \zeta_\alpha^l \eta_\alpha dx_3 \text{ for all } \eta_\alpha \in H_0^2(0, 1). \tag{4.20}$$

Step 5 The strong convergence of $\boldsymbol{u}^l(\varepsilon)$ in $\boldsymbol{V}$ follows once again as in [12]. □

Remark 1 As already mentioned in the introduction, there is an important difference between the limit equations obtained in Le Dret [12] and the equations obtained above. In the former, the right-hand side of the second equation is a function of the vertical components of the forces whereas in the latter we get zero. This is because the horizontal and vertical components of the displacement and forces have been scaled in different ways in Le Dret [12] to balance the different powers of ε occurring on both sides of the equation. In the case of eigenvalue problem, we can scale only the displacements and the eigenvalues and hence the powers do not get balanced. This leads to vanishing of the vertical component of the displacement and we get a one-dimensional fourth order elliptic eigenvalue problem in which the eigenvector is the horizontal components of the vector $\boldsymbol{u}^l$. □

Lemma 4.3 The problem (4.5) is a well posed eigenvalue problem which has a sequence of eigenvalues and the corresponding eigenvectors form an orthonormal basis for $(L^2(0, 1))^2$ and a basis for $(H_0^2(0, 1))^2$.

Proof The result follows from the ellipticity of the bilinear form appearing in the left-hand side of the equation (4.5) over $(H_0^2(0, 1))^2$. □

Though we have proved that each subsequence $(\boldsymbol{u}^l(\varepsilon), \xi^l(\varepsilon))_{\varepsilon>0}$, $l \geq 1$ strongly converges in $(H_0^1(\Omega))^3 \times \mathbb{R}$ to a solution $(\boldsymbol{u}^l, \xi^l)$ of the limit eigenvalue problem (4.5) (cf. Theorem 4.2), nothing tells us so far whether ξ^l is precisely the l-th eigenvalue of (4.5). We shall answer this question by the affirmative in the next lemma using the ideas developed by Kesavan [9].

Lemma 4.4 The sequence $\{\xi^l\}_{l=1}^\infty$ comprises all the eigenvalues of the problem (4.5) and the corresponding eigenvectors form an orthogonal basis for $(L^2(0,1))^2$ and a basis for $(H_0^2(0,1))^2$.

Proof Since we already know that

$$0 < \xi^1(\varepsilon) \le \xi^2(\varepsilon) \le \ldots \le \xi^l(\varepsilon) \le \xi^{l+1}(\varepsilon) \ldots \to \infty \tag{4.21}$$

and since the Green's operator associated with the limit problem is compact, it follows that

$$0 < \xi^1 \le \xi^2 \ldots \le \xi^l \le \xi^{l+1} \ldots \to \infty \tag{4.22}$$

Passing to the limit in the orthogonality relation (3.7), we get

$$\int_0^1 u_\alpha^l u_\alpha^m \, dx = \frac{1}{a} \delta_{lm}. \tag{4.23}$$

Suppose that there exists $\xi \in R$ such that $\xi \neq \xi^l$ for all l and ξ is an eigenvalue of the problem (4.5). Then there exists an eigenfunction ζ_α such that

$$\int_0^1 \zeta_\alpha \zeta_\alpha \, dx = \frac{1}{a}, \int_0^1 \zeta_\alpha \zeta_\alpha^l \, dx = 0, \text{ for all } l. \tag{4.24}$$

For each $\varepsilon > 0$, let $w(\varepsilon) \in V$ be the unique solution of

$$b(\varepsilon)(\boldsymbol{w}(\varepsilon), \boldsymbol{v}) = \xi \int_\Omega \zeta_\alpha v_\alpha \, dx, \text{ for all } \boldsymbol{v} \in V. \tag{4.25}$$

Then proceeding as in Theorem 4.2, we can show that $\boldsymbol{w}(\varepsilon) \to \boldsymbol{w} \in V$ and $w_\alpha = z_\alpha(x_3)$, $w_3 = -\ x_0 z'_\alpha(x_3)$ for some $z_\alpha \in\ (H_0^2(0,1)^2$. Further (z_α) will satisfy

$$\int_0^1 \frac{E}{a} I_{\alpha\beta} z''_\alpha \eta''_\alpha \, dx_3 = \xi \int_0^1 \zeta_\alpha \eta_\alpha \, dx_3, \text{ for all } \eta \in (H^2(0,1))^2. \tag{4.26}$$

By the uniqueness of the solution, it follows that $z_\alpha = \zeta_\alpha$.
Since the sequence ξ^l is unbounded, we can choose l such that

$$\xi < \xi^l \tag{4.27}$$

For $\boldsymbol{u}, \boldsymbol{v} \in V$, define

$$D(\varepsilon)\,(\boldsymbol{u}, \boldsymbol{v}) = \int u_\alpha v_\alpha \, dx + \varepsilon^2 \int u_3 v_3 \, dx. \tag{4.28}$$

Consider the vector

$$\boldsymbol{v}(\varepsilon) = \boldsymbol{w}(\varepsilon) - \sum_{k=1}^{l} D(\varepsilon)\,(\omega(\varepsilon), \boldsymbol{u}^k(\varepsilon))\, \boldsymbol{u}^k(\varepsilon). \tag{4.29}$$

Then

$$D(\varepsilon)(\boldsymbol{v}(\varepsilon), \boldsymbol{u}^k(\varepsilon)) = 0 \text{ for all } 1 \leq k \leq l. \tag{4.30}$$

Therefore it follows from the variational characterization of the eigenvalues that

$$\xi^{l+1}(\varepsilon) \leq \frac{b(\varepsilon)(\boldsymbol{v}(\varepsilon), \boldsymbol{v}(\varepsilon))}{D(\varepsilon)(\boldsymbol{v}(\varepsilon), \boldsymbol{v}(\varepsilon))}. \tag{4.31}$$

Passing to the limit in the above inequality, it can be shown that

$$\xi^{l+1} \leq \xi \tag{4.32}$$

which contradicts (4.27) and the proof is complete. □

Remark 2 We can also show that if ξ^l is a simple eigenvalue of the limit problem, then $\xi^l(\varepsilon)$ is a simple eigenvalue of (3.5) for sufficiently small ε. In this case we can choose the eigenvector $\boldsymbol{u}^l(\varepsilon)$ such that the entire family $\boldsymbol{u}^l(\varepsilon)$ converges, instead of just a subsequence. □

Lemma 4.5 A smooth enough solution (ζ^l_α, ξ^l) of the variational equation (4.5) solves the following equation.

$$\frac{E}{a}(I_{\alpha\beta}\zeta^{l''}_\alpha(x_3))'' = \xi^l \zeta^l_\alpha \text{ in } (0, 1)$$

$$\zeta^l_\alpha(0) = \zeta^{l'}_\alpha(0) = \zeta^l_\alpha(1) = \zeta^{l'}_\alpha(1) = 0. \tag{4.33}$$

Proof It follows easily from integration by parts in equation (4.5). □

Remark 3 The convergence of the rescaled shear and cross sectional stress components (i.e., $\sigma^l_{\alpha 3}(\varepsilon)$ and $\sigma^l_{\alpha\beta}(\varepsilon)$) can be proved in the same way as in Section 5 and 6 in Le Dret [12], for the same proof holds when the load f is replaced by a family $f(\varepsilon)$ that is uniformly bounded in $(L^2(\Omega, R^3))$. □

5. Conclusion

By rescaling the variables and posing the three dimensional eigenvalue problem for thin rods over a fixed domain, we have been able to show that the eigensolutions converge towards those of a one-dimensional fourth order eigenvalue problem. It is possible to now effect a descaling and obtain a one-dimensional model approximating the original problem.

The main difference between the model obtained for the eigenvalue problem studied here and the stationary problem studied by Le Dret [12] is that in the former we get a one-dimensional eigenvalue problem involving only the horizontal components of the limit eigenvectors and the vertical component converges to zero, while in the latter, we have a coupled fourth order system involving all three components. This has been commented upon in detail earlier (cf. Remark 1 in Section 4) and, in some sense, our model is more intrinsic since it does not involve special kinds of scalings of the force components which the stationary problem needed.

Another minor difference is in the presence of the coefficient $\frac{1}{a}$, a being the area of the cross section ω, in the bilinear form of the one-dimensional model. Again this is natural. Even the stationary problem should have this coefficient (cf. Equation (35) of Le Dret [12]). The right-hand side of this equation would then have the average of the forces over a cross-section rather than just the integral.

Both these phenomena were also observed in the case of shallow shells by Kesavan and Sabu [10]. In that case it showed that shallow vibrating shells, in the limit, behaved in a manner similar to vibrating plates.

References

1. Bourquin F. and Ciarlet P.G. Modelling and justification of eigenvalue problem for junctions between elastic structures, J. Functional Analysis, 87, 1989, pp. 392–427.
2. Ciarlet P.G. and Destuynder P. A justification of the two-dimensional plate model, J. Mécanique, 18, 1979, pp. 315–344.
3. Ciarlet P.G. and Kesavan S. Two-dimensional approximation of three dimensional eigenvalue problem in plate theory, Comp. Methods in Appl. Mech. Engrg., 26, 1981, pp. 145–172.
4. Ciarlet P.G. and Lods V. Asymptotic analysis of linearly elastic shells. I. Justification of membrane shell equation, Arch. Rational Mech. Anal., 136, 1996, pp. 119–161.
5. Ciarlet P.G. and Lods V. Asymptotic analysis of linearly elastic shells. III. Justification of Koiter's shell equations, Arch. Rational Mech. Anal., 136, 1996, pp. 191–200.
6. Ciarlet P.G., and Lods V. Asymptotic analysis of linearly elastic shells. "Generalized membrane shells", J. Elasticity, 43, 1996, pp. 147–188.
7. Ciarlet P.G. Lods V. and Miara B. Asymptotic analysis of linearly elastic shells. II. Justification of flexural shell equations, Arch. Rational. Mech. Anal., 136, 1996, pp. 162–190.
8. Ciarlet P.G. and Miara B. Justification of the two-dimensional equations of a linearly elastic shallow shells, Comm. Pure. Appl. Math. XLV, 1992, pp. 327–360.
9. Kesavan S. Homogenization of elliptic eigenvalue problems, Part I, Appl. Math. Optim. 5, 1979, pp. 153–167.
10. Kesavan S. and Sabu N. Two dimensional approximation of eigenvalue problem in shallow shell theory (to appear in Mathematics and Mechanics of Solids).
11. Le Dret H. Vibration of a folded plate, Math. Modelling and Numerical Anal. 24, 1990, pp. 501–521.
12. Le Dret H. Convergence of displacement and stress in linearly elastic slender rods as thickness of the rods goes to zero.-Asymptotic Analysis, 10, 1995, pp. 365–402.

Function Spaces and Applications
D.E. Edmunds et al (Eds)

12. Some Comments to the Hardy Inequality

Alois Kufner

Institute of Mathematics, Academy of Sciences of the Czech Republic
Zitná 25, 115 67 Prague, Czech Republic

This article aims to fill some gaps connected with the validity of Hardy's inequality

$$\left(\int_a^b | g(x) |^q u(x) \, dx \right)^{1/q} \leq C \left(\int_a^b | g^{(k)}(x) |^p v(x) \, dx \right)^{1/q} \tag{0.1}$$

for functions $g \in AC^{(k-1)}(a, b)$ satisfying certain additional conditions. Here we assume that $k \geq 1$, $-\infty \leq a < b \leq +\infty$, $0 < q < \infty$, $1 < p < \infty$, u and v are (given) weight functions, i.e. functions measurable and positive a.e. in (a, b).

If we denote, for $0 < r < \infty$ and w a weight function defined on the interval (α, β),

$$\| g \|_{r,u,(\alpha,\beta)} = \left(\int_\alpha^\beta | g(t) |^r w(t) \, dt \right)^{1/r}, \tag{0.2}$$

then we can rewrite (0.1) in the form

$$\| g \|_{q,u,(a,b)} \leq C \| g^{(k)} \|_{p,v,(a,b)}. \tag{0.3}$$

Section 1 will deal with the case $k = 1$, while Section 2 will give some examples for the case $k = 2$ to illustrate some (open) problems.

1. The "First Order" Hardy Inequality

Inequality (0.1) for $k = 1$ was considered for functions $g \in AC(a, b)$ satisfying one of the following conditions:

$$g(a) = 0 \quad \text{or } g(b) = 0 \quad \text{or} \quad g(a) = g(b) = 0. \tag{1.1}$$

[If $a = -\infty$ then $g(a) = 0$ means that $g(t) \to 0$ for $t \to -\infty$, and analogously for $b = \infty$.] These additional "boundary conditions" are natural since they

1991 Mathematics Subject Classification, 26D10.

*This work has been supported by the Grant Agency of Czech Republic, Grant No. 201/97/0744.

avoid the case when g is a (nonzero) constant for which inequality (0.1) makes no sense.

In this sections we will give necessary and sufficient conditions under which (0.1) holds (for $k = 1$) if $g \in AC(a, b)$ satisfies

$$g(c) = 0 \text{ with } c \text{ fixed, } a < c < b, \tag{1.2}$$

or

$$\lambda_1 g(a) + \lambda_2 g(b) = 0 \quad \text{with} \quad \lambda_1 \neq 0, \lambda_2 \neq 0. \tag{1.3}$$

But first, let us recall the necessary and sufficient conditions on the parameters p, q and on the weight functions u, v under which (0.1) holds (for $k = 1$) provided $g \in AC(a, b)$ satisfies $g(a) = 0$ or $g(b) = 0$.

Proposition 1.1 Let $1 < p < \infty$, $0 < q < \infty$, and denote

$$A_L = A_L(a, b) = \sup_{a<x<b} \left(\int_x^b u(t)\, dt \right)^{1/q} \left(\int_a^x v^{1-p'}(t)\, dt \right)^{1/p'} \tag{1.4}$$

with $p' = \dfrac{p}{p-1}$, if $1 < p \leq q < \infty$, and

$$A_L = A_L(a, b) = \left(\int_a^b \left(\int_x^b u(t)\, dt \right)^{r/p} \left(\int_a^x v^{1-p'}(t)\, dt \right)^{r/p'} u(x)\, dx \right)^{1/r} \tag{1.5}$$

with $\dfrac{1}{r} = \dfrac{1}{q} - \dfrac{1}{p}$, if $0 < q < p < \infty$, $1 < p < \infty$.

Then the first order Hardy inequality

$$\| g \|_{q,u,(a,b)} \leq C \| g' \|_{p,v,(a,b)} \tag{1.6}$$

holds for $g \in AC(a, b)$ satisfying $g(a) = 0$ if and only if

$$A_L < \infty. \tag{1.7}$$

Proposition 1.2 Let $1 < p < \infty$, $0 < q < \infty$, and denote

$$A_R = A_R(a, b) = \sup_{a<x<b} \left(\int_a^x u(t)\, dt \right)^{1/q} \left(\int_x^b v^{1-p'}(t)\, dt \right)^{1/p'} \tag{1.8}$$

if $1 < p < q < \infty$, and

$$A_R = A_R(a, b) = \left(\int_a^b \left(\int_a^x u(t)\, dt \right)^{r/p} \left(\int_x^b v^{1-p'}(t)\, dt \right)^{r/p'} u(x)\, dx \right)^{1/r} \tag{1.9}$$

if $0 < q < p < \infty$, $1 < p < \infty$.

Then the first order Hardy inequality (1.6) holds for $g \in AC(a, b)$ satisfying $g(b) = 0$ if and only if

$$A_R < \infty. \tag{1.10}$$

Remark 1.1 (i) The proof of the foregoing assertions can be found e.g. in [2], where also estimates for the constant C in (1.6) are given. In general, it is $C \approx A_R$ or $C \approx A_L$, respectively.

(ii) Inequality (1.6) is equivalent to the weighted norm inequality

$$\| Tf \|_{q,u,(a,b)} \leq C \| f \|_{p,v,(a,b)} \tag{1.11}$$

for nonnegative measurable functions f, where

$$(Tf)(x) = \int_a^x f(t)\, dt \tag{1.12}$$

for the case considered in Proposition 1.1 (with $g(a) = 0$) and

$$(Tf)(x) = \int_x^b f(t)\, dt \tag{1.13}$$

for the case considered in Proposition 1.2 (with $g(b) = 0$). See again [2].

Now, let us consider inequality (1.6) for functions g satisfying (1.2).

Theorem 1.1 Let $c \in (a, b)$ be fixed. Let $1 < p < \infty$, $0 < q < \infty$. Then the first order Hardy inequality (1.6) holds for all functions $g \in AC(a, b)$ satisfying the condition $g(c) = 0$ if and only if

$$A_R(a, c) + A_L(c, b) < \infty. \tag{1.14}$$

Proof (i) Necessity of (1.14). Suppose that (1.6) holds for every g *satisfying* $g(c) = 0$ and choose $g \in AC(a, b)$ such that $g(x) \equiv 0$ in $[c, b)$. Then inequality (1.6) attains the form

$$\left(\int_a^c | g(x) |^q u(x)\, dx \right)^{1/q} \leq C \left(\int_a^c | g'(x) |^p v(x)\, dx \right)^{1/p}, \tag{1.15}$$

i.e., we have the first order Hardy inequality on the interval (a, c) with $g(c) = 0$. Thus, according to Proposition 1.2, $A_R(a, c) < \infty$. Similarly, for g such that $g(x) \equiv 0$ in $(a, c]$, (1.6) attains the form

$$\| g \|_{q,u,(c,b)} \leq C \| g' \|_{p,v,(c,b)}, \tag{1.16}$$

i.e., we have the first order Hardy inequality on the interval (c, b) with $g(c) = 0$. Thus, according to Proposition 1.1., $A_L(c, b) < \infty$.

(ii) Sufficiency of (1.14). Suppose that condition (1.14) is satisfied. Then $A_R(a, c) < \infty$ and $A_L(c, b) < \infty$ and according to Propositions 1.2 and 1.1, inequality (1.15) holds for every $g_1 \in AC(a, c)$ with $g_1(c) = 0$ and inequality (1.16) holds for every $g_2 \in AC(a, b)$ with $g_2(c) = 0$. Taking $g \in AC(a, b)$ such that

$$g(x) = \begin{cases} g_1(x) & \text{for} \quad x \in (a, c] \\ g_2(x) & \text{for} \quad x \in (c, b], \end{cases}$$

we have that

$$\int_a^b | g(x) |^q u(x)\,dx = \int_a^c | g_1(x) |^q u(x)\,dx + \int_c^b | g_2(x) |^q u(x)\,dx$$

$$\leq C^q \left[\left(\int_a^c | g_1'(x) |^p v(x)\,dx \right)^{q/p} \right.$$

$$\left. + \left(\int_c^b | g_2'(x) |^q v(x)\,dx \right)^{q/p} \right]$$

$$\leq 2C^q \left(\int_a^b | g'(x) |^p v(x)\,dx \right)^{q/p}$$

Now, we will consider inequality (1.6) for functions satisfying (1.3). Obviously, this condition can be rewritten in the form

$$g(a) + \lambda g(b) = 0,\ \lambda \neq 0. \tag{1.17}$$

Remark 1.2 We have to consider also $\lambda \neq -1$, since for $\lambda = -1$, condition (1.17) has the form

$$g(a) - g(b) = 0$$

and is satisfied by the function $g(x) \equiv \text{const.}\ (\neq 0)$, but for this function, inequality (1.6) is meaningless.

Theorem 1.2 Let $\lambda \in \mathbb{R}$ be fixed, $\lambda \neq 0, \lambda \neq -1$. Let $1 < p < \infty, 0 < q < \infty$. Then the first order Hardy inequality (1.6) holds for all functions $g \in AC(a, b)$ satisfying the condition (1.17) if and only if

$$A_L(a, b) + A_R(a, b) < \infty. \tag{1.18}$$

Proof If we define the operator G as

$$(Gf)(x) = \frac{1}{\lambda + 1} \int_a^x f(t)\,dt - \frac{\lambda}{\lambda + 1} \int_x^b f(t)\,dt, \tag{1.19}$$

then the function $g(x) = (Gf)(x)$ satisfies condition (1.17) and it is $g' = f$. Consequently, instead of the Hardy inequality (1.6), we can consider the weighted norm inequality

$$\| Gf \|_{q,u,(a,b)} \leq C \| f \|_{p,v,(a,b)}. \tag{1.20}$$

(i) Sufficiency of (1.18). Suppose that condition (1.18) is satisfied. Then $A_L(a, b) < \infty$ and $A_R(a, b) < \infty$, and, according to Remark 1.1 (ii), the inequalities

$$\| T_L f \|_{q,u,(a,b)} \leq C \| f \|_{p,v,(a,b)},$$
$$\| T_R f \|_{q,u,(a,b)} \leq C \| f \|_{p,v,(a,b)} \tag{1.21}$$

hold, with

$$(T_L f)(x) = \int_a^x f(t)\,dt \quad \text{and} \quad (T_R f)(x) = \int_x^b f(t)\,dt. \tag{1.22}$$

Since, due to (1.19), $Gf = \frac{1}{\lambda+1} T_L - \frac{\lambda}{\lambda+1} T_R$, we have

$$\| Gf \|_{q,u,(a,b)} \leq \left| \frac{1}{\lambda+1} \right| \| T_L f \|_{q,u,(a,b)} + \left| \frac{\lambda}{\lambda+1} \right| \| T_R f \|_{q,u,(a,b)}$$

$$\leq C_1(\lambda) \, \| f \|_{p,v,(a,b)},$$

and this is inequality (1.6) since $Gf = g$ and $f = g'$.

(ii) Necessity of (1.18). We have to show that (1.18) is necessary for (1.20) to hold. The proof will be given for $f \geq 0$ and for $\lambda \in (-1, 0)$; for the general case, the assertion follows from a more general result due to Zharov (see [5]).

Hence, suppose that (1.20) holds for every nonnegative measurable function f and suppose that $-1 < \lambda < 0$. Then $\frac{1}{\lambda+1} > 0$ and $-\frac{\lambda}{\lambda+1} > 0$, and for $f \geq 0$ we have

$$T_L f \leq (\lambda + 1)\, Gf, \; T_R f \leq \frac{\lambda + 1}{|\lambda|} Gf.$$

From (1.20), it follows that $\| T_L f \|_{q,u,(a,b)} \leq (\lambda + 1)\, C \, \| f \|_{p,v,(a,b)}$ which implies, in view of Remark 1.1 (ii), that $A_L(a, b) < \infty$. Similarly, it is $\| T_R f \|_{q,u,(a,b)} \leq \frac{\lambda+1}{|\lambda|} C \, \| f \|_{p,v,(a,b)}$ and consequently, $A_R(a, b) < \infty$.

2. The Higher Order Hardy Inequality

Now, let us consider inequality (0.1) for $k > 1$. Again, in order to make this inequality meaningful, we have to consider functions $g \in AC^{(k-1)}(a, b)$ satisfying certain "boundary conditions". In the general case, which corresponds to conditions (1.3) for $k = 1$, they have the form

$$\sum_{j=0}^{k-1} [\alpha_{ij} g^{(j)}(a) + \beta_{ij} g^{(j)}(b)] = 0, \quad i = 0, 1, \ldots, k-1 \tag{2.1}$$

with given coefficients α_{ij} and β_{ij}.

Till now, the problem is solved in general only for the special case of conditions

$$\begin{aligned} g^{(i)}(a) &= 0, \quad i \in M_0, \\ g^{(j)}(b) &= 0, \quad j \in M_1, \end{aligned} \tag{2.2}$$

where M_0, M_1 are subsets of the k-tuple $\{0, 1, \ldots, k-1\}$ and the number of elements of M_0 and M_1 is equal to k,

$$\# M_0 + \# M_1 = k. \tag{2.3}$$

For details, see [1] and [3]; the method is based on the reduction of Hardy's inequality (0.1) to the weighted norm inequality

$$\| Gf \|_{q,u,(a,b)} \le C \| f \|_{p,v,(a,b)} \tag{2.4}$$

for nonnegative functions f, where G is the Green operator of the boundary value problem

$$\{g^{(k)} = f \text{ in } (a, b) \text{ and boundary conditions (2.2)}\}. \tag{2.5}$$

The general case is still open; here we will consider for illustration some special cases. For simplicity, we will take $k = 2$ and $(a, b) = (0, 1)$.

Hence, we are interested in necessary and sufficient conditions of the validity of inequality

$$\left(\int_0^1 | g(x) |^q u(x)\, dx \right)^{1/q} \le C \left(\int_0^1 | g''(x) |^p v(x)\, dx \right)^{1/p} \tag{2.6}$$

for functions $g \in AC^{(1)}(0, 1)$ satisfying the boundary conditions

$$\begin{aligned} \alpha_{00} g(0) + \alpha_{01} g'(0) + \beta_{00} g(1) + \beta_{01} g'(1) &= 0 \\ \alpha_{10} g(0) + \alpha_{11} g'(0) + \beta_{10} g(1) + \beta_{11} g'(1) &= 0 \end{aligned} \tag{2.7}$$

with given coefficients α_{ij}, β_{ij} $(i, j = 0, 1)$.

Example 2.1 (Robin-type boundary conditions). Let us consider inequality (2.6) under the conditions

$$\alpha g(0) + \beta g'(0) = 0, \quad \gamma g(1) + \delta g'(1) = 0. \tag{2.8}$$

Suppose that

$$\Delta := \beta\gamma - \alpha(\gamma + \delta) \ne 0, \tag{2.9}$$

and denote

$$\begin{aligned} K_1(x, t) &= \frac{1}{\Delta} (\alpha t - \beta)(\gamma + \delta - \gamma x),\ 0 < t < x < 1, \\ K_2(x, t) &= \frac{1}{\Delta} (\alpha x - \beta)(\gamma + \delta - \gamma t),\ 0 < x < t < 1. \end{aligned} \tag{2.10}$$

The function

$$g(x) = (Gf)(x) = \int_0^x K_1(x, t) f(t)\, dt + \int_x^1 K_2(x, t) f(t)\, dt \tag{2.11}$$

satisfies the differential equation $g'' = f$ in $(0, 1)$ and the boundary conditions (2.8). Thus, we can reduce the Hardy inequality (2.6) for functions g satisfying (2.8) to the inequality (2.4) for G given by (2.10).

Remark 2.1 Let us mention that condition (2.9) is substantial. Indeed: for $\alpha = \gamma = 1$, $\beta = 0$ and $\delta = -1$, i.e., for the boundary conditions

$$g(0) = 0,\ g(1) - g'(1) = 0,$$

we have $\Delta = \beta\gamma - \alpha(\gamma + \delta) = 0$ and the function $g(x) = x$ satisfies these conditions, but $g''(x) \equiv 0$ so that inequality (2.6) becomes meaningless.

Remark 2.2 The operator G from (2.11) is of the form

$$G = S_L + S_R$$

with

$$(S_L f)(x) = \varphi(x) T_L(f\psi)(x),\ (S_R f)(x) = \psi(x) T_R(f\varphi)(x)$$

where $\varphi(x) = \frac{1}{\Delta}(\gamma + \delta - \gamma x)$, $\psi(x) = \alpha x - \beta$ and T_L, T_R are defined by (1.22).

Using now the results mentioned in Remark 1.1 (ii) (see also (1.21)) it is an easy exercise to show that the weighted norm inequality

$$\| S_L f \|_{q,u,(0,1)} \leq C \| f \|_{p,v,(0,1)}$$

holds if and only if

$$\sup_{0<x<1} \left(\int_x^1 |\varphi(t)|^q u(t)\, dt \right)^{1/q} \left(\int_0^x |\psi(t)|^{p'} v^{1-p'}(t)\, dt \right)^{1/p'} < \infty$$

for $1 < p \leq q < \infty$ and

$$\left(\int_0^1 \left(\int_x^1 |\varphi(t)|^q u(t)\, dt \right)^{r/p} \right.$$

$$\left. \times \left(\int_0^x |\psi(t)|^{p'} v^{1-p'}(t)\, dt \right)^{r/p'} |\varphi(x)|^q u(x)\, dx \right)^{1/r} < \infty$$

with $\frac{1}{r} = \frac{1}{q} - \frac{1}{p}$ for $0 < q < p < \infty$, $1 < p < \infty$, and analogously for the operator S_R.

Hence, using the same arguments as in the proof of Theorem 1.2 (together with the results of [5] mentioned there) we immediately obtain the following assertion.

Theorem 2.1 The second order Hardy inequality (2.6) holds for functions $g \in AC^{(1)}(0, 1)$ satisfying (2.8) (with $\Delta \neq 0$) if and only if

$$\sup_{0<x<1} \left(\int_x^1 |\gamma + \delta - \gamma t|^q u(t)\, dt \right)^{1/q} \left(\int_0^x |\alpha t - \beta|^{p'} v^{1-p'}(t)\, dt \right)^{1/p'} < \infty,$$

$$\sup_{0<x<1} \left(\int_0^x |\alpha t - \beta|^q u(t)\, dt \right)^{1/q} \left(\int_x^1 |\gamma + \delta - \gamma t|^{p'} v^{1-p'}(t)\, dt \right)^{1/p'} < \infty$$

for $1 < p \le q < \infty$, and

$$\left(\int_0^1 \left(\int_x^1 |\gamma + \delta - \gamma t|^q\, u(t)\, dt\right)\right)^{r/p} \left(\int_0^x |\alpha t - \beta|^{p'}\, v^{1-p'}(t)\, dt\right)^{r/p'}$$

$$\times |\gamma + \delta - \gamma x|^q u(x)\, dx)^{1/r} < \infty,$$

$$\left(\int_0^1 \left(\int_0^x |\alpha t - \beta|^q\, u(t)\, dt\right)^{r/p} \left(\int_x^1 |\gamma + \delta - \gamma t|^{p'}\, v^{1-p'}(t)\, dt\right)^{r/p'}\right.$$

$$\left. \times |\alpha t - \beta|^q\, u(x)\, dx\right)^{1/r} < \infty$$

with $\frac{1}{r} = \frac{1}{q} - \frac{1}{p}$ for $0 < q < p < \infty$, $1 < p < \infty$.

Example 2.2 Consider again inequality (2.6), but now under the conditions

$$\alpha g(0) + \beta g(1) = 0, \quad \gamma g'(0) + \delta g'(1) = 0. \tag{2.12}$$

Suppose that

$$\Delta := (\alpha + \beta)(\gamma + \delta) \neq 0 \tag{2.13}$$

and denote

$$K_1(x, t) = Cx + (B - 1)\, t - BC, \; 0 < t < x < 1,$$

$$K_2(x, t) = (C - 1)\, x + Bt - BC, \; 0 < x < t < 1 \tag{2.14}$$

with $B = \frac{\beta}{\alpha + \beta}$, $C = \frac{\gamma}{\gamma + \delta}$. Then the function $g = Gf$ defined by (2.11) satisfies the differential equation $g'' = f$ in (0, 1) and the boundary conditions (2.12). Thus, again we can reduce the Hardy inequality (2.6) for functions g satisfying (2.12) to the inequality (2.4) for G given by (2.11) with kernels K_i from (2.14).

Remark 2.3 Again, Hardy's inequality (2.6) becomes meaningless if condition (2.13) is violated. As a counterexample we can use the function $g(x) = (\alpha + \beta)\, x - \beta$ with $\alpha + \beta \neq 0$, which satisfies (2.12) (the second condition with $\gamma = 1$, $\delta = -1$, i.e., $\gamma + \delta = 0$), and again $g''(x) \equiv 0$.

Remark 2.4 In the case mentioned in Example 2.1, we cannot use the approach described in Remark 2.2, since the kernels K_i from (2.14) have *not* the form of a product $\varphi(t)\psi(x)$ as in (2.10). Nevertheless, in some special cases we can give necessary and sufficient conditions for the validity of (2.6). For example, let us suppose that $\beta = 0$, i.e., $B = 0$. Then we have

$$K_1(x, t) = Cx - t, \quad K_2(x, t) = (C - 1)\, x$$

and hence

$$(Gf)(x) = \int_0^x (Cx - t)f(t)\,dt + (C-1)\,x\int_x^1 f(t)\,dt. \tag{2.15}$$

If $C > 1$, then $K_1(x, t) > 0$ for $0 < t < x < 1$ and we obtain the following *triple* of necessary and sufficient conditions for the validity of the second order Hardy inequality (2.6) under the boundary conditions

$$g(0) = 0, \quad \gamma g'(0) + (1-\gamma)g'(1) = 0, \quad \gamma > 1:$$

In the case $1 < p \le q < \infty$,

$$\sup_{0<x<1}\left(\int_x^1 (\gamma x - t)^q u(t)\,dt\right)^{1/q}\left(\int_0^x v^{1-p'}(t)\,dt\right)^{1/p'} < \infty,$$

$$\sup_{0<x<1}\left(\int_x^1 u(t)\,dt\right)^{1/q}\left(\int_0^x (\gamma t - x)^{p'} v^{1-p'}(t)\,dt\right)^{1/p'} < \infty, \tag{2.16}$$

$$\sup_{0<x<1}\left(\int_0^x t^q u(t)\,dt\right)^{1/q}\left(\int_x^1 v^{1-p'}(t)\,dt\right)^{1/p'} < \infty;$$

In the case $1 \le q < p < \infty$,

$$\left(\int_0^1\left(\int_x^1 (\gamma x - t)^q u(t)\,dt\right)^{r/q}\left(\int_0^x v^{1-p'}(t)\,dt\right)^{r/q'} v^{1-p'}(x)\,dx\right)^{1/r} < \infty,$$

$$\left(\int_0^1\left(\int_x^1 u(t)\,dt\right)^{r/p}\left(\int_0^x (\gamma t - x)^{p'} v^{1-p'}(t)\,dt\right)^{r/p'} u(x)\,dx\right)^{1/r} < \infty, \tag{2.17}$$

$$\left(\int_0^1\left(\int_0^x t^q u(t)\,dt\right)^{r/p}\left(\int_x^1 v^{1-p'}(t)\,dt\right)^{r/p'} x^q u(x)\,dx\right)^{1/r} < \infty$$

with $\dfrac{1}{r} = \dfrac{1}{q} - \dfrac{1}{p}$.

Notice that the first two conditions in (2.16) or (2.17) are necessary and sufficient for the validity of the inequality

$$\left(\int_0^1\left(\int_x^1 (\gamma x - t)\,f(t)\,dt\right)^q u(x)\,dx\right)^{1/q} \le C\left(\int_0^1 f^p(x)v(x)\,dx\right)^{1/p}$$

for $f \ge 0$ and cover the first integral in formula (2.15) for G (see e.g., [4]) while the third condition in (2.16) or (2.17) is necessary and sufficient for

$$\left(\int_0^1 \left(x\int_0^x f(t)\,dt\right)^q u(x)\,dx\right)^{1/q} \le C\left(\int_0^1 f^p(x)v(x)\,dx\right)^{1/p}$$

and covers the second integral in (2.15).

References

1. A. Kufner. Higher order Hardy inequalities. Bayreuth. Math. Schriften 44 (1993), 105–146.
2. B. Opic, A. Kufner. Hardy-type inequalities. Pitman Research Notes 219. Longman, Harlow 1990.
3. G. Sinnamon. Kufner's conjecture for higher order Hardy inequalities. Real Analysis Exchange 21 (2), 1995/6, 590–603.
4. V.D. Stepanov. Weighted inequalities for a class of Volterra convolution operators. J. London Math. Soc. 45 (1992), 232–242.
5. P.A. Zharov. On a two-weights inequality. Generalizatoin of the inequalities of Hardy and Poincaré. Trudy Mat. Inst. RAN 194 (1992), 97–110 (Russian).

Function Spaces and Applications
D.E. Edmunds et al (Eds)

13. On Asymptotic Behaviour of the Approximation Numbers and Estimates of Schatten-von Neumann Norms of the Hardy-type Integral Operators

Elena N. Lomakina*[1] and Valdimir D. Stepanov*[2]

[1]Khabarovsk State University, Tichookeanskaya, 136, Khabarovsk, 680035, Russia

[2]Computer Center, Far-Eastern Branch of the Russian Academy of Sciences, Shelest 118-205, Khabarovsk, 680042, Russia

1. Introduction

In 1907 E. Schmidt [14] obtained an analog of the spectral representation formulae for a non-symmetric integral operator A in L^2 with a continuous kernel, using the eigenvalues of positive operator $(A^*A)^{1/2}$, which were later called *s-numbers* of A. In 1957 D.E. Allakhverdiev proved the approximation Property of s-numbers (see [2], Chapter 2, Theorem 2.1), which made it possible to extend the notion of s-numbers to operators acting on Banach spaces. This was performed by A. Pietsch [11] and, thus, given $\mathbf{T} : X \to Y$ *the* m-th *approximation number* $a_m(\mathbf{T})$ is defined by

$$a_m(\mathbf{T}) = \inf \{ \| \mathbf{T} - P \|; P : X \to Y, \text{rank } P < m\}, m = 1, 2, \ldots .$$

Basic information on the approximation numbers (*a-numbers)* can be found in monographs [3, 6, 12].

Let $1 < p < \infty$ and $L^p = L^p(\mathbf{R}^+)$ be the Lebesgue space with the norm

$$\|f\|_{L^p} = \left(\int_0^\infty |f(x)|^p \, dx \right)^{1/p}$$

For the space of functions restricted to an interval $I \subset \mathbf{R}^+$ we use the notations $L^p(I)$ and $\|f\|_{L^p(I)}$. Put $p' = \dfrac{p}{(p-1)}$. Let $u(y)$ and $v(x)$ be real-valued functions

---

1991 Mathematics Subject Classification: 47G10.

*The research work of both the authors was partially supported by the INTAS project 94–881 and by the Russian Fund of Basic Researches grant 97-01-00604 and by the Ministry of Education grant 10.98GR. The research work of the first author was also partially supported by the INTAS Fellowship grant YSF 99-4005.

(weights). We study the behaviour of a-numbers of the Hardy-type integral operator $T : L^p(\mathbf{R}^+) \to L^q(\mathbf{R}^+)$ of the form

$$Tf(x) = v(x) \int_0^x u(y) f(y)\, dy \tag{1.1}$$

which is merely supposed to be bounded. It implies, in particular, that $u \in L^{p'}(0, t)$ and $v \in L^q(t, \infty)$ for all $t > 0$. This topic was initiated by D.E. Edmunds, W.D. Evans and D.J. Harris in [4], where implicit lower and upper estimates of $\{a_n(T)\}$ were found for $T : L^p \to L^q$, $1 < p \leq q < \infty$. Recently it was continued in [5], where the authors invented the new powerful method, which allowed to turn the implicit relations into precise asymptotic or Schatten-von Neumann type norm estimates of $\{a_n(T)\}$. The main results of [5] are the following. Let the sequence $\{\xi_n\}$ be defined from the equation

$$U(\xi_n) = \int_0^{\xi_n} | u(t) |^{p'} dt = 2^{n+1}. \tag{1.2}$$

We suppose for simplicity, that $\| u \|_{L^{p'}} = \infty$, and consequently, $\{\xi_n\}$ does exist for every $n \in \mathbf{Z}$. Put

$$\sigma_n = \left(\int_{\xi_{n-1}}^{\xi_n} | u(t) |^{p'} dt \right)^{1/p'} \left(\int_{\xi_n}^{\xi_{n+1}} | v(t) |^q dt \right)^{1/q} \tag{1.3}$$

If $T : L^p \to L^p$, $1 < p < \infty$ is compact, then

$$\frac{1}{4} \alpha_p \int_0^\infty | uv | \leq \liminf_{n\to\infty} n a_n(T) \leq \limsup_{n\to\infty} n a_n(T) \leq \alpha_p \int_0^\infty | uv |, p \neq 2, \tag{1.4}$$

where

$$\alpha_p = \sup_{\|f\|_{L^p_{[0,1]}} \leq 1} \left(\int_0^1 \int_0^1 | \int_x^y f(\tau)\, d\tau |^p \, dx\, dy \right)^{1/p}, \tag{1.5}$$

and

$$\lim_{n\to\infty} n a_n(T) = \frac{1}{\pi} \int_0^\infty | uv |, p = 2,$$

provided $\sum_n \sigma_n < \infty$. Secondly, if $1 < s < \infty$, then

$$\sum_n a_n^s(T) \asymp \sum_n \sigma_n^s \tag{1.6}$$

and

$$\sup_{t>0} t(\# \{n \in \mathbf{N} : a_n(T) > t\})^{1/s} \asymp \sup_{t>0} t(\# \{n \in \mathbf{Z} : \sigma_n > t\})^{1/s}. \tag{1.7}$$

The aim of the present paper is twofold. The first is to extend (1.4), (1.6) and (1.7) for $T : L^p \to L^q$, when $1 < p, q < \infty$, and the second purpose is to replace the right-hand side of (1.6) by an integral expression in terms of u and v. The main results of the paper are the following two theorems.

Theorem 1 *Suppose operator* $T : L^p(\mathbf{R}^+) \to L^q(\mathbf{R}^+)$ *of the form* (1.1) *be compact.* (1) *If* $1 < p \le q < \infty$, $r = \dfrac{qp'}{q+p'} \ge 1$ *and* $\sum_{k\in\mathbf{Z}} \sigma_k^r < \infty$, *then*

$$\frac{1}{2} 3^{\frac{1-r}{r}} \gamma_{pq} \left(\int_0^\infty | u(t)v(t) |^r \, dt \right)^{1/r} \le \liminf_{N\to\infty} N a_N(T) \tag{1.8}$$

and

$$\limsup_{N\to\infty} N^{1/r} a_N(T) \le \gamma_{pq} \left(\int_0^\infty | u(t) |^r | v(t) |^r \, dt \right)^{1/r} \tag{1.9}$$

(2) *If* $1 < q < p < \infty$, $\dfrac{1}{r} = \dfrac{q+p'}{qp'} > 1$ *and* $\sum_{k\in\mathbf{Z}} \sigma_k^r < \infty$, *then*

$$\frac{1}{2} \gamma_{pq} \left(\int_0^\infty | u(t)v(t) |^r \, dt \right)^{1/r} \le \liminf_{N\to\infty} N^{1/r} a_N(T) \tag{1.10}$$

and

$$\limsup_{N\to\infty} N a_N(T) \le \gamma_{pq} \left(\int_0^\infty | u(t)v(t) |^r \, dt \right)^{1/r}, \tag{1.11}$$

where

$$\gamma_{pq} = \sup_{\|f\|_{L^p_{[0,1]}} \le 1} \left(\int_0^1 | \int_0^1 \left(\int_y^x f(\tau)\, d\tau \right) dy |^q \, dx \right)^{1/q}$$

Let $1 < p, q < \infty$, $0 < s < \infty$ and put

$$J = \left(\int_0^\infty \left(\int_0^x | u |^{p'} \right)^{s/p'} \left(\int_x^\infty | v |^q \right)^{\frac{s}{q}-1} | v(x) |^q \, dx \right)^{1/s}$$

Theorem 2 *Let operator* $T : L^p \to L^q$ *be compact. Then the following estimates hold*

$$J^s \le 6 \cdot 2^s \sum_{k\in\mathbf{N}} \left[a_k(T)\, k^{\frac{1}{p}-\frac{1}{q}} \right]^s, \; 1 < p \le q < \infty$$

and

$$J^s \le 6 \cdot 2^s \left[\frac{q(p-1)}{p-q} \right]^{\frac{s(p-q)}{pq}} \sum_{k\in\mathbf{N}} a_k^s(T), \; 1 < q < p < \infty$$

for all $s \in (0, \infty)$. If $s > r = \dfrac{qp'}{q + p'}$, *then*

$$\sum_{k \in \mathbf{N}} a_k^s(T) \le C_1(p, q, s)\, J^s, \quad 1 < p \le q < \infty$$

and

$$\sum_{k \in \mathbf{N}} \left[a_k(T)\, k^{\frac{1}{p} - \frac{1}{q}} \right]^s \le C_2(p, q, s)\, J^s, \quad 1 < q < p < \infty.$$

Throughout the paper the expressions of the form $0. \infty$, $0/0$, ∞/∞ are taken to zero, the inequality $A << B$ means $A \le cB$ with a constant, depending on p, q, s only, however the relationship $A \asymp B$ is interpreted as $A << B << A$ or $A = cB$. χ_E denotes the characteristic function of a set $E \subset \mathbf{R}^+$ and $\#\,\{k \in \Omega\}$ is established for the number of integers with prescribed property.

2. Auxiliary Results

The mapping properties of the Hardy-type integral operator (1.1) in Lebesgue space are well-known. We summarize the results on the boundedness and compactness due to B. Muckenhoupt [10], V.G. Mazja and A.L. Rozin [9], J.S. Bradley [1], S.D. Riemenschneider [13] in the following statement. Notate

$$A(t) = \left(\int_0^t |u(y)|^{p'} dy \right)^{1/p'} \left(\int_t^\infty |v(y)|^q dy \right)^{1/q}$$

and suppose without a loss of generality that

$$\inf\{t > 0; \;\; A(t) > 0\} = 0, \quad \sup\{t > 0; A(t) > 0\} = \infty.$$

Theorem 2.1 *Let the integral operator* $T : L^p(\mathbf{R}^+) \to L^q(\mathbf{R}^+)$ *be given by* (1.1).

(*a*1) *If* $1 < p \le q < \infty$, *then T is bounded if, and only if,*

$$A = \sup_{t>0} A(t) = \sup_{t>0} \left(\int_0^t |u(y)|^{p'} dy \right)^{1/p'} \left(\int_t^\infty |v(y)|^q dy \right)^{1/q} < \infty.$$

Moreover, $\| T \| \asymp A$.

(*a*2) *For* $1 < p \le q < \infty$ *the operator T is compact if, and only if,*

$$A < \infty \text{ and } \lim_{t \to 0} A(t) = \lim_{t \to \infty} A(t) = 0.$$

(*b*1) *Let* $1 < q < p < \infty$. *The integral operator T is bounded if, and only if,*

$$B = \left\{ \int_0^\infty \left[\left(\int_0^x |u(y)|^{p'} dy \right)^{q-1} \int_x^\infty |v(y)|^q dy \right]^{\frac{p}{p-q}} |u(x)|^{p'} dx \right\}^{\frac{p-q}{pq}} < \infty$$

and $\| T \| \asymp B$.

(b2) If $1 < q < p < \infty$, then T is compact if, and only if, $B < \infty$.

Remark 2.1 Theorem 2.1 has an obvious analog for a finite interval. For the boundedness and compactness of more general Hardy-type integral operators see [8], [15].

As we mentioned above the first *implicit estimates* of a-numbers for $T : L^p \to L^q$, $1 < p \le q < \infty$, were found in [4]. Below we remind the scheme with a slight simplification and amendment given in [8].

For an interval $I = [a, b] \subset \mathbf{R}^+$, $0 < a < b < \infty$ we define the measure μ by

$$d\mu(x) = |\, v(x) \,|^q \, dx$$

and suppose, that

$$\mu(I) = \int_I |\, v(x) \,|^q \, dx \neq 0.$$

For

$$F(x) = \int_a^x u(t) \, f(t) \, dt$$

we put

$$F_I = \frac{1}{\mu(I)} \int_I F(x) \, d\mu(x).$$

Now, for the operator

$$\mathcal{J}_f(x) = v(x) \, (F(x) - F_I), \; x \in I$$

we find the two-sided estimate of the norm of $\mathcal{J}_I : L^p(I) \to L^q(I)$. To this end let $c \in I$ be chosen so that

$$\int_a^c |\, v(x) \,|^q \, dx = \frac{1}{2} \int_a^b |\, v(x) \,|^q \, dx,$$

and set

$$A^*(a, c) = \sup_{a<t<c} \left(\int_a^t |\, v(y) \,|^q \, dy \right)^{1/q} \left(\int_t^c |\, u(y) \,|^{p'} \, dy \right)^{1/p'},$$

$$A(c, b) = \sup_{c<t<b} \left(\int_t^b |\, v(y) \,|^q \, dy \right)^{1/q} \left(\int_c^t |\, u(y) \,|^{p'} \, dy \right)^{1/p'},$$

$$B^*(a, c) = \left\{ \int_a^c \left[\left(\int_a^t |\, v(x) \,|^q \, dx \right)^{1/q} \right. \right.$$

$$\left. \left. \times \left(\int_t^c |\, u(x) \,|^{p'} \, dx \right)^{1/q'} \right]^{\frac{pq}{p-q}} |\, u(t) \,|^{p'} \, dt \right\}^{\frac{p-q}{pq}},$$

$$B(c,b) = \left\{ \int_c^b \left[\left(\int_c^t |u(x)|^{p'} dx \right)^{1/q'} \right. \right.$$

$$\left. \left. \times \left(\int_t^b |v(x)|^q dx \right)^{1/q} \right]^{\frac{pq}{p-q}} |u(t)|^{p'} dt \right\}^{\frac{p-q}{pq}}$$

Theorem 2.2 [4], [7].

(1) *Let* $1 \le p \le q < \infty$, *then* $\| \mathcal{T}_I \|_{L^p(I) \to L^q(I)} \approx \max\,(A^*(a,c), A(c,b))$

(2) *Let* $1 < q < p < \infty$, *then* $\| \mathcal{T}_I \|_{L^p(I) \to L^q(I)} \approx \max\,(B^*(a,c), B(c,b))$.

Assume that the operator T is compact. Given $\varepsilon > 0$ such that $0 < \varepsilon < \| T \|$, by Theorem 2.1 we choose the numbers $0 = c_0 < c_1 < c_N < c_{N+1} = \infty$ such, that

$$A[c_1] = \sup_{0<t<c_1} A(t) = \varepsilon,$$

$$A[c_N] = \sup_{c_N<t<c_{N+1}} A(t) \le \varepsilon,$$

in the case $1 < p \le q < \infty$, and

$$B(0, c_1) = \varepsilon, \quad B(c_N, \infty) \le \varepsilon,$$

when $1 < q < p < \infty$.

By Theorem 2.2 the norm of operator $\mathcal{T}_I$ continuously depends on interval I and therefore for sufficiently small $\varepsilon > 0$ we may and shall find the points $0 = c_0 < c_1 < c_2 < \ldots < c_{N-1} < c_N < c_{N+1} = \infty$ and the intervals $J_k = [c_k, c_{k+1}]$, $k = 1, \ldots, N-1$, so, that

$$\| \mathcal{T}_{J_k} \| = \varepsilon, \quad k = 1, 2, \ldots, N-1,$$

where

$$\mathcal{T}_{J_k} f(x) = \chi_{J_k}(x) v(x) \left(F(x) - F_I \right).$$

Note, that

$$N = N(\varepsilon), \tag{2.1}$$

that is the number N depends on $\varepsilon > 0$. If we take

$$P_k f(x) = \chi_{J_k}(x) \{ Tf(x) - \mathcal{T}_{J_k} f(x) \}, k = 1, 2, \ldots, N-1$$

and define $P : L^p(\mathbf{R}^+) \to L^q(\mathbf{R}^+)$ by

$$P = \sum_{k=1}^{N-1} P_k,$$

then rank $P \le N - 1$. Using the Jensen inequality and Theorem 2.2 we obtain

$$\| Tf - Pf \|_{L^q}^q \leq \varepsilon^q \| f \|_{L^p}^q, \; 1 < p \leq q < \infty$$

and

$$\| Tf - Pf \|_{L^q}^q \leq \varepsilon^q N^{1-\frac{q}{p}} \| f \|_{L^p}^q, \; 1 < q < p < \infty$$

which imply the upper bounds for the approximation numbers

$$\begin{aligned} a_N(T) &\leq \varepsilon, \quad 1 < p \leq q < \infty, \\ a_N(T) &\leq \varepsilon N^{\frac{1}{q}-\frac{1}{p}}, \quad 1 < q < p < \infty. \end{aligned} \tag{2.2}$$

For the lower bound similar to the work [4] we use the estimate

$$\inf_{c \in \mathbf{R}} \left(\int_I | F(x) - c |^q \, d\mu(x) \right)^{1/q} \geq \frac{1}{2} \left(\int_I | F(x) - F_I |^q \, d\mu(x) \right)^{1/q},$$

$$2 \neq q \in (1, \infty) \tag{2.3}$$

or

$$\inf_{c \in \mathbf{R}} \left(\int_I | F(x) - c |^2 \, d\mu(x) \right)^{1/2} = \left(\int_I | F(x) - F_I |^2 \, d\mu(x) \right)^{1/2}.$$

Let $0 < \delta < 1$ and $\{ f_k \}$, $k = 0, 1, \ldots N - 1$ be the functions such, that

$$\begin{aligned} \| Tf_0 \|_{L^q_{[0,c_1]}} &= \delta \varepsilon \| f_0 \|_{L^p_{[0,c_1]}}, \\ \| \mathcal{T}_{J_k} f_k \|_{L^q_{(J_k)}} &= \delta \varepsilon \| f_k \|_{L^p_{(J_k)}} \end{aligned} \tag{2.4}$$

For any operator P with rank $P \leq N - 1$, there exist constants $\{\lambda_k\}$, $k = 0, 1, \ldots, N - 1$ not all equal to zero such, that $P(\sum_{k=0}^{N-1} \lambda_k f_k) = 0$. Putting $f = \sum_{k=0}^{N-1} \lambda_k f_k$, we write

$$\| Tf - Pf \|_q^q = \| Tf \|_q^q \geq (\delta \varepsilon)^q \, | \lambda_0 |^q \, \| f_0 \|_{L^p_{[0,c_1]}}^q$$

$$+ \sum_{k=1}^{N-1} \int_{c_k}^{c_{k+1}} \left| \int_0^x u(y) f(y) \, dy \right|^q \, d\mu(x).$$

Obviously, for $x \in (c_k, c_{k+1})$

$$\int_0^x u(y) f(y) \, dy = \lambda_k \int_{c_k}^x u(y) f_k(y) \, dy + \int_0^{c_k} u(y) f(y) \, dy := \lambda_k F_k(x) + \nu_k$$

and, using (2.3) and (2.4), the above estimate, say for $q \neq 2$, is continued by

$$\geq (\delta\varepsilon)^q \, | \lambda_0 |^q \, \| f_0 \|^q_{L^p_{[0,c_1]}} + \sum_{k=1}^{N-1} \int_{J_k} | \lambda_k F_k(x) + \nu_k |^q \, d\mu(x)$$

$$\geq (\delta\varepsilon)^q \, | \lambda_0 |^q \, \| f_0 \|^q_{L^p_{[0,c_1]}} + \frac{1}{2^q} \sum_{k=1}^{N-1} \int_{J_k} | \lambda_k |^q \, | F_k - F_{k,I_k} |^q \, d\mu(x)$$

$$\geq \left(\frac{\delta\varepsilon}{2} \right)^q \sum_{k=0}^{N-1} | \lambda_k |^q \, \| f_k \|^q_{L^p_{(J_k)}}$$

Letting $\delta \to 1$ the Jensen inequality yields

$$\| T - P \| \geq \varepsilon, \quad q = 2 \leq p < \infty$$

$$\| T - P \| \geq \frac{\varepsilon}{2}, \quad 1 < q < p < \infty, \; q \neq 2,$$

$$\| T - P \| \geq \frac{1}{2} \, \varepsilon N^{1/q - 1/p}, \quad 1 < p \leq q < \infty, \; q \neq 2,$$

which implies the lower bounds for the approximation numbers complemented the upper bounds (2.2)

$$a_N(T) \geq \varepsilon, \; q = 2 < p < \infty,$$

$$a_N(T) \geq \frac{1}{2} \, \varepsilon, \quad 1 < q < p < \infty, \; q \neq 2 \tag{2.5}$$

$$a_N(T) \geq \frac{1}{2} \, \varepsilon N^{1/q - 1/p}, \quad 1 < p \leq q < \infty, \; q \neq 2.$$

3. Asymptotic Behavior (Proof of Theorem 1)

Let $1 < p, q < \infty$, $\frac{1}{p} + \frac{1}{p'} = 1$, $\frac{1}{q} + \frac{1}{q'} = 1$; $\mathbf{R}^+ = [0, \infty)$, $I = [a, b] \subset \mathbf{R}^+$, where $0 \leq a < b \leq \infty$. We consider the operator $\mathcal{T}_I : L^p(I) \to L^q(I)$ of the form

$$\mathcal{T}_I f(x) = \mathcal{T}_{u,v;I} f(x)$$

$$= v(x) \left\{ \int_a^x f(\tau) u(\tau) \, d\tau - \frac{\displaystyle\int_a^b \left(\int_a^y f(\tau) u(\tau) \, d\tau \right) | v(y) |^q \, dy}{\displaystyle\int_a^b | v(y) |^q \, dy} \right\}, x \in I, \tag{3.1}$$

provided $0 < \int_a^b | v(y) |^q \, dy < \infty$.

In the case $u = v = 1$, $I = [0, 1]$, we put

$$\mathscr{T}_{1,1;[0,1]}f(x) = \int_0^x f(\tau)\,d\tau - \int_0^1 \left(\int_0^y f(\tau)\,d\tau \right) dy = \int_0^1 \left(\int_y^x f(\tau)\,d\tau \right) dy$$

and

$$\gamma_{pq} \equiv \| \mathscr{T}_{1,1;[0,1]} \|_{L^p_{[0,1]} \to L^q_{[0,1]}} = \sup_{\|f\|_{L^p_{[0,1]}} \le 1} \left(\int_0^1 \left| \int_0^1 \left(\int_y^x f(\tau)\,d\tau \right) dy \right|^q dx \right)^{1/q} \tag{3.2}$$

Note, that $\gamma_{22} = \frac{1}{\pi}$, and

$$\gamma_{pp} < \alpha_p < 2\gamma_{pp}, \quad 1 < p < \infty, \tag{3.3}$$

where α_p is given by (1.5) and, thus, Theorem 1 gives better result for $1 < p = q < \infty$, $p \neq 2$ than (1.4).

We begin with the following simple lemma.

Lemma 3.1 *Let $1 < p, q < \infty$ and $I = [a, b] \subset \mathbf{R}^+$, $u = u_0$, and $v = v_0$ are the constant weights. Then*

$$\| \mathscr{T}_{u_0,v_0;I} \|_{L^p(I) \to L^q(I)} = \gamma_{pq} \mid u_0 || v_0 || b - a |^{1-\frac{1}{p}+\frac{1}{q}}, \tag{3.4}$$

where γ_{pq} is determined by (3.2).

Proof Follows by change of variables.

Lemma 3.2 *Let $1 < p, q < \infty$ $u_1, u_2 \in L^{p'}(I)$ and $v \in L^q(I)$. Then*

$$\| \mathscr{T}_{u_1,v;I} f - \mathscr{T}_{u_2,v;I} f \|_{L^q(I)} \le \| u_1 - u_2 \|_{L^{p'}(I)} \| v \|_{L^q(I)} \| f \|_{L^p(I)}.$$

Proof Since

$$\mathscr{T}_{u_1,v;I} f - \mathscr{T}_{u_2,v;I} f = \mathscr{T}_{u_1-u_2,v;I} f$$

$$= \frac{v(x)}{\| v \|^q_{L^q(I)}} \int_a^b | \int_y^x f(u_1 - u_2) | | v(y) |^q \,dy,$$

then using Hölder's inequality, we have

$$\| \mathscr{T}_{u_1,v;I} f - \mathscr{T}_{u_2,v;I} f \|_{L^q(I)} \le \| u_1 - u_2 \|_{L^{p'}(I)} \| v \|_{L^q(I)} \| f \|_{L^p(I)}$$

Lemma 3.3 *Let $1 < p, q < \infty$, $u \in L^{p'}(I)$ and $v_1, v_2 \in L^q(I)$. Then*

$$\| \mathscr{T}_{u,v_1;I} f - \mathscr{T}_{u,v_2;I} f \|_{L^q(I)} \le (2 + q2^q) \| v_1 - v_2 \|_{L^q(I)} \| u \|_{L^{p'}(I)} \| f \|_{L^p(I)}.$$

Proof Write

$$\mathscr{T}_1 f(x) \equiv \mathscr{T}_{u,v_1;I} f(x), \mathscr{T}_2 f(x) \equiv \mathscr{T}_{u,v_2;I} f(x)$$

and consider the difference

$$\mathcal{T}_1 f(x) - \mathcal{T}_2 f(x) = [v_1(x) - v_2(x)] \int_a^x f(\tau)u(\tau)\, d\tau$$

$$+ \frac{[v_2(x) - v_1(x)]}{\int_a^b |v_1(y)|^q\, dy} \int_I \left(\int_a^y f(\tau)\, u(\tau) d\tau \right) |v_1(y)|^q\, dy$$

$$+ v_2(x) \int_I \left(\int_a^y fu \right) \left[\frac{|v_2|^q}{\| v_2 \|^q_{L^q(I)}} - \frac{|v_1|^q}{\| v_1 \|^q_{L^q(I)}} \right] dy \equiv J_1 + J_2 + J_3.$$

Then

$$\| \mathcal{T}_1 f - \mathcal{T}_2 f \|_{L^q(I)} \le \| J_1 \|_{L^q(I)} + \| J_2 \|_{L^q(I)} + \| J_3 \|_{L^q(I)}.$$

For the first term we write

$$\| J_1 \|_{L^q(I)} = \| [v_1(x) - v_2(x)] \int_a^x f(\tau)u(\tau)\, d\tau \|_{L^q(I)}$$

$$\le \int_I |f(\tau)u(\tau)|\, d\tau \| v_1(x) - v_2(x) \|_{L^q(I)}.$$

Applying Hölder's inequality, we obtain

$$\| J_1 \|_{L^q(I)} \le \| v_1 - v_2 \|_{L^q(I)} \| u \|_{L^{p'}(I)} \| f \|_{L^p(I)}.$$

The second term is estimated similarly.

$$\| J_2 \|_{L^q(I)} = \left\| \frac{[v_2(x) - v_1(x)]}{\int_I |v_1(y)|^q\, dy} \int_I \left(\int_a^y f(\tau)u(\tau)\, d\tau \right) |v_1(y)|^q\, dy \right\|_{L^q(I)}$$

$$\le \frac{\| v_1 - v_2 \|_{L^p(I)} \left| \int_I \left(\int_I f(\tau)u(\tau)\, d\tau \right) |v_1(y)|^q\, dy \right|}{\| v_1 \|^q_{L^q(I)}}$$

$$\le \frac{\| v_1 - v_2 \|_{L^q(I)} \| u \|_{L^{p'}(I)} \| f \|_{L^p(I)} \| v_1 \|^q_{L^q(I)}}{\| v_1 \|^q_{L^q(I)}}$$

$$= \| v_1 - v_2 \|_{L^q(I)} \| u \|_{L^{p'}(I)} \| f \|_{L^p(I)}.$$

$$\| J_2 \|_{L^q(I)} \le \| v_1 - v_2 \|_{L^q(I)} \| u \|_{L^{p'}(I)} \| f \|_{L^p(I)}.$$

For the estimate of J_3 we need the following inequality

$$||x|^s - |y|^s| \leq s\,(|x| + |y|)^{s-1}\,|x - y|, \quad 1 \leq s < \infty. \tag{3.5}$$

Really, if $|y| < z < |x|$ and $1 \leq s < \infty$, then

$$||x|^s - |y|^s| = s\left|\int_{|y|}^{|x|} z^{s-1}\,dz\right|$$

$$\leq s\left|\int_{|y|}^{|x|} (|x| + |y|)^{s-1}\,dz\right| \leq s\,(|x| + |y|)^{s-1}\,|x - y|.$$

Thus, using Hölder's inequality with p, p' and q, q', we find

$$\| J_3 \|_{L^q(I)}$$

$$\leq \| v_2 \|_{L^q(I)} \int_I |f(\tau)u(\tau)|\,d\tau \left|\int_I \left[\left(\frac{|v_2|}{\| v_2 \|_{L^q(I)}}\right)^q - \left(\frac{|v_1|}{\| v_1 \|_{L^q(I)}}\right)^q\right] dy\right|$$

$$\leq q\,\| v_2 \|_{L^q(I)}\, \| u \|_{L^{p'}(I)}\, \| f \|_{L^p(I)} \int_a^b \left|\frac{|v_1|}{\| v_1 \|_{L^q(I)}} + \frac{|v_2|}{\| v_2 \|_{L^q(I)}}\right|^{q-1}$$

$$\times \left|\frac{|v_2|}{\| v_2 \|_{L^q(I)}} - \frac{|v_1|}{\| v_1 \|_{L^q(I)}}\right| dy$$

$$\leq q\,\| v_2 \|_{L^q(I)}\, \| u \|_{L^{p'}(I)}\, \| f \|_{L^p(I)} \left(\int_I \left|\frac{|v_2|}{\| v_2 \|_{L^q(I)}} - \frac{|v_1|}{\| v_1 \|_{L^q(I)}}\right|^q dy\right)^{1/q}$$

$$\times \left(\int_I \left|\frac{|v_1|}{\| v_1 \|_{L^q(I)}} + \frac{|v_2|}{\| v_2 \|_{L^q(I)}}\right|^{q'(q-1)} dy\right)^{1/q'}$$

(we apply the triangle inequality)

$$\leq q2^{q-1}\,\| v_2 \|_{L^q(I)}\, \| u \|_{L^{p'}(I)}\, \| f \|_{L^p(I)} \left(\int_I \left|\frac{|v_2|}{\| v_2 \|_{L^q(I)}} - \frac{|v_1|}{\| v_1 \|_{L^q(I)}}\right|^q dy\right)^{1/q}$$

$$= q2^{q-1}\,\| u \|_{L^{p'}(I)}\, \| f \|_{L^p(I)} \left(\int_I \left| |v_2| - |v_1| + |v_1| \left|1 - \frac{\| v_2 \|_{L^q(I)}}{\| v_1 \|_{L^q(I)}}\right|\right|^q dy\right)^{1/q}$$

$$\left(\text{using } \left|1 - \frac{\| v_2 \|_{L^q(I)}}{\| v_1 \|_{L^q(I)}}\right| = \left|\frac{\| v_1 \|_{L^q(I)} - \| v_2 \|_{L^q(I)}}{\| v_1 \|_{L^q(I)}}\right| \le \frac{\| v_1 - v_2 \|_{L^q(I)}}{\| v_1 \|_{L^q(I)}}\right.$$

$$\left.\text{and } \| v_2 | - | v_1 \| \le | v_1 - v_2 |\right)$$

$$\le q2^q \| u \|_{L^{p'}(I)} \| f \|_{L^p(I)} \| v_1 - v_2 \|_{L^q(I)}$$

Hence

$$\| \mathcal{T}_1 f - \mathcal{T}_2 f \|_{L^q(I)} \le (2 + q2^q) \| u \|_{L^{p'}(I)} \| f \|_{L^p(I)} \| v_1 - v_2 \|_{L^q(I)}$$

and Lemma 3.3 follows.

Summarizing the results of Lemmas 3.2 and 3.3 we obtain

$$\| \mathcal{T}_{u_1, v_1; I} - \mathcal{T}_{u_2, v_2; I} \|_{L^q(I)} \le (2 + q2^q) \| v_1 - v_2 \|_{L^q(I)} \| u_1 \|_{L^{p'}(I)} + \| u_1 - u_2 \|_{L^{p'}(I)} \| v_2 \|_{L^q(I)} \tag{3.6}$$

Proof of Theorem 1 In the proof we essentially use the ideas of the work [5]. Fix a sufficiently large $x \in \mathbf{R}^+$. Since the set of all step functions is dense in Lebesgue space, then for each $\eta > 0$ there exist simple functions u_η, v_η on $[1/x, x] = \bigcup_{k=1}^{m} I_k$ such that

$$u_\eta(t) = \sum_{k=1}^{m} u_k \chi_{I_k}(t), \quad v_\eta(t) = \sum_{k=1}^{m} v_k \chi_{I_k}(t)$$

and the inequalities

$$\| u(t) - u_\eta(t) \|_{L^{p'}_{[1/x,x]}} \le \eta, \quad \| v(t) - v_\eta(t) \|_{L^q_{[1/x,x]}} \le \eta \tag{3.7}$$

hold. Given $\varepsilon > 0$, $0 < \varepsilon < \| T \|$ we consider the finite sequence $0 = c_0 < c_1 < c_2 < c_3 < \cdots < c_{N-1} < c_N < c_{N+1} = \infty$, such that

$$\| \chi_{J_j} T \| \le \varepsilon, \quad j = 0,$$

$$N \| \mathcal{T}_{J_j} \| = \varepsilon, \; j = 1, 2, \ldots, N - 1$$

Put

$$N(x, \varepsilon) = \max \{ j \in \mathbf{N} : 1/x < c_j \le x \},$$

$$N\left(\frac{1}{x}, \varepsilon\right) = \min \{ j \in \mathbf{N} : 1/x < c_j \le x \},$$

$$J_j(\varepsilon) = [c_j, c_{j+1}], \text{ if } N\left(\frac{1}{x}, \varepsilon\right) < j < N(x, \varepsilon),$$

$$J_j(\varepsilon) = [c_j, x], \text{ if } j = N(x, \varepsilon) \text{ or } \left[\frac{1}{x}, c_j\right], \text{ if } j = N\left(\frac{1}{x}, \varepsilon\right).$$

If I_k is contained in some $J_j(\varepsilon)$ for all $\varepsilon > 0$, then $\| \mathcal{T}_{I_k,u,v} \| = 0$. By Theorem 2.2 it follows that $u(t)v(t) = 0$ a.e. on I_k.

Let $\varepsilon_k = \inf \{\varepsilon > 0$, there exists $j : J_j(\varepsilon) \supset I_k\}$ and put $\delta = \min \{\varepsilon_k : \varepsilon_k > 0\}$. Then if $0 < \varepsilon < \delta$, it follows for all j and k that $I_k \not\subseteq J_j(\varepsilon)$. We estimate the difference

$$\left| \int_{1/x}^{x} | u(t) |^r \, | v(t) |^r \, dt - \int_{1/x}^{x} \sum_{k=1}^{m} | u_k |^r \, | v_k |^r \, \chi_{I_k}(t) \, dt \right|$$

where the summation is taken over those $k \in \{1, 2, \dots, m\}$ for which $\int_{I_k} u(t)v(t)\, dt \neq 0$ and when $1 < p < q < \infty$, $r = \dfrac{qp'}{q + p'} > 1$. We have

$$\left| \int_{1/x}^{x} | u(t) |^r \, | v(t) |^r \, dt - \int_{1/x}^{x} \sum_{k=1}^{m} | u_k |^r \, | v_k |^r \, \chi_{I_k}(t) \, dt \right|$$

$$= \left| \int_{1/x}^{x} (| u |^r | v |^r - | u_\eta |^r | v_\eta |^r) \right|$$

(using inequality (3.5))

$$\leq r \left| \int_{1/x}^{x} | v |^{\frac{qp'}{q+p'}} (| u | + | u_\eta |)^{\frac{qp'-p'-q}{q+p'}} | u - u_\eta | \right|$$

$$+ \, r \left| \int_{1/x}^{x} | u_\eta |^{\frac{qp'}{q+p'}} (| v | + | v_\eta |)^{\frac{qp'-p'-q}{q+p'}} | v - v_\eta | \right|$$

(applying Hölder's inequality for the three factors with the exponents $\dfrac{p' + q}{p'}, \dfrac{p'(p' + q)}{qp' - p' - q}, p'$ in the first integral and with the exponents $\dfrac{p' + q}{q}, \dfrac{q(p' + q)}{qp' - p' - q}, q$ in the second)

$$\leq r \| v \|^r_{L^q_{[1/x,x]}} \| \, | u | + | u_\eta | \, \|^{r-1}_{L^{p'}_{[1/x,x]}} \| u - u_\eta \|_{L^{p'}_{[1/x,x]}}$$

$$+ \, r \| u_\eta \|^r_{L^{p'}_{[1/x,x]}} \| \, | v | + | v_\eta | \, \|^{r-1}_{L^q_{[1/x,x]}} \| v - v_\eta \|_{L^q_{[1/x,x]}}$$

$$\leq r \| v \|^r_{L^q_{[1/x,x]}} \| \, | u | + | u_\eta | \, \|^{r-1}_{L^{p'}_{[1/x,x]}} \eta + r \| u_\eta \|^r_{L^{p'}_{[1/x,x]}} \| \, | v | + | v_\eta | \, \|^{r-1}_{L^q_{[1/x,x]}} \eta$$

$$= \left| r \| v \|^r_{L^q_{[1/x,x]}} \| \, |u| + |u_\eta| \, \|^{r-1}_{L^{p'}_{[1/x,x]}} + r \| u_\eta \|^r_{L^{p'}_{[1/x,x]}} \| \, |v| + |v_\eta| \, \|^{r-1}_{L^q_{[1/x,x]}} \right| \eta$$

$$= O(\eta).$$

Thus

$$\left| \int_{1/x}^{x} |u(t)|^r |v(t)|^r \, dt - \int_{1/x}^{x} \sum_{k=1}^{m} |u_k|^r |v_k|^r \chi_{I_k}(t) \, dt \right| = O(\eta), \quad (3.8)$$

when $1 < p < q < \infty$.

Now we begin the proof of the estimate (1.8). Let $0 < \varepsilon < \delta$, and in the case $1 < p < q < \infty$, using Lemma 3.1 with $r = \dfrac{qp'}{q + p'} > 1$, we obtain

$$\gamma^r_{pq} |u_k|^r |v_k|^r |I_k| = \sum_{j:J_j \subset I_k} \| \mathcal{T}_{u_k,v_k;J_j} \|^r + \| \mathcal{T}_{u_k,v_k;J_{j_1}(\varepsilon)\cap I_k} \|^r + \| \mathcal{T}_{u_k,v_k;J_{j_2}(\varepsilon)\cap I_k} \|^r,$$

where the numbers $j_1 = j_1(k)$ and $j_2 = j_2(k)$ are such, that the intervals $J_{j1}(\varepsilon)$ and $J_{j2}(\varepsilon)$ contain the left and right ends of I_k, respectively. Summing over all the intervals I_k, we get

$$\gamma^r_{pq} \sum_{k=1}^{m} |u_k|^r |v_k|^r |I_k| = \sum_{k=1}^{m} \sum_{j:J_j(\varepsilon) \subset I_k} \| \mathcal{T}_{u_k,v_k;J_j(\varepsilon)} \|^r$$
$$+ \sum_{k=1}^{m} \left(\| \mathcal{T}_{u_k,v_k;J_{j_1}(\varepsilon)\cap I_k} \|^r + \| \mathcal{T}_{u_k,v_k;J_{j_2}(\varepsilon)\cap I_k} \|^r \right)$$
$$\le \sum_{k=1}^{m} \sum_{j:J_j(\varepsilon) \subset I_k} \left(\| \mathcal{T}_{u,v;J_j(\varepsilon)} \| + \| \mathcal{T}_{u_k,v_k;J_j(\varepsilon)} - \mathcal{T}_{u,v;J_j(\varepsilon)} \| \right)^r$$
$$+ \sum_{l=1}^{2} \sum_{k=1}^{m} \left(\| \mathcal{T}_{u,v;J_{j_l}(\varepsilon)\cap I_k} \| + \| \mathcal{T}_{u_k,v_k;J_{j_l}(\varepsilon)\cap I_k} - \mathcal{T}_{u,v;J_{j_1}(\varepsilon)\cap I_k} \| \right)^r$$

(by (3.6))

$$\le \sum_{k=1}^{m} \sum_{j:J_j(\varepsilon) \subset I_k} \Big(\varepsilon + (2 + q2^q) \| v - v_k \|_{L^q(J_j(\varepsilon))} \| u \|_{L^{p'}(J_j(\varepsilon))}$$
$$+ \| u - u_k \|_{L^{p'}(J_j(\varepsilon))} \| v_k \|_{L^q(J_j(\varepsilon))} \Big)^r$$
$$+ \sum_{l=1}^{2} \sum_{k=1}^{m} \Big(\varepsilon + (2 + q2^q) \| v - v_k \|_{L^q(J_{j_l}(\varepsilon))\cap I_k)} \| u \|_{L^{p'}(J_{j_l}(\varepsilon)\cap I_k)}$$
$$+ \| u - u_k \|_{L^{p'}(J_{j_l}(\varepsilon)\cap I_k)} \| v_k \|_{L^q(J_{j_l}(\varepsilon)\cap I_k)} \Big)^r$$

(applying the inequality $(a+b+c)^r \le 3^{r-1}(a^r+b^r+c^r)$)

$$\le 3^{r-1}\Bigg[\sum_{k=1}^{m}\sum_{j;J_j(\varepsilon)\subset I_k}\varepsilon^r$$

$$+(2+q2^q)^r\sum_{k=1}^{m}\sum_{j;J_j(\varepsilon)\subset I_k}\|v-v_k\|^r_{L^q(J_j(\varepsilon))}\|u\|^r_{L^{p'}(J_j(\varepsilon))}$$

$$+\sum_{k=1}^{m}\sum_{j;J_j(\varepsilon)\subset I_k}\|u-u_k\|^r_{L^{p'}(J_j(\varepsilon))}\|v_k\|^r_{L^q(J_j(\varepsilon))}$$

$$+\sum_{l=1}^{2}\sum_{k=1}^{m}\Bigg(\varepsilon^r+(2+q2^q)^r\sum_{k=1}^{m}\|v-v_k\|^r_{L^q(J_{j_l}(\varepsilon)\cap I_k)}\|u\|^r_{L^{p'}(J_{j_l}(\varepsilon)\cap I_k)}$$

$$+\sum_{k=1}^{m}\|u-u_k\|^r_{L^{p'}(J_{j_l}(\varepsilon)\cap I_k)}\|v_k\|^r_{L^q(J_{j_l}(\varepsilon)\cap I_k)}\Bigg)\Bigg]$$

Applying twice Hölder's inequality with $\frac{p'+q}{p'}, \frac{p'+q}{q}$ and (3.7)

$$\le 3^{r-1}[\varepsilon^r(N(x,\varepsilon)+2m+1)+(2+q2^q)^r\|v-v_k\|^r_{L^q[1/x,x]}\|u\|^r_{L^{p'}[1/x,x]}$$

$$+\|u-u_k\|^r_{L^{p'}[1/x,x]}\|v_k\|^r_{L^q[1/x,x]}$$

$$+2\{(2+q2^q)^r\eta^r\|u\|^r_{L^{p'}[1/x,x]}+\eta^r\|v_k\|^r_{L^q[1/x,x]}\}]$$

$$\le 3^{r-1}\varepsilon^r(N(x,\varepsilon)+2m+1)+O(\eta^r). \tag{3.9}$$

Thus, by (3.8) we see that

$$\gamma^r_{pq}\left(\int_{1/x}^{x}|u(t)|^r\,|v(t)|^r\,dt\right)\le 3^{r-1}\varepsilon^r(N(x,\varepsilon)+1+2m)+O(\eta).$$

Hence,

$$\gamma^r_{pq}\left(\int_{1/x}^{x}|u(t)|^r\,|v(t)|^r\,dt\right)^{1/r}\le 3^{(r-1)/r}\liminf_{\varepsilon\to 0+}\varepsilon N^{1/r}(x,\varepsilon)+O(\eta)$$

and letting $\eta\to 0$, we conclude

$$\gamma_{pq}\left(\int_{1/x}^{x}|u(t)|^r\,|v(t)|^r\,dt\right)^{1/r}\le 3^{(r-1)/r}\liminf_{\varepsilon\to 0+}\varepsilon N^{1/r}(x,\varepsilon).$$

Now assume that $\liminf_{N\to\infty} N\,a_N(T)<\infty$ and show that $u(t)v(t)\in L^r(\mathbf{R}^+)$. Indeed, using (2.5) we find, say for $q\ne 2$,

$$\gamma_{pq}\left(\int_{1/x}^{x} |u(t)|^r |v(t)|^r\,dt\right)^{1/r} \le 3^{(r-1)/r} \liminf_{\varepsilon\to 0+} \varepsilon N^{1/r}(x,\varepsilon)$$

$$\le 3^{(r-1)/r} \liminf_{\varepsilon\to 0+} \varepsilon N^{1/r}(\infty,\varepsilon) \le 2\cdot 3^{(r-1)/r} \liminf_{N\to\infty} N a_N(T).$$

Letting $x \to \infty$, we have

$$\gamma_{pq}\left(\int_{0}^{\infty} |u(t)|^r |v(t)|^r\,dt\right)^{1/r} \le 2\cdot 3^{(r-1)/r} \liminf_{N\to\infty} N a_N(T)$$

and the proof of first upper estimate of Theorem 1 is complete.

For the proof of the lower bound (1.9) we denote $\mathscr{K} = \{j$: there exists k: $J_j(\varepsilon) \subset I_k\}$. Then $\#\mathscr{K} \ge N(x,\varepsilon) - N\left(\frac{1}{x},\varepsilon\right) - 2m$. Using the arguments which have led us to (3.9), we find

$$\left(N(x,\varepsilon) - N\left(\frac{1}{x},\varepsilon\right) - 2m\right)\varepsilon^r \le \sum_{j\in\mathscr{K}} \| \mathscr{T}_{u,v,J_j} \|^r$$

$$\le \sum_{j\in\mathscr{K}} \| \mathscr{T}_{u_\eta,v_\eta,J_j} \|^r + \sum_{k=1}^{m} \sum_{j:J_j(\varepsilon)\in I_k} \| \mathscr{T}_{u_k,v_k,J_j(\varepsilon)} - \mathscr{T}_{u,v,J_j(\varepsilon)} \|^r$$

$$\le \gamma_{pq}^r \int_{1/x}^{x} |u_\eta(t) v_\eta(t)|^r\,dt + O(\eta^r),$$

and applying (3.8) we write

$$\left(N(x,\varepsilon) - N\left(\frac{1}{x},\varepsilon\right) - 2m\right)\varepsilon^r \le \gamma_{pq}^r \int_{1/x}^{x} |u(t)v(t)|^r\,dt + O(\eta^r). \tag{3.10}$$

Now, set

$$n = \max\{k : \xi_k < N(x,\varepsilon)\}$$

and

$$m = \min\left\{k : \xi_k > N\left(\frac{1}{x},\varepsilon\right)\right\},$$

where $\{\xi_k\}$ defined by (1.2) and let $N = N(\varepsilon) = N(\infty,\varepsilon)$ be the total number of intervals $\{I_k\}$. Note, that $n \to \infty$, $m \to -\infty$ if $x \to \infty$.

For the proof of the reverse estimates we need the following two lemmas. Recall, that $\{\xi_n\}$, $n \in \mathbf{Z}$ are defined by

$$U(\xi_n) = \int_0^{\xi_n} |u(t)|^{p'}\,dt = 2^{n+1}$$

and without a loss of generality we assume, that ξ exist for all $n \in \mathbf{Z}$. Put

$$\Delta_n = (\xi_{n-1}, \xi_n)$$

and

$$\sigma_n = \left(\int_{\xi_{n-1}}^{\xi_n} | u(t) |^{p'} dt \right)^{1/p'} \left(\int_{\xi_n}^{\xi_{n+1}} | v(t) |^q dt \right)^{1/q}$$

Lemma 3.4 *Let* $n_1, n_2, n_3 \in \mathbf{Z}$, $n_1 < n_2 < n_3$ *and* $c_0 \in \Delta_{n_1}$, $x_0 \in \Delta_{n_2}$, $c_1 \in \Delta_{n_3}$. *Then*

$$\left(\int_{c_0}^{x_0} | u(t) |^{p'} dt \right)^{1/p'} \left(\int_{x_0}^{c_1} | v(t) |^q dt \right)^{1/q} \leq 2^{\frac{2}{p'}+\frac{1}{q}} \max_{\substack{n_2-1\leq n \\ \leq n_3-1}} \sigma_n .$$

Proof of Lemma 3.4

$$\left(\int_{c_0}^{x_0} | u(t) |^{p'} dt \right)^{1/p'} \left(\int_{x_0}^{c_1} | v(t) |^q dt \right)^{1/q}$$

$$\leq \left(\int_{\xi_{n_1-1}}^{\xi_{n_2}} | u(t) |^{p'} dt \right)^{1/p'} \left(\int_{\xi_{n_2-1}}^{\xi_{n_3}} | v(t) |^q dt \right)^{1/q}$$

$$\leq 2^{(n_2+1)/p'} \left(\frac{\sigma^q_{n_2-1}}{2^{(n_2-1)q/p'}} + \frac{\sigma^q_{n_3-1}}{2^{(n_3-1)q/p'}} \right)^{1/q}$$

$$\leq 2^{\frac{2}{p'}+\frac{1}{q}} \max_{\substack{n_2-1\leq n \\ \leq n_3-1}} \sigma_n .$$

Lemma 3.4 is proved.

Lemma 3.5 *Let* $1 < p \leq q < \infty$, $r = \dfrac{p'q}{p'+q} > 1$, $I_k = (c_k, c_{k+1})$, $x_k \in I_k$ *and* $\xi_{n-1} < c_1 < c_2 < \ldots < c_l < \xi_n$, *Then*

$$\sum_{k=1}^{l} \left(\int_{c_k}^{x_k} | u(t) |^{p'} dt \right)^{r/p'} \left(\int_{x_k}^{c_{k+1}} | v(t) |^q dt \right)^{r/q} \leq 2^{r/p'} \sigma^r_{n-1} .$$

Proof of Lemma 3.5 Applying Hölder's inequality with $\dfrac{p'+q}{q}$ and $\dfrac{p'+q}{p'}$, we obtain

$$\sum_{k=1}^{l} \left(\int_{c_k}^{x_k} | u(t) |^{p'} dt \right)^{r/p'} \left(\int_{x_k}^{c_{k+1}} | v(t) |^q dt \right)^{r/q}$$

$$\leq \sum_{k=1}^{l} \left(\int_{I_k} |u(t)|^{p'} dt \right)^{q/(p'+q)} \left(\int_{I_k} |v(t)|^{q} dt \right)^{p'/(q+p')}$$

$$\leq \left(\sum_{k=1}^{l} \int_{I_k} |u(t)|^{p'} dt \right)^{r/p'} \left(\sum_{k=1}^{l} \int_{I_k} |v(t)|^{q} dt \right)^{r/q}$$

$$\leq \left(\int_{\xi_{n-1}}^{\xi_n} |u(t)|^{p'} dt \right)^{r/p'} \left(\int_{\xi_{n-1}}^{\xi_n} |v(t)|^{q} dt \right)^{r/q} = 2^{r/p'} \sigma_{n-1}^{r}.$$

Lemma 3.5 is proved.

Now we continue the proof of the reverse estimate (1.9) of Theorem 1. Using Theorem 2.2, Lemmas 3.4 and 3.5 we find

$$(N(\infty, \varepsilon) - N(x, \varepsilon))\, \varepsilon^r << \sum_{k \geq n} \sigma_k^r, \tag{3.11}$$

$$N\left(\frac{1}{x}, \varepsilon\right) \varepsilon^r << \sum_{k \leq m} \sigma_k^r. \tag{3.12}$$

Thus,

$$\limsup_{\varepsilon \to 0} \varepsilon^r N(\infty, \varepsilon) \leq \gamma_{pq}^r \int_{1/x}^{x} |u_\eta(t) v_\eta(t)|^r dt + \sum_{\substack{|k| \geq \min \\ \{|n|; |m|\}}} \sigma_k^r + O(\eta^r).$$

Note, that the middle term on the right hand side vanishes, if $\sum_{k \in \mathbf{Z}} \sigma_k^r < \infty$, and $x \to \infty$. Letting $\eta \to 0$ and $x \to \infty$ we establish (1.9).

For the proof of the second part of Theorem 1, let $1 < q < p < \infty$ and $\frac{1}{r} = \frac{q + p'}{qp'} > 1$. Then summing over those $k \in \{1, 2, \ldots, m\}$ for which $\int_{I_k} u(t)v(t)\, dt \neq 0$, we prove (3.8) as follows.

$$\left| \int_{1/x}^{x} |u(t)|^r |v(t)|^r dt - \int_{1/x}^{x} \sum_{k=1}^{m} |u_k|^r |v_k|^r \chi_{I_k}(t)\, dt \right|$$

$$= \left| \int_{1/x}^{x} (|u|^r |v|^r - |u_\eta|^r |v_\eta|^r) \right|$$

$$\leq \left| \int_{1/x}^{x} |v|^r (|u|^r - |u_\eta|^r) \right| + \left| \int_{1/x}^{x} |u_\eta|^r (|v|^r - |v_\eta|^r) \right|$$

$$\leq \int_{\{t\in[1/x,x]:|u(t)|>|u_\eta(t)|\}} |v|^r(|u|^r - |u_\eta|^r)$$

$$+ \int_{\{t\in[1/x,x]:|u(t)|<|u_\eta(t)|\}} |v|^r(|u_\eta|^r - |u|^r)$$

$$+ \int_{\{t\in[1/x,x]:|v(t)|>|v_\eta(t)|\}} |u_\eta|^r(|v|^r - |v_\eta|^r)$$

$$+ \int_{\{t\in[1/x,x]:|v(t)|<|v_\eta(t)|\}} |u_\eta|^r(|v_\eta|^r - |v|^r)$$

$$\leq \int_{\{t\in[1/x,x]:|u(t)|>|u_\eta(t)|\}} |v|^r(|u| - |u_\eta|)^r$$

$$+ \int_{\{t\in[1/x,x]:|u(t)|<|u_\eta(t)|\}} |v|^r(|u_\eta| - |u|)^r$$

$$+ \int_{\{t\in[1/x,x]:|v(t)|>|v_\eta(t)|\}} |u_\eta|^r(|v| - |v_\eta|)^r$$

$$+ \int_{\{t\in[1/x,x]:|v(t)|<|v_\eta(t)|\}} |u_\eta|^r(|v_\eta| - |v|)^r$$

(Applying Hölder's inequality for each above integrals with exponents $\frac{p'+q}{p'}$ and $\frac{p'+q}{q}$.)

$$\leq 2\,\|v\|^r_{L^q_{[1/x,x]}}\,\|u-u_\eta\|^r_{L^{p'}_{[1/x,x]}} + 2\,\|u_\eta\|^r_{L^{p'}_{[1/x,x]}}\,\|v-v_\eta\|^r_{L^q_{[1/x,x]}}$$

$$\leq 2\,\|v\|^r_{L^q_{[1/x,x]}}\,\eta + 2\,\|u_\eta\|^r_{L^{p'}_{[1/x,x]}}\,\eta = O(\eta).$$

Lemma 3.1 yields

$$\gamma^r_{pq}|u_k|^r|v_k|^r|I_k| = \sum_{j:J_j\subset I_k}\left(\|\mathcal{T}_{u_k,v_k;J_j}\|^r + \|\mathcal{T}_{u_k,v_k;J_{j_1(\varepsilon)}\cap I_k}\|^r + \|\mathcal{T}_{u_k,v_k;J_{j_2(\varepsilon)}\cap I_k}\|^r\right)$$

and since $0 < r < 1$, we find

$$\gamma^r_{pq}\sum_{k=1}^m |u_k|^r|v_k|^r|I_k| = \sum_{k=1}^m \sum_{j:J_j(\varepsilon)\subset I_k}\|\mathcal{T}_{u_k,v_k;J_j(\varepsilon)}\|^r + \sum_{k=1}^m\left(\|\mathcal{T}_{u_k,v_k;J_{j_1}(\varepsilon)\cap I_k}\|^r + \|\mathcal{T}_{u_k,v_k;J_{j_2}(\varepsilon)\cap I_k}\|^r\right)$$

$$\leq \sum_{k=1}^{m} \sum_{j:J_j(\varepsilon)\subset I_k} \left(\| \mathcal{T}_{u,v,J_j(\varepsilon)} \| + \| \mathcal{T}_{u_k,v_k;J_j(\varepsilon)} - \mathcal{T}_{u,v,J_j(\varepsilon)} \| \right)^r$$

$$+ \sum_{l=1}^{2} \sum_{k=1}^{m} \left(\| \mathcal{T}_{u,v,J_{j_1}(\varepsilon)\cap I_k} \| + \| \mathcal{T}_{u_k,v_k;J_{j_l}(\varepsilon)\cap I_k} - \mathcal{T}_{u,v,J_{j_l}(\varepsilon)\cap I_k} \| \right)^r$$

(Using (3.6))

$$\leq \sum_{k=1}^{m} \sum_{j:J_j(\varepsilon)\subset I_k} \Bigg(\varepsilon^r + (2 + q2^q)^r \| v - v_k \|^r_{L^q(J_j(\varepsilon))} \| u \|^r_{L^{p'}(J_j(\varepsilon))}$$

$$+ \| u - u_k \|^r_{L^{p'}(J_j(\varepsilon))} \| v_k \|^r_{L^q(J_j(\varepsilon))} \Bigg)$$

$$+ \sum_{l=1}^{2} \sum_{k=1}^{m} \Bigg(\varepsilon^r + (2 + q2^q)^r \| v - v_k \|^r_{L^q(J_{j_l}(\varepsilon)\cap I_k)} \| u \|^r_{L^{p'}(J_{j_l}(\varepsilon)\cap I_k)}$$

$$+ \| u - u_k \|^r_{L^{p'}(J_{j_l}(\varepsilon)\cap I_k)} \| v_k \|^r_{L^q(J_{j_l}(\varepsilon)\cap I_k)} \Bigg)$$

(Applying Hölder's inequality with $\frac{p'+q}{p'}, \frac{p'+q}{q}$ twice and (3.7))

$$\leq \varepsilon^r (N(x,\varepsilon) + 2m + 1) + (2 + 2^q)^r \sum_{k=1}^{m} \| v - v_k \|^{\frac{qp'}{p'+q}}_{L^q(I_k)} \| u \|^{\frac{qp'}{p'+q}}_{L^{p'}(I_k)}$$

$$+ \sum_{k=1}^{m} \| u - u_k \|^{\frac{qp'}{p'+q}}_{L^{p'}(I_k)} \| v_k \|^{\frac{qp'}{p'+q}}_{L^q(I_k)}$$

$$+ 2\{(2 + q2^q)^r \| v - v_k \|^r_{L^q[1/x,x]} \| u \|^r_{L^{p'}[1/x,x]}$$

$$+ \| u - u_k \|^r_{L^{p'}[1/x,x]} \| v_k \|^r_{L^q[1/x,x]}\}$$

$$\leq \varepsilon^r(N(x,\varepsilon) + 2m+1) + (2+q2^q)^r \eta^r \| u \|^r_{L^{p'}[1/x,x]} + \eta^r \| v_k \|^r_{L^q[1/x,x]}$$

$$+ 2\{(2 + q2^q)^r \eta^r \| u \|^r_{L^{p'}[1/x,x]} + \eta^r \| v_k \|^r_{L^q[1/x,x]}\}$$

$$= \varepsilon^r(N(x,\varepsilon) + 2m + 1) + C_3\eta^r,$$

where

$$C_3 = 3\{(2 + q2^q)^r \| u \|^r_{L^{p'}[1/x,x]} + \| v_k \|^r_{L^q[1/x,x]}\}$$

So,

$$\gamma^r_{pq} \int_{1/x}^{x} | u(t) |^r | v(t) |^r \, dt \leq \varepsilon^r(N(x,\varepsilon) + 1 + 2m) + C_3\eta^r + O(\eta).$$

When $\eta \to 0$, we have

$$\gamma_{pq}^{r} \int_{1/x}^{x} | u(t) |^r | v(t) |^r \, dt \leq \varepsilon^r (N(x, \varepsilon) + 1 + 2m).$$

$$\gamma_{pq} \left(\int_{1/x}^{x} | u(t) |^r | v(t) |^r \, dt \right)^{1/r} \leq \lim_{\varepsilon \to 0+} \inf \varepsilon N^{1/r}(x, \varepsilon).$$

Assume that $\liminf\limits_{N \to \infty} N^{1/r} a_N(T) < \infty$, then $u(t)v(t) \in L^r(\mathbf{R}^+)$. Indeed,

$$\gamma_{pq} \left(\int_{1/x}^{x} | u(t) |^r | v(t) |^r \, dt \right)^{1/r} \leq \lim_{\varepsilon \to 0+} \inf \varepsilon N^{1/r}(x, \varepsilon)$$

$$\leq \lim_{\varepsilon \to 0+} \inf \varepsilon N^{1/r}(\infty, \varepsilon) \leq 2 \liminf_{N \to \infty} N^{1/r}(\varepsilon) a_N(T)$$

and for $x \to \infty$ we have

$$\gamma_{pq} \left(\int_{0}^{\infty} | u(t) |^r | v(t) |^r \, dt \right)^{1/r} \leq 2 \liminf_{N \to \infty} N^{1/r} a_N(T)$$

and the upper estimate (1.10) is established.

As in the proof of (1.9) for the lower bound (1.11) we need Lemma 3.4 and the following.

Lemma 3.6 Let $1 < q < p < \infty$, $\dfrac{1}{r} = \dfrac{p' + q}{p'q} > 1$, $\dfrac{1}{\gamma} = \dfrac{1}{q} - \dfrac{1}{p}$, $I_k = (c_k, c_{k+1})$, $x \in I_k$ *and* $\xi_{n-1} < c_1 < c_2 < \ldots < c_l < \xi_n$. *Then*

$$\sum_{k=1}^{l} \left\{ \int_{I_k} \left(\int_{c_k}^{x} | u(t) |^{p'} \, dt \right)^{\gamma/q'} \left(\int_{x}^{c_{k+1}} | v(t) |^q \, dt \right)^{\gamma/q'} | u(x) |^{p'} \, dx \right\}^{r/\gamma}$$

$$\leq \left(\frac{p'}{\gamma} \right)^{r/\gamma} 2^{r/p'} \sigma_{n-1}^{r}.$$

Proof We have

$$\sum_{k=1}^{l} \left\{ \int_{I_k} \left(\int_{c_k}^{x} | u(t) |^{p'} \, dt \right)^{\gamma/q'} \left(\int_{x}^{c_{k+1}} | v(t) |^q \, dt \right)^{\gamma/q} | u(x) |^{p'} \, dx \right\}^{(1/\gamma)+1}$$

$$\leq \sum_{k=1}^{l} \left(\int_{I_k} | v(t) |^q \, dt \right)^{r/q} \left\{ \int_{I_k} \left(\int_{c_k}^{x} | u(t) |^{p'} \, dt \right)^{\gamma/q'} | u(x) |^{p'} \, dx) \right\}^{(1/\gamma)+1}$$

$$= \left(\frac{p'}{\gamma} \right)^{(1/\gamma)+1} \sum_{k=1}^{l} \left(\int_{I_k} | v(t) |^q \, dt \right)^{r/q} \left(\int_{I_k} | u(t) |^{p'} \, dt \right)^{r/p'}$$

$$\left(\text{using Hölder's inequality with } \frac{p'+q}{q} \text{ and } \frac{p'+q}{p'}\right)$$

$$\leq \left(\frac{p'}{\gamma}\right)^{1/(\gamma+1)} \left(\int_{\xi_{n-1}}^{\xi_n} |u(t)|^{p'}\,dt\right)^{r/p'} \left(\int_{\xi_{n-1}}^{\xi_n} |v(t)|^{q}\,dt\right)^{r/q}$$

$$= \left(\frac{p'}{\gamma}\right)^{r/\gamma} 2^{r/p'} \sigma_{n-1}^{r}.$$

Lemma 3.6 is proved.

Now, arguing similar to the proof of (1.9) we obtain (3.10) and, using Theorem 2.2, Lemmas 3.4 and 3.6, we prove (3.4) and (3.12), which imply (1.11).

Proof of the Theorem 1 is completed.

4. Schatten-von Neumann Norm Estimates

In this section using the ideas of remarkable work [5] we estimate norms given by

$$\| \{a_n\} \|_{l^s} = (\sum_{n=1}^{\infty} a_n^s)^{1/s}$$

and

$$\| \{a_n\} \|_{l^s_{\text{weak}}} = \sup t\, (\#\{n : a_n > t\})^{1/s}$$

of the sequence of the approximation numbers of $\{a_n(T)\}$. As in the work [5] this goal will be achieved by comparison of distribution functions of sequences $\{a_n(T)\}$ and $\{\sigma_k\}$. To this and we extend on the cases $1 < p < q < \infty$ and $1 < q < p < \infty$ the sequence of lemmas from [5], where the case $1 < p = q < \infty$ was treated. Then we prove equivalence of the l^s norms of $\{\sigma_k\}$ to the integral functionals depending on weights u and v. Let $\{\xi_n\}$ and $\{\sigma_n\}$ be given by (1.2), (1.3), respectively, and interval $\Delta_n = (\xi_{n-1}, \xi_n)$.

Lemma 4.1 Let $1 < p \leq q < \infty$, $I = (a, b)$, $I \subset \mathbf{R}^+$,

$$A_0(I) = \sup_{a<x<b} \left(\int_x^b |u(t)|^{p'}\,dt\right)^{1/p'} \left(\int_a^x |v(t)|^{q}\,dt\right)^{1/q},$$

$$A_1(I) = \sup_{a<x<b} \left(\int_a^x |u(t)|^{p'}\,dt\right)^{1/p'} \left(\int_x^b |v(t)|^{q}\,dt\right)^{1/q}$$

Then

$$A_0(\overline{\Delta_n} \cup \overline{\Delta_{n+1}}) \geq 4^{1/p'} \sigma_{n-1}, \quad A_1(\overline{\Delta_n} \cup \overline{\Delta_{n+1}}) \geq \sigma_n.$$

Proof

$$A_0(\overline{\Delta_n} \cup \overline{\Delta_{n+1}}) = \sup_{\substack{\xi_{n-1}<x \\ <\xi_{n+1}}} \left(\int_x^{\xi_{n+1}} |u(t)|^{p'} dt \right)^{1/p'} \left(\int_{\xi_{n-1}}^x |v(t)|^q dt \right)^{1/q}$$

$$\geq \left(\int_{\xi_n}^{\xi_{n+1}} |u(t)|^{p'} dt \right)^{1/p'} \left(\int_{\xi_{n-1}}^{\xi_n} |v(t)|^q dt \right)^{1/q} = 4^{1/p'} \sigma_{n-1}.$$

$$A_1(\overline{\Delta_n} \cup \overline{\Delta_{n+1}}) = \sup_{\substack{\xi_{n-1}<x \\ <\xi_{n+1}}} \left(\int_{\xi_{n-1}}^x |u(t)|^{p'} dt \right)^{1/p'} \left(\int_x^{\xi_{n+1}} |v(t)|^q dt \right)^{1/q}$$

$$\geq \left(\int_{\xi_{n-1}}^{\xi_n} |u(t)|^{p'} dt \right)^{1/p'} \left(\int_{\xi_n}^{\xi_{n+1}} |v(t)|^q dt \right)^{1/q} = \sigma_{n_1}.$$

Lemma 4.2 *Let* $1 < q < p < \infty$, $\frac{1}{\gamma} = \frac{1}{q} - \frac{1}{p}$, $I = (a, b)$, $I \subset \mathbf{R}^+$, *and*

$$B_0(I) = \left\{ \int_a^b \left(\int_a^x |v(t)|^q dt \right)^{\gamma/q} \left(\int_x^b |u(t)|^{p'} dt \right)^{\gamma/q'} |u(x)|^{p'} dx \right\}^{1/\gamma},$$

$$B_1(I) = \left\{ \int_a^b \left(\int_a^x |u(t)|^{p'} dt \right)^{\gamma/q'} \left(\int_x^b |v(t)|^q dt \right)^{\gamma/q} |u(x)|^{p'} dx \right\}^{1/\gamma}$$

Then

$$B_0(\overline{\Delta_n} \cup \overline{\Delta_{n+1}}) \geq 4^{1/p'} \left(\frac{p'}{\gamma} \right)^{1/\gamma} \sigma_{n-1}, \quad B_1(\overline{\Delta_n} \cup \overline{\Delta_{n+1}}) \geq \left(\frac{p'}{\gamma} \right)^{1/\gamma} \sigma_n.$$

Proof

$$B_0(I) = \left(\int_a^b \left(\int_a^x |v(t)|^q dt \right)^{\gamma/q} \left(\int_x^b |u(t)|^{p'} dt \right)^{\gamma/q'} |u(x)|^{p'} dx \right)^{1/\gamma}$$

$$= \left\{ \int_a^b \left(\int_a^x |v(t)|^q dt \right)^{\gamma/q} \left(\int_x^b |u(t)|^{p'} dt \right)^{\gamma/q'} d\left(-\int_x^b |u(t)|^{p'} dt \right) \right\}^{1/\gamma}$$

$$= \left\{ \int_a^b \left(\int_a^x |v(t)|^q dt \right)^{\gamma/q} \left(\int_x^b |u(t)|^{p'} dt \right)^{\frac{\gamma}{p'}-1} d\left(-\int_x^b |u(t)|^{p'} dt \right) \right\}^{1/\gamma}$$

$$\geq \left(\frac{p'}{\gamma} \right)^{1/\gamma} \left\{ \int_x^b \left(\int_a^x |v(t)|^q dt \right)^{\gamma/q} d\left(-\int_x^b |u(t)|^{p'} dt \right)^{\gamma/p'} \right\}^{1/\gamma}$$

$$\geq \left(\frac{p'}{\gamma} \right)^{1/\gamma} \left(\int_a^x |v(t)|^q dt \right)^{1/q} \left(\int_x^b |u(t)|^{p'} dt \right)^{1/p'}$$

It implies

$$B_0(\overline{\Delta_n} \cup \overline{\Delta_{n+1}}) \geq \left(\frac{p'}{\gamma}\right)^{1/\gamma} \left(\int_{\xi_{n-1}}^{x} |v(t)|^q \, dt\right)^{1/q} \left(\int_{x}^{\xi_{n+1}} |u(t)|^{p'} \, dt\right)^{1/p'}$$

$$\geq \left(\frac{p'}{\gamma}\right)^{1/\gamma} \left(\int_{\xi_{n-1}}^{\xi_n} |v(t)|^q \, dt\right)^{1/q} \left(\int_{\xi_n}^{\xi_{n+1}} |u(t)|^{p'} \, dt\right)^{1/p'} = 4^{1/p'} \left(\frac{p'}{\gamma}\right)^{1/\gamma} \sigma_{n-1}.$$

Similary we get

$$B_1(\overline{\Delta_n} \cup \overline{\Delta_{n+1}}) \geq \left(\frac{p'}{\gamma}\right)^{1/\gamma} \sigma_n.$$

Lemma 4.3 *Let* $1 < p \leq q < \infty$, $0 < a < b < \infty$, $I = (a, b) \subset \mathbf{R}^+$. *We define*

$$D(I) = \max\left(A_0(a, c), A_1(c, b)\right),$$

where

$$A_0(a, c) = \sup_{a<s<c} \left(\int_s^c |u(t)|^{p'} \, dt\right)^{1/p'} \left(\int_a^s |v(t)|^q \, dt\right)^{1/q},$$

$$A_1(c, b) = \sup_{c<s<b} \left(\int_c^s |u(t)|^{p'} \, dt\right)^{1/p'} \left(\int_s^b |v(t)|^q \, dt\right)^{1/q}$$

and the point c in (a, b) *is chosen so that*

$$\int_a^c |v(t)|^q \, dt = \frac{1}{2} \int_a^b |v(t)|^q \, dt.$$

Let $0 < \varepsilon < \| T \|$, *where* ε *is taken sufficiently small, and assume*

$$S_I(\varepsilon) = \{n \in \mathbf{Z} : \overline{\Delta_{n+1}} \subset I, \sigma_n > \varepsilon\} \text{ and } \# S_I(\varepsilon) \geq 4.$$

Then $D(I) > \varepsilon$.

Proof If $n_1 = \min\{n : n \in S_I(\varepsilon)\}$, $n_2 = \max\{n : n \in S_I(\varepsilon)\}$, then $\overline{\Delta_{n_1}} \cup \overline{\Delta_{n_1+1}} \subset (a, c)$, and by Lemma 2.4

$$A_0(a, c) \geq A_0(\overline{\Delta_{n_1}} \cup \overline{\Delta_{n_1+1}}) \geq 4^{1/p'} \sigma_{n_1} > 4^{1/p'} \varepsilon > \varepsilon.$$

Analogously, $\overline{\Delta_{n_2-1}} \cup \overline{\Delta_{n_2}} \subset (c, b)$ and

$$A_1(c, b) \geq A_1(\overline{\Delta_{n_2-1}} \cup \overline{\Delta_{n_2}}) \geq \sigma_{n_2-1} > \varepsilon.$$

Hence, $D(I) = \max\left(A_0(a, c), A_1(c, b)\right) > \varepsilon$.

Lemma 4.4 *Let* $1 < q < p < \infty$, $\frac{1}{\gamma} = \frac{1}{q} - \frac{1}{p}$, $0 < a < b < \infty$, $I = (a, b) \subset \mathbf{R}^+$. *We define*

$$D(I) = \max\,(B_0(a, c), B_1(c, b)),$$

where

$$B_0(a, c) = \left\{ \int_a^c \left(\int_a^x |v(t)|^q \, dt \right)^{\gamma/q} \left(\int_x^c |u(t)|^{p'} \, dt \right)^{\gamma/q'} |u(x)|^{p'} \, dx \right\}^{1/\gamma},$$

$$B_1(c, b) = \left\{ \int_c^b \left(\int_c^x |u(t)|^{p'} \, dt \right)^{\gamma/q'} \left(\int_x^b |v(t)|^q \, dt \right)^{\gamma/q} |u(x)|^{p'} \, dx \right\}^{1/\gamma}$$

The point c in (a, b) *is chosen so that*

$$\int_a^c |v(t)|^q \, dt = \frac{1}{2} \int_a^b |v(t)|^q \, dt.$$

Let $0 < \varepsilon < \| T \|$, *where* ε *is taken sufficiently small, and assume*

$$S_I(\varepsilon) = \left\{ n \in \mathbf{Z} : \overline{\Delta_{n+1}} \subset I, \sigma_n > \left(\frac{\gamma}{p'} \right)^{1/\gamma} \varepsilon \right\} \text{ and } \# \, S_I\left(\left(\frac{\gamma}{p'} \right)^{1/\gamma} \varepsilon \right) \geq 4.$$

Then

$$D(I) > \varepsilon.$$

Proof of Lemma 4.4 is similar to Lemma 4.3.

Lemma 4.5 *Let* $0 < \varepsilon < \| T \|$, $\# S_I(C_q \varepsilon) \geq 4$, *where*

$$C_q = \begin{cases} 1, & 1 < p \leq q < \infty, \\ \left(\frac{\gamma}{p'} \right)^{1/\gamma} & 1 < q < p < \infty. \end{cases} \tag{4.1}$$

Then

$$\| \mathcal{T}_I \| > \varepsilon.$$

Proof Since $\| \mathcal{T}_I \| \asymp D(I)$, then the result follows from Lemma 4.4.

Lemma 4.6 *Let* $0 < \varepsilon < \| T \|$, *an integer* $N = N(\varepsilon)$ *is defined by* (2.1). *Then*

$$\# \, \{ k \in \mathbf{Z} : \sigma_k > C_q \varepsilon \} \leq 6N(\varepsilon).$$

Proof Note, that $\# \, \{ k \in \mathbf{Z} : c_i \in \overline{\Delta_k}$ for some i, $1 \leq i \leq N \} \leq 2N$. For every $k \in \mathbf{Z}$ outside of the above set we have $\overline{\Delta_k} \subset I_i = (c_i, c_{i+1})$ for some $1 \leq i \leq N$,

$$\# \, \{ k \in \mathbf{Z} : \overline{\Delta_k} \subset I_i, \sigma_k > C_q \varepsilon \} \leq 3.$$

Hence,

$$\# \{k \in \mathbf{Z} : \sigma_k > C_q\varepsilon\} = \sum_{i=0}^{N} \# \{k \in \mathbf{Z} : \overline{\Delta_k} \subset I_i, \sigma_k > C_q\varepsilon\} + 2N$$

$$\leq 3(N+1) + 2N \leq 6N.$$

Lemma 4.7 *Let* $1 < p \leq q < \infty$, *then for all* $t > 0$

$$\# \{k \in \mathbf{Z} : \sigma_k > t\} \leq 6\, \# \left\{k \in \mathbf{N} : a_k(T)\, k^{\frac{1}{p}-\frac{1}{q}} \geq \frac{t}{4}\right\}.$$

Let $1 < q < p < \infty$, *then for all* $t > 0$

$$\# \{k \in \mathbf{Z} : \sigma_k > t\} \leq 6\, \# \left\{k \in \mathbf{N} : a_k(T) \geq \frac{t}{2C_q}\right\},$$

where C_q *is determinate by* (4.1).

Proof We use (2.5), when $1 < p \leq q < \infty$ and see that

$$\# \left\{k \in \mathbf{N} : a_k(T)\, k^{\frac{1}{p}-\frac{1}{q}} \geq \frac{1}{2}\,\varepsilon\right\} \geq N(\varepsilon),$$

By Lemma 4.6 we have

$$\# \{k \in \mathbf{Z} : \sigma_k > t\} \leq 6N(t) \leq 6\# \left\{k \in \mathbf{N} : a_k(T)\, k^{\frac{1}{p}-\frac{1}{q}} \geq \frac{t}{2}\right\}.$$

If $1 < q < p < \infty$, then we use (2.5) and Lemma 4.6 we obtain

$$\# \left\{k \in \mathbf{N} : a_k(T) \geq \frac{1}{2}\,\varepsilon\right\} \geq N(\varepsilon)$$

and

$$\# \{k \in \mathbf{Z} : \sigma_k > t\} \leq 6N(t/C_q) \leq 6\# \left\{k \in \mathbf{N} : a_k(T) \geq \frac{t}{2C_q}\right\}.$$

The lemma follows.

Let $l^s_\omega(\mathbf{Z})$, $s > 1$ be the space of all the sequences $\{x_k\}$ such that $\| \{x_k\} \|_{l^s_\omega(\mathbf{Z})} < \infty$, where

$$\| \{x_k\} \|_{l^s_\omega(\mathbf{Z})} = \sup_{t>0} t(\#\{k \in \mathrm{Z} : |x_k| > t\})^{1/s}.$$

Theorem 4.1 *Let* $s \in (1, \infty)$. *Then*

$$\| \{\sigma_k\} \|^s_{l^s_\omega(\mathbf{Z})} \leq 6 \cdot 2^s \| \{a_k(T)\, k^{\frac{1}{p}-\frac{1}{q}}\} \|^s_{l^s_\omega(\mathbf{N})}, \text{ if } 1 < p \leq q < \infty,$$

and

$$\| \{\sigma_k\} \|^s_{l^s_\omega(\mathbf{Z})} \le 6 \cdot (2C_q)^s \| \{a_k(T)\} \|^s_{l^s_\omega(\mathbf{N})}, \quad \text{if } 1 < q < p < \infty.$$

Proof Let $1 < p \le q < \infty$ and $\left\{ a_k(T)\, k^{\frac{1}{p}-\frac{1}{q}} \right\} \in l^s_\omega(\mathbf{N})$. By Lemma 4.7 we see

$$\# \{k \in \mathbf{Z} : \sigma_k > t\} \le 6\# \left\{ k \in \mathbf{N} : a_k(T)\, k^{\frac{1}{p}-\frac{1}{q}} \ge \frac{t}{2} \right\}.$$

Hence,

$$\| \{\sigma_k\} \|^s_{l^s_\omega(\mathbf{Z})} \le 6 \cdot 2^s \| \{a_k(T)\, k^{\frac{1}{p}-\frac{1}{q}}\} \|^s_{l^s_\omega(\mathbf{N})}.$$

Let $1 < q < p < \infty$ and $\{a_k(T)\} \in l^s_\omega(\mathbf{N})$. Using Lemma 4.7 we obtain

$$\# \{k \in \mathbf{Z} : \sigma_k > t\} \le 6\# \left\{ k \in \mathbf{N} : a_k(T) \ge \frac{t}{2C_q} \right\}.$$

Hence,

$$\sup_{t>0} t^s \#\{k \in \mathbf{Z} : \sigma_k > t\} \le 6 \sup_{t>0} t^s \# \left\{ k \in \mathbf{N} : a_k(T) \ge \frac{t}{2C_q} \right\}$$

Theorem 4.2 *Let* $s \in (0, \infty)$. *Then*

$$\| \{\sigma_k\} \|^s_{l^s(\mathbf{Z})} \le 6 \cdot 2^s \| \{a_k(T)\, k^{\frac{1}{p}-\frac{1}{q}}\} \|^s_{l^s(\mathbf{N})}, \quad \text{if } 1 < p \le q < \infty$$

and

$$\| \{\sigma_k\} \|^s_{l^s(\mathbf{Z})} \le 6 \cdot (2C_q)^s \| \{a_k(T)\} \|^s_{l^s(\mathbf{N})}, \quad \text{if } 1 < q < p < \infty.$$

Proof We begin with the case $1 < p \le q < \infty$.

$$\begin{aligned} \| \{\sigma_k\} \|^s_{l^s(\mathbf{Z})} &= s \int_0^\infty t^{s-1} \#\{k \in \mathrm{Z} : \sigma_k > t\}\, dt \\ &\le 6s \int_0^\infty t^{s-1} \# \left\{ k \in \mathrm{N} : a_k(T) k^{\frac{1}{p}-\frac{1}{q}} \ge \frac{t}{2} \right\} dt \\ &= 6(2C_q)^s \| \{a_k(T)\, k^{\frac{1}{p}-\frac{1}{q}}\} \|^s_{l^s(\mathbf{N})}. \end{aligned}$$

Let $1 < q < p < \infty$, and again using Lemma 4.7, we have

$$\| \{\sigma_k\} \|^s_{l^s(\mathbf{Z})} = s\int_0^\infty t^{s-1}\#\{k \in \mathbf{Z} : \sigma_k > t\}\, dt$$

$$\leq 6s\int_0^\infty t^{s-1}\#\left\{k \in \mathbf{N} : a_k(T) \geq \frac{t}{2C_q}\right\} dt$$

$$\leq 6 \cdot (2C_q)^s \| \{a_k(T)\} \|^s_{l^s(\mathbf{N})}$$

This completes the proof of Theorem 4.2.

Theorem 4.3 *Let $T{:}L^p(\mathbf{R}^+) \to L^q(\mathbf{R}^+)$ be compact operator given by* (1.1), *and* $r = \dfrac{qp'}{q+p'}$, $s > r$.

If $1 < p \leq q < \infty$, then

$$\| \{a_k(T)\} \|^s_{l^s_\omega(\mathbf{N})} \leq C^s(p,q)\beta(s/r) \| \{\sigma_k\} \|^s_{l^s_\omega(\mathbf{Z})},$$

$$\| \{a_k(T)\} \|^s_{l^s(\mathbf{N})} \leq C^s(p,q)\beta(s/r) \| \{\sigma_k\} \|^s_{l^s(\mathbf{Z})}.$$

If $1 < q < p < \infty$, then

$$\| \{a_{k-1}(T)\, k^{\frac{1}{p}-\frac{1}{q}}\} \|^s_{l^s_\omega(\mathbf{N})} \leq C^s(p,q)\beta(s/r) \| \{\sigma_k\} \|^s_{l^s_\omega(\mathbf{Z})}$$

$$\| \{a_k(T)\, k^{\frac{1}{p}-\frac{1}{q}}\} \|^s_{l^s(\mathbf{N})} \leq C^s(p,q)\beta(s/r) \| \{\sigma_k\} \|^s_{l^s(\mathbf{Z})}.$$

Proof Let $1 < p \leq q < \infty$, $0 < \varepsilon < \| T \|$ and $N = N(\varepsilon)$ given by (2.1). Then for all k there exists number j_k such that $c_k \subset \bar{J}_{jk}$, and there are two choices.

(1) $j_{k_0} < j_{k_0+1}$,

(2) $j_k = j_{k+1} = \ldots = j_{k+m_k}$, $I_i \subset J_{j_k}$, $k \leq i \leq k + m_k$, $m_k > 1$.

Using Lemma 3.4 and Lemma 3.5, we obtain

$$\varepsilon = \| \mathcal{T}_{I_{k_0}} \| \leq C_1 D(I_{k_0}) \leq C_1(A_0(I_{k_0}) + A_1(I_{k_0})) \leq C \sup_{\substack{j_{k_0} \leq j \\ \leq j_{k_0}+1}} \sigma_j \equiv C\sigma_{jk}, \tag{1}$$

for some $j_k \in [j_{k_0}, j_{k_0+1}]$.

$$\varepsilon^r m_k = \sum_{i=k}^{k+m_k} \| \mathcal{T}_{I_i} \|^r \leq C^r \sigma^r_{jk}, \text{ where } r = \frac{qp'}{q+p'}. \tag{2}$$

$$N(\varepsilon) = \#\left\{k : \sigma_{j_k} \geq \frac{\varepsilon}{C}\right\} + \sum_{k:m_k>1} \#\left\{k : \sigma_{j_k} \geq \frac{\varepsilon m_k^{1/r}}{C}\right\}$$

$$\leq \sum_{n=1}^{\infty} \#\left\{k : \sigma_k \geq \frac{n^{1/r}\varepsilon}{C}\right\}.$$

Let $1 < p \leq q < \infty$, then by (2.2), we see

$$\#\{k \in \mathbf{N} : a_k(T) > \varepsilon\} \leq N(\varepsilon) + 1 \leq 2N(\varepsilon).$$

$$\| (a_k(T)\} \|^s_{l^s_\omega(\mathbf{N})} = 2\sup_{t>0} t^s N(t)$$

$$\leq 2 \sup_{t>0} t^s \sum_{n=1}^{\infty} \#\left\{k \in \mathbf{Z} : \sigma_k \geq \frac{n^{1/r}t}{C}\right\} = 2C^s\left(\sum_{n=1}^{\infty} \frac{1}{n^{s/r}}\right) \| \{\sigma_k\} \|^s_{l^s_\omega(\mathbf{Z})}.$$

A similar result can be obtained for the spaces l^s.

$$\| \{a_k(T)\} \|^s_{l^s(\mathbf{N})} = s \int_0^\infty t^{s-1} \#\{k \in \mathbf{N} : a_k(T) > t\}\, dt$$

$$\leq s \int_0^\infty t^{s-1} 2N(t)\, dt \leq 2s \int_0^\infty \sum_{n=1}^{\infty} t^{s-1} \#\left\{k \in \mathbf{Z} : \sigma_k \geq \frac{n^{1/r}t}{C}\right\} dt$$

$$\leq 2sC^s \int_0^\infty \sum_{n=1}^{\infty} \frac{1}{n^{s/r}} \left(\frac{tn^{s/r}}{C}\right)^{s-1} \#\left\{k : \sigma_k \geq \frac{n^{1/r}t}{C}\right\} d\left(\frac{n^{s/r}t}{C}\right) dt$$

$$= 2C^s \left(\sum_{n=1}^{\infty} \frac{1}{n^{s/r}}\right) \| \{\sigma_k\} \|^s_{l^s(\mathbf{Z})} \equiv C^s(p,q)\beta(s/r) \| \{\sigma_k\} \|^s_{l^s(\mathbf{Z})}.$$

Let $1 < q < p < \infty$, then by (2.2)

$$\# \{k \in \mathrm{N} : a_{k-1}(T)\, k^{\frac{1}{p}-\frac{1}{q}} > \varepsilon\} \leq N(\varepsilon).$$

Hence,

$$\| \{a_{k-1}(T)\, k^{\frac{1}{p}-\frac{1}{q}}\} \|^s_{l^s_\omega(\mathrm{N})} = \sup_{t>0} t^s N(t)$$

$$\leq \sup_{t>0} t^s \sum_{n=1}^{\infty} \#\left\{k : \sigma_k \geq \frac{n^{1/r}t}{C}\right\} = C^s\left(\sum_{n=1}^{\infty} \frac{1}{n^{s/r}}\right) \| \{\sigma_k\} \|^s_{l^s_\omega(\mathbf{Z})}.$$

$$\| \{a_{k-1}(T)\, k^{\frac{1}{p}-\frac{1}{q}}\} \|^s_{l^s(\mathbf{N})} = s \int_0^\infty t^{s-1} \#\{k \in \mathbf{N} : a_{k-1}(T) k^{\frac{1}{p}-\frac{1}{q}} > t\}\, dt$$

$$\le s\int_0^\infty t^{s-1}N(T)\,dt \le s\int_0^\infty \sum_{n=1}^{\infty} t^{s-1}\#\left\{k:\sigma_k \ge \frac{n^{1/r}t}{C}\right\}dt$$

$$= sC^s\int_0^\infty \sum_{n=1}^{\infty}\frac{1}{n^{s/r}}\left(\frac{tn^{s/r}}{C}\right)^{s-1}\#\left\{k:\sigma_k\ge\frac{n^{1/r}t}{C}\right\}d\left(\frac{n^{s/r}t}{C}\right)$$

$$= C^s\left(\sum_{n=1}^{\infty}\frac{1}{n^{s/r}}\right)\|\{\sigma_k\}\|^s_{l^s(\mathbf{Z})} \equiv C^s(p,q)\beta(s/r)\|\{\sigma_k\}\|^s_{l^s(\mathbf{Z})}.$$

Let $\{\eta_k\}$, $k\in\mathbf{Z}$ be given by

$$V(\eta_k) = \int_{\eta_k}^{\infty}|v(t)|^q\,dt = 2^{-k+1}.$$

Put

$$\delta_k = \left(\int_{\eta_{k-1}}^{\eta_k}|u(t)|^{p'}\,dt\right)^{1/p'}\left(\int_{\eta_k}^{\eta_{k+1}}|v(t)|^q\,dt\right)^{1/q}$$

$$= 2^{-k/q}\left(\int_{\eta_{k-1}}^{\eta_k}|u(t)|^{p'}\,dt\right)^{1/p'},$$

$$J_s = \left(\int_0^\infty\left(\int_0^x|u|^{p'}\right)^{s/p'}\left(\int_x^\infty|v|^q\right)^{\frac{s}{q}-1}|v(x)|^q\,dx\right)^{1/s},$$

$$J_s' = \left(\int_0^\infty\left(\int_0^x|u|^{p'}\right)^{\frac{s}{p'}-1}\left(\int_x^\infty|v|^q\right)^{s/q}|u(x)|^{p'}\,dx\right)^{1/s}$$

Lemma 4.8 *Let $0<s<\infty$, $1<p,q<\infty$, and assume $J_s<\infty$ ($J_s'<\infty$), then $J_s'<\infty$ ($J_s<\infty$) and in this case $J_s = \left(\frac{q}{p'}\right)^{1/s}J_s'$.*

Proof Suppose that $0\le J_s<\infty$, then

$$\lim_{t\to\infty}\int_0^\infty\left(\int_0^x|u|^{p'}\right)^{s/p'}\left(\int_x^\infty|v|^q\right)^{\frac{s}{q}-1}|v(x)|^q\,dx = 0.$$

It implies

$$\lim_{t\to\infty}\left(\int_0^t|u|^{p'}\right)^{s/p'}\left(\int_t^\infty|v|^q\right)^{s/q} = 0.$$

Integrating by parts, we find

$$\infty > J_s^s = \frac{q}{s}\int_0^\infty\left(\int_0^x |u|^{p'}\right)^{s/p'} \frac{s}{q}\left(\int_x^\infty |v|^q\right)^{\frac{s}{q}-1} |v(x)|^q\,dx$$

$$= \frac{q}{s}\int_0^\infty\left(\int_0^x |u|^{p'}\right)^{s/p'} d\left(-\int_x^\infty |v|^q\right)^{s/q}$$

$$\geq \frac{q}{s}\int_0^\infty\left(\int_x^\infty |v|^q\right)^{s/q} d\left(\int_0^x |u|^{p'}\right)^{s/p'}$$

$$= \frac{q}{p'}\int_0^\infty\left(\int_0^x |u|^{p'}\right)^{\frac{s}{p'}-1}\left(\int_x^\infty |v|^q\right)^{s/q} |u(x)|^{p'}\,dx = \frac{q}{p'} J_s'^{\,s}.$$

So $J_s \geq \left(\frac{q}{p'}\right)^{1/s} J_s'$, and therefore $J_s' < \infty$. Conversely, suppose $J_s' < \infty$, a similar argument shows that $\lim_{t\to 0}\left(\int_0^t |u|^{p'}\right)^{s/p'}\left(\int_t^\infty |v|^q\right)^{s/q} = 0$, and $J_s' \geq \left(\frac{p'}{q}\right)^{1/s} J_s$.

Let

$$A_s = (\sum_k \sigma_k^s)^{1/s}, \quad B_s = (\sum_k \delta_k^s)^{1/s}$$

Theorem 4.4 *If* $0 < s < \infty$, $1 < p, q < \infty$, *then*

$$A_s \asymp B_s \asymp J_s \asymp J_s'.$$

Proof The last equivalence follows from Lemma 4.8. Now, let $0 < s < \infty$, $1 < p, q < \infty$. We have

$$A_s^s = \sum_k \sigma_k^s = \sum_k 2^{ks/p'}\left(\int_{\xi_k}^{\xi_{k+1}} |v(t)|^q\,dt\right)^{s/q} \leq \sum_k 2^{ks/p'}\left(\int_{\xi_k}^\infty |v(t)|^q\,dt\right)^{s/q}$$

Put

$$\sum_k 2^{ks/p'}\left(\int_{\xi_k}^\infty |v(t)|^q\,dt\right)^{s/q} := \mathscr{A}_s^s.$$

$$J_s^s = \sum_k \int_{\xi_k}^{\xi_{k+1}}\left(\int_0^x |u|^{p'}\right)^{s/p'}\left(\int_x^\infty |v|^q\right)^{\frac{s}{q}-1} |v|^q\,dx$$

$$\geq \sum_k \int_{\xi_k}^{\xi_{k+1}}\left(\int_0^{\xi_k} |u|^{p'}\right)^{s/p'}\left(\int_x^\infty |v|^q\right)^{\frac{s}{q}-1} |v|^q\,dx$$

$$= \frac{q}{s} 2^{s/p'} \sum_k 2^{ks/p'} \left[\left(\int_{\xi_k}^{\infty} |v|^q \right)^{s/q} - \left(\int_{\xi_{k+1}}^{\infty} |v|^q \right)^{s/q} \right]$$

$$= \frac{q}{s} 2^{s/p'} [A_s^s - 2^{-s/p'} A_s^s] = \frac{q}{s} [2^{s/p'} - 1] A_s^s \geq \frac{q}{s} [2^{s/p'} - 1] A_s^s.$$

$$J_s^s \geq \frac{q}{s} [2^{s/p'} - 1] A_s^s.$$

Hence,

$$A_s = (\sum_k \sigma_k^s)^{1/s} \leq [\frac{q}{s} (2^{s/p'} - 1)]^{-1/s} J_s, \ 0 < s < \infty, \ 1 < p, q < \infty.$$

To prove the reverse inequality we suppose first, that $0 < s < \infty$, $1 < p, q < \infty$ and $s \leq q$. Then $\frac{s}{q} \leq 1$, $\frac{s}{q} - 1 \leq 0$, and if

$$\int_x^{\xi_{k+1}} |v|^q \leq \int_x^{\infty} |v|^q,$$

it follows

$$\left(\int_x^{\infty} |v|^q \right)^{\frac{s}{q}-1} \leq \left(\int_x^{\xi_{k+1}} |v|^q \right)^{\frac{s}{q}-1}$$

$$\left(\int_{\xi_k}^{\xi_{k+1}} |v|^q \right)^{s/q} = \frac{s}{q} \int_{\xi_k}^{\xi_{k+1}} \left(\int_x^{\xi_{k+1}} |v|^q \right)^{\frac{s}{q}-1} |v(x)|^q \, dx$$

$$\geq \frac{s}{q} \int_{\xi_k}^{\xi_{k+1}} \left(\int_x^{\infty} |v|^q \right)^{\frac{s}{q}-1} |v(x)|^q \, dx.$$

Thus,

$$\left(\int_{\xi_k}^{\xi_{k+1}} |v|^q \right)^{s/q} \geq \frac{s}{q} \int_{\xi_k}^{\xi_{k+1}} \left(\int_x^{\infty} |v|^q \right)^{\frac{s}{q}-1} |v(x)|^q \, dx.$$

$$\sum_k 2^{ks/p'} \left(\int_{\xi_k}^{\xi_{k+1}} |v|^q \right)^{s/q} \geq \frac{s}{q} \sum_k 2^{ks/p'} \int_{\xi_k}^{\xi_{k+1}} \left(\int_x^{\infty} |v|^q \right)^{\frac{s}{q}-1} |v(x)|^q \, dx$$

$$A_s^s \geq \frac{s}{q} \sum_k \int_{\xi_k}^{\xi_{k+1}} 2^{ks/p'} 2^{2s/p'} 2^{-2s/p'} \left(\int_x^{\infty} |v|^q \right)^{\frac{s}{q}-1} |v(x)|^q \, dx$$

$$= \frac{s}{q} \sum_k 2^{-2s/p'} \int_{\xi_k}^{\xi_{k+1}} 2^{(k+2)s/p'} \left(\int_x^{\infty} |v|^q \right)^{\frac{s}{q}-1} |v(x)|^q \, dx$$

$$\left(2^{(k+2)s/p'} = \left(\int_0^{\xi_{k+1}} |u|^{p'} \right)^{s/p'} \geq \left(\int_0^x |u|^{p'} \right)^{s/p'}, \xi_k \leq x \leq \xi_{k+1} \right)$$

$$\geq \frac{s}{q} 2^{-2s/p'} \sum_k \int_{\xi_k}^{\xi_{k+1}} \left(\int_0^x |u|^{p'} \right)^{s/p'} \left(\int_x^\infty |v|^q \right)^{\frac{s}{q}-1} |v(x)|^q \, dx$$

$$= \frac{s}{q} 2^{-2s/p'} \int_0^\infty \left(\int_0^x |u|^{p'} \right)^{s/p'} \left(\int_x^\infty |v|^q \right)^{\frac{s}{q}-1} |v(x)|^q \, dx = \frac{s}{q} 2^{-2s/p'} J_s^s.$$

We obtain

$$\left(\frac{s}{q} \right)^{1/s} 2^{-2/p'} J_s \leq \left(\sum_k \sigma_k^s\right)^{1/s}, \; 0 < s < \infty, \; 1 < p, q < \infty \text{ and } s \leq q.$$

Now, let $1 < q < s < \infty$.

$$J_s^s = \sum_k \int_{\xi_k}^{\xi_{k+1}} \left(\int_0^x |u|^{p'} \right)^{s/p'} \left(\int_x^\infty |v|^q \right)^{\frac{s}{q}-1} |v|^q \, dx$$

$$\leq \sum_k \int_{\xi_k}^{\xi_{k+1}} 2^{(k+2)s/p'} \left(\int_x^\infty |v|^q \right)^{s/q-1} |v|^q \, dx$$

$$= \frac{q}{s} 2^{2s/p'} \sum_k 2^{ks/p'} \left[\left(\int_{\xi_k}^\infty |v|^q \right)^{s/q} - \left(\int_{\xi_{k+1}}^\infty |v|^q \right)^{s/q} \right]$$

$$\leq \frac{q}{s} 2^{2s/p'} \sum_k 2^{ks/p'} \left(\int_{\xi_k}^\infty |v|^q \right)^{s/q}$$

Let $\alpha = \dfrac{q}{2p'}$, then applying Hölder's inequality we find

$$\int_{\xi_k}^\infty |v|^q = \sum_{m \geq k} \int_{\xi_m}^{\xi_{m+1}} |v|^q = \sum_{m \geq k} \left[2^{\alpha m} \int_{\xi_m}^{\xi_{m+1}} |v|^q \right] 2^{-\alpha m}$$

$$\leq \left[\sum_{m \geq k} 2^{\frac{\alpha m s}{q}} \left(\int_{\xi_m}^{\xi_{m+1}} |v|^q \right)^{s/q} \right]^{q/s} \left[\sum_{m \geq k} 2^{-\alpha m \cdot \frac{s}{s-q}} \right]^{1-\frac{q}{s}}$$

$$= C_1 2^{-k\alpha} \left[\sum_{m \geq k} 2^{\frac{\alpha m s}{q}} \left(\int_{\xi_m}^{\xi_{m+1}} |v|^q \right)^{s/q} \right]^{q/s},$$

where $C_1 = \dfrac{2^\alpha}{\left(2^{\frac{\alpha s}{s-q}} - 1\right)^{\frac{s-q}{s}}}$. Consequently,

$$J_s^s \le \frac{q}{s} 2^{2s/p'} \sum_k 2^{ks/p'} C_1^{s/q} 2^{\frac{-k\alpha s}{q}} \sum_{m\ge k} 2^{\frac{\alpha m s}{q}} \left(\int_{\xi_m}^{\xi_{m+1}} |v|^q \right)^{s/q}$$

$$= C_2 \sum_k 2^{\frac{ks}{2p'}} \sum_{m\ge k} 2^{\frac{ms}{2p'}} \left(\int_{\xi_m}^{\xi_{m+1}} |v|^q \right)^{s/q}, \text{ where } C_2 = \frac{q}{s} \cdot \frac{2^{5s/2p'}}{(2^{\frac{qs}{2p'(s-q)}} - 1)^{\frac{s}{q}-1}}.$$

$$= C_2 \sum_k 2^{\frac{ks}{2p'}} \sum_{m\ge k} 2^{\frac{ms}{2p'}} \left(\int_{\xi_m}^{\xi_{m+1}} |v|^q \right)^{s/q}$$

$$= C_2 \sum_m 2^{\frac{ms}{2p'}} \left(\int_{\xi_m}^{\xi_{m+1}} |v|^q \right)^{s/q} \sum_{k\le m} 2^{\frac{ks}{2p'}} \le C_3 A_s^s,$$

where
$$C_3 = \frac{q}{s} \cdot \frac{2^{3s/2p'}}{(2^{\frac{qs}{2p'(s-q)}} - 1)^{\frac{s}{q}-1} (2^{\frac{s}{2p'}} - 1)}.$$

So,
$$J_s \le C_3^{1/s} A_s, \quad 1 < q < s < \infty.$$

Consequently,

$$\left(\sum_k \sigma_k^s \right)^{1/s} \asymp \left(\int_0^\infty \left(\int_0^x |u|^{p'} \right)^{s/p'} \left(\int_x^\infty |v|^q \right)^{\frac{s}{q}-1} |v(x)|^q \, dx \right)^{1/s},$$

when $0 < s < \infty$, $1 < p, q < \infty$.

The equivalence $B_s \asymp J_s'$ follows by the similar arguments.

$$J_s' \ge \left(\frac{p'}{s} [2^{s/q} - 1] \right)^{1/s} B_s.$$

$$B_s \ge \left[\frac{s}{p'} 2^{\frac{-2s}{q}} \right]^{1/s} J_s', \qquad 0 < s \le p'.$$

$$J_s' \le C_5^{1/s} B_s, \qquad 1 < p' < s < \infty,$$

where

$$C_5 = \frac{p'}{s} \cdot \frac{2^{2s/q}}{(2^{\frac{p's}{2q(s-p')}} - 1)^{\frac{s-p'}{p'}} (2^{s/2q} - 1)}.$$

Theorem 4.4 is proved.

Corollary 4.1 *Let* $1 < p < \infty$, $1 < s < \infty$ *and* $T : L^p(\mathbf{R}^+) \to L^p(\mathbf{R}^+)$ *be compact operator given by* (1.1). *Then*

$$\left(\sum_n a_n^s(T) \right)^{1/s} \asymp \left(\int_0^\infty \left(\int_0^x |u(y)|^{p'} dy \right)^{s/p'} \left(\int_x^\infty |v(y)|^p \right)^{\frac{s}{p}-1} |v(x)|^p \, dx \right)^{1/s}$$

References

1. J.S. Bradley, Hardy's inequalities with mixed normes, *Canad. Math. Bull.* **21** (1978), 405–408.
2. I.C. Gohberg, M.G. Krein, *Introduction to the theory of linear non-selfadjont operators,* AMS Transl. Math. Monographs, 18, Providence, R.I., 1969.
3. D.E. Edmunds, W.D. Evans. *Spectral Theory and Differential Operators.* Oxford: Univ. Press, Oxford, 1987.
4. D.E. Edmunds, W.D. Evans and D.J. Harris Approximation numbers of certain Volterra integral operators. *J. London Math. Soc.* (2) **38** (1988), 471–489.
5. D.E. Edmunds, W.D. Evans and D.J. Harris. Two-sided estimates of the approximation numbers of certain Volterra integral operators. *Studia Math.* **124** (1997), 59–80.
6. H. König. *Eigenvalue distribution of compact operators.* Birkhäuser, Boston, 1986.
7. E.N. Lomakina, On estimates of the approximation numbers of certain integral operators, *Report of Computer Center,* Khabarovsk, 1993.
8. E. Lomakina, V. Stepanov. On the compactness and approximation numbers of Hardy type integral operators in Lorentz spases. *J. London Math. Soc.* (2) **53** (1996), 369–382.
9. V.G. Maz'ya, *Sobolev spaces,* Springer-Verlag, 1985.
10. B. Muckenhoupt. Hardy's inequality with weghts. *Studia Math.* **44** (1972), 31–38.
11. A. Pietsch. Einige neue Klassen von Kompacten linearen Abbildungen, *Rev. Roumaine Math. Pures Appl.* **8** (1963), 427–447.
12. A. Pietsch *Eigenvalues and s-numbers.* Geest Portig, Leipzig, 1987.
13. S.D. Riemenschneider Compactness of a class of Volterra operators, *Tohoku Math. Jorn.,* 1974, 26, 3, 385–387.
14. E. Schmidt. Entwicklung willkürlicher Functionen nach Systemen vorgeschriebener. *Math. Ann.* **63** (1907), 433–476.
15. V.D. Stepanov, Weighted inequalities for a class of Volterra convolution operators. *J. London Math. Soc.,* **45** (1992), 232–242.

Function Spaces and Applications
D.E. Edmunds et al (Eds)

14. Expansions in Series of Legendre Functions

E.R. Love and M.N. Hunter

Department of Mathematics, The University of Melbourne.
Parkville, Victoria 3052, Australia

1. Introduction

The Legendre functions $P_\nu^\mu(x)$ occurring in this paper are sometimes called modified Legendre functions, or Legendre functions on the cut. They are defined in [**2**: 3.4 (6) on p. 143] as

$$P_\nu^\mu(x) = \frac{1}{\Gamma(1-\mu)}\left(\frac{1+x}{1-x}\right)^{\frac{1}{2}\mu} F\left(\begin{matrix}-\nu, 1+\nu; & \frac{1-x}{2} \\ 1-\mu; & \end{matrix}\right) \quad \text{for } -1 < x < 1, \tag{1}$$

where F is Gauss's hypergeometric function and μ and ν are real or complex parameters. They satisfy the recurrence relation

$$(\nu - \mu + 1)\, P_{\nu+1}^\mu(x) + (\nu + \mu)\, P_{\nu-1}^\mu(x) = (2\nu + 1)\, x\, P_\nu^\mu(x). \tag{2}$$

The Legendre polynomials are $P_n^0(x)$ (usually written $P_n(x)$), the case $\mu = 0$ and $\nu = n$ a positive integer or zero; they will not often appear in this paper.

The main result obtained in [5], Theorem 8, runs as follows.

If $(1-t^2)^{-1/4} f(t) \in L(-1,1)$, f is Dini (see below) at a certain $x \in (-1, 1)$, $|\,\mathrm{re}\mu\,| < \frac{1}{2}$ and ν is not half an odd integer, then

$$f(x) = \sum_{n=-\infty}^{\infty} a_n\, P_{\nu+n}^\mu(x),$$

where
$$a_n = (-1)^n \frac{\nu + n + \frac{1}{2}}{2\cos\nu\pi} \int_{-1}^{1} f(t) P_{\nu+n}^{-\mu}(-t)\, dt;$$

and the two "halves" of the series are separately convergent.

The Dini condition on f at x is that for some $\delta > 0$

$$\frac{f(x) - f(t)}{x - t} \;\varepsilon\; L(x - \delta, x + \delta).$$

It allows f to be differentiable at x, and more generally Hölder-continuous of any order in (0, 1] at x; but it does not allow f to be discontinuous at x.

The purpose of this paper is to present several improvements and tidyings of the work in [5] that have since become apparent. Perhaps the most notable of these is one of M.N. Hunter's identities; this appears in Lemma II.

Sections, theorems and lemmas in this paper are given the same numbers as their counterparts in [5], when they have such counterparts.

2. Christoffel Summation Formula

Theorem 1 If n is a positive integer, x and t are in $(-1, 1)$ and $x \neq t$, then

$$\sum_{r=0}^{n-1} (-1)^r \left(\nu + r + \frac{1}{2}\right) P^{\mu}_{\nu+r}(x)\, P^{-\mu}_{\nu+r}(-t)$$

$$= \frac{(-1)^{n-1} D(\nu+n, t) + D(\nu, t)}{x - t} \tag{3}$$

where $D(\nu, t) = D(\mu, \nu; x, t) = \frac{1}{2}\{(\nu - \mu)\, P^{\mu}_{\nu}(x)\, P^{-\mu}_{\nu-1}(-t)$

$$+ (\nu + \mu)\, P^{\mu}_{\nu-1}(x) P^{-\mu}_{\nu}(-t)\}. \tag{4}$$

Proof is essentially the same as in [5: p. 581], with m replaced by 1 and n by $n - 1$. It involves nothing more than elementary algebra from (2).

Christoffel's Formula was the case $\mu = 0 = \nu$; it involved only Legendre polynomials.

3. Stieltjes's Inequality

As in [5: pp. 582–585]. This includes:

Theorem 2 (Existence of the coefficients a_n). If $(1 - t^2)^{-1/4} f(t) \in L(-1, 1)$ and $|re\mu| < 1/2$,

then
$$\int_{-1}^{1} f(t)\, P^{\mu}_{\nu}(\pm t)\, dt$$

exist as L-integrals.

Theorem 3 (Asymptotic estimate). If μ and Im ν are fixed with $|re\mu| < \frac{1}{2}$, then

$$P^{\mu}_{\nu}(\cos\theta) = 0\left(\frac{(re\nu)^{re\mu - 1/2}}{(\sin\theta)^{1/2}}\right)$$

as $re\nu \to +\infty$, uniformly on $0 < \theta < \pi$.

4. Riemann-Lebesgue-Type Theorem

Theorem 4 If $(1 - t^2)^{-1/4}(t) \in L(-1, 1)$, μ and Im ν are fixed and $|re\mu| < \frac{1}{2}$, then

$$\int_{-1}^{1} f(t)\, P_\nu^\mu(\pm t)\, dt = 0((re\nu)^{re\mu - 1/2}) \text{ as } re\nu \to +\infty.$$

Proof as in [5: pp. 585–586].

5. Singular Integrals

Lemma 5 If $-\infty < a < b < \infty$, $\psi \in L(a, b)$ and ψ is Dini at a certain $x \in (a, b)$, then the singular (that is, Cauchy principal value) integral

$$^{*}\!\int_a^b \frac{\psi(t)}{x - t}\, dt := \lim_{y \to 0+} \left(\int_a^{x-y} + \int_{x+y}^b \right) \frac{\psi(t)}{x - t}\, dt$$

exists.

Proof as in [5: pp. 587–588].

Lemma 7 If ϕ is differentiable at x and ψ is Dini at x, then the product $\phi\psi$ is Dini at x.

Proof as in [5: p. 588].

Theorem 6 (Dirichlet's Integral subdivided) If $(1 - t^2)^{-1/4} f(t) \in L(-1, 1)$, f is Dini at a certain $x \in (-1, 1)$ and $|re\mu| < \frac{1}{2}$, then

$$\sum_{r=0}^{n-1} b_r\, P_{\nu+r}^\mu(x) = (-1)^{n-1} I_n + I_0, \tag{5}$$

where
$$b_r = (-1)^r \left(\nu + r + \frac{1}{2} \right) \int_{-1}^{1} f(t)\, P_{\nu+r}^{-\mu}(-t)\, dt \tag{6}$$

and
$$I_n = {}^{*}\!\int_{-1}^{1} \frac{f(t)}{x - t} D(\nu + n, t)\, dt. \tag{7}$$

Proof Integrating (3) in Theorem 1 with respect to $f(t)\, dt$ over $(-1, 1)$,

$$\sum_{r=0}^{n-1} (-1)^r \left(\nu + r + \frac{1}{2} \right) P_{\nu+r}^\mu(x) \int_{-1}^{1} P_{\nu+r}^{-\mu}(-t)\, f(t)\, dt$$

$$= \int_{-1}^{1} \frac{(-1)^{n-1} D(\nu + n, t) + D(\nu, t)}{x - t} f(t)\, dt,$$

the integrals on the left existing by Theorem 2, and consequently that on the right by linearity.

By (4) $D(\nu + n, t)$ is differentiable at x, and by hypothesis f is Dini at x. So by Lemma 7 $D(\nu + n, t)\, f(t)$ is Dini at x. By (4) and Theorem 2, $D(\nu + n, t)\, f(t)$ is in $L(-1, 1)$. So by Lemma 5 the singular integral

$$^*\int_{-1}^{1} \frac{D(\nu + n, t) f(t)}{x - t} dt$$

exists; in particular it exists for $n = 0$. By linearity of singular integrals, and (7),

$$(-1)^{n-1} I_n + I_0 = {}^*\int_{-1}^{1} \frac{(-1)^{n-1} D(\nu + n, t) + D(\nu, t)}{x - t} f(t) dt. \qquad (8)$$

The integral on the right is equal to the corresponding L-integral, since this exists as remarked in connection with (8). With (8) and (6), this proves (5).

6. Neumann's Integral

Franz Neumann proved [2: 3.6 (29) on p. 154] that Q_n, a Legendre function of the second kind, is related to the corresponding Legendre polynomial P_n by

$$Q_n(z) = \frac{1}{2} \int_{-1}^{1} \frac{P_n(t)}{z - t} dt$$

for all z in the complex plane cut along the real axis from –1 to 1. A generalization of this was given by Love [3 pp. 450–453]; and limit processes indicated by [2: 3.4 (5) on p. 143] applied to the generalization give

Theorem 7 If $re\mu < 1$, $re\nu > -1$, $re\,(\mu + \nu) > -1$, $n \geq 0$ and

$$\theta(t) = \frac{(1 + t)^{\nu + \frac{1}{2}\mu}}{(1 - t)^{\frac{1}{2}\mu}} \quad \text{for } -1 < t < 1,$$

then $\theta(t)\ P^{\mu}_{\nu+n}(t) \in L(-1, 1)$ and

$$\theta(x) Q^{\mu}_{\nu+n}(x) = \frac{1}{2}\ {}^*\int_{-1}^{1} \frac{\theta(t) P^{\mu}_{\nu+n}(t)}{x - t} dt \quad \text{for } -1 < x < 1,$$

where $Q^{\mu}_{\nu}(x)$ may be defined by [2: 3.4 (10) on p. 144].

Proof as in [5: pp. 589–590], with $k = 0$ and $\rho(t) = 1$.

7. Consequences of Neumann's Integral

Lemma 9 If $(1 - t^2)^{-1/4} f(t) \in L(-1, 1)$, f is Dini at a certain $x \in (-1, 1)$, $|\,re\mu\,| < \frac{1}{2}$, $re\nu \geq -\frac{1}{2}$ and $n \geq 0$, then

$$\frac{1}{2}\ {}^*\int_{-1}^{1} \frac{f(t)}{x - t} P^{\mu}_{\nu+n}(t) dt = f(x) Q^{\mu}_{\nu+n}(x)$$

$$- \frac{1}{2} \int_{-1}^{1} \frac{h(x) - h(t)}{x - t} \frac{(1 + t)^{\nu + \frac{1}{2}\mu}}{(1 - t)^{\frac{1}{2}\mu}} P^{\mu}_{\nu+n}(t) dt,$$

where $$h(t) = \frac{(1-t)^{\frac{1}{2}\mu}}{(1+t)^{\nu+\frac{1}{2}\mu}} f(t)$$

and the last integral is Lebesgue.

Proof Let $\theta(t) = \dfrac{(1+t)^{\nu+\frac{1}{2}\mu}}{(1-t)^{\frac{1}{2}\mu}}$ and $\rho(t) = \dfrac{h(x)-h(t)}{x-t}$. (9)

Then $f(t) = h(t)\ \theta(t)$ and

$$\begin{aligned} f(x)Q^{r}_{\nu+n}(x) &= h(x)\theta(x)Q^{\mu}_{\nu+n}(x) \\ &= h(x)\frac{1}{2}\ {}^{*}\!\!\int_{-1}^{1} \frac{\theta(t)\,P^{\mu}_{\nu+n}(t)}{x-t}\,dt \text{ (by Theorem 7)} \\ &= \frac{1}{2}\ {}^{*}\!\!\int_{-1}^{1} \frac{h(x)-h(t)+h(t)}{x-t}\,\theta(t)\,P^{\mu}_{\nu+n}(t)\,dt \\ &= \frac{1}{2}\ {}^{*}\!\!\int_{-1}^{1} \rho(t)\theta(t)\,P^{\mu}_{\nu+n}(t)\,dt + \frac{1}{2}\ {}^{*}\!\!\int_{-1}^{1} \frac{f(t)}{x-t}\,P^{\mu}_{\nu+n}(t)\,dt, \quad (10)\end{aligned}$$

which gives the required equation, provided that the separation into two integrals in the last step is correct and the former of these integrals is Lebesgue.

To justify these provisions, observe first that h is Dini at x, by its definition and Lemma 7. So there is $\delta > 0$ such that $\rho \in L(x-\delta, x+\delta)$, and δ can be chosen small enough for $[x-\delta, x+\delta] \subset (-1, 1)$. Thus $p(t)\ \theta(t)\ (1-t^2)^{-1/2} \in L(x-\delta, x+\delta)$. Outside $(x-\delta, x+\delta)$ but inside $(-1, 1)$,

$$\begin{aligned} \delta\,|\,(1-t^2)^{-1/4}\ \theta(t)\ \rho(t)\,| &\le (1-t^2)^{-1/4}\,|\,\theta(t)\,|\ \{|\,h(x)\,| + |\,h(t)\,|\} \\ &= |\,h(x)\,|\,(1+t)^{re\left(\nu+\frac{1}{2}\left(\mu-\frac{1}{2}\right)\right)}\ (1-t)^{-\frac{1}{2}\left(re\,\mu+\frac{1}{2}\right)} \\ &\quad + |\,(1-t^2)^{-1/4} f(t)\,|. \end{aligned}$$

The last two terms are in $L(-1, 1)$. Consequently $(1-t^2)^{-1/4}\ \theta(t)\ \rho(t)$ is in $L\{(-1, x-\delta) \cup (x+\delta, 1)\}$

Having thus established that $(1-t^2)^{-1/4}\ \theta(t)\ \rho(t) \in L(-1, 1)$, it follows from Theorem 2 that the first integral on the right of (10) exists; this justifies the separation of the previous line into the sum of the two integrals in (10), and completes the proof of Lemma 9.

Lemma 10 If $(1-t^2)^{-1/4} f(t) \in L(-1, 1)$, f is Dini at a certain $x \in (-1, 1)$, μ and ν are fixed with $|\,re\mu\,| < 1/2$, then as $n \to \infty$

$$\frac{1}{2}\ {}^{*}\!\!\int_{-1}^{1} \frac{f(t)}{x-t}\,P^{\mu}_{\nu+n}(-t)\,dt = -f(x)Q^{\mu}_{\nu+n}(-x) + o(n^{re\mu-\frac{1}{2}}).$$

Proof Suppose temporarily that $re\nu \geq -\frac{1}{2}$. For $n \geq 0$. Lemma 9 gives

$$\frac{1}{2}\,{}^{*}\!\!\int_{-1}^{1} \frac{f(t)}{x-t} P_{\nu+n}^{\mu}(t)\,dt = f(x)Q_{\nu+n}^{\mu}(x) - \frac{1}{2}\int_{-1}^{1} \theta(t)\rho(t)P_{\nu+n}^{\mu}(t)\,dt$$

with θ and ρ defined as in (9). Since by Lemma 9 $(1-t^2)^{-1/4}\,\theta(t)\,\rho(t) \in L(-1, 1)$, Theorem 4 gives that the last integral is $o((re\nu + n)^{re\mu-1/2})$ as $n \to \infty$, and therefore $0(n^{re\mu-\frac{1}{2}})$.

The temporary hypothesis that $re\nu \geq -\frac{1}{2}$ can now be omitted. For, any given ν, there is positive m such that $re(\nu + m) \geq -\frac{1}{2}$. Replacing ν by $\nu + m$ and n by $n - m$,

$$\frac{1}{2}\,{}^{*}\!\!\int_{-1}^{1} \frac{f(t)}{x-t} P_{\nu+n}^{\mu}(t)\,dt = f(x)Q_{\nu+n}^{\mu}(x) + o(n^{re\mu-\frac{1}{2}}) \text{ as } n \to \infty. \quad (11)$$

Now let $y = -x$, $u = -t$ and $g(t) = f(u) = f(-t)$. Then

$$(1-u^2)^{-1/4}\,g(u) = (1-t^2)^{-1/4}\,f(t) \in L(-1.1),$$

and also g is Dini at y. So the left side of the desired equation is equal, by (11), to

$$-\frac{1}{2}\,{}^{*}\!\!\int_{-1}^{1} \frac{g(u)}{y-u} P_{\nu+n}^{\mu}(u)\,du = -g(y)Q_{\nu+n}^{\mu}(y) + o(n^{re\mu-\frac{1}{2}})$$

$$= -f(x)\,Q_{\nu+n}^{\mu}(-x) + o(n^{re\mu-\frac{1}{2}}),$$

as desired.

8. Hunter's Identities

A typical one of several identities due to M.N. Hunter, and the only one that we require in this paper, is as follows. It replaces the asymptotic result given in [5: Lemma 11, on p. 592].

Lemma 11 For all real or complex μ and ν for which the functions are defined, and for $-1 < x < 1$,

$$(\nu-\mu)\,P_{\nu}^{\mu}(x)Q_{\nu-1}^{-\mu}(-x) + (\nu+\mu)\,P_{\nu-1}^{\mu}(x)Q_{\nu}^{-\mu}(-x) = \cos\nu\pi.$$

Proof Using [2: 3.4 (15), (18) and (17) on p. 144], the left side is equal to

$$(\nu-\mu)\,P_{\nu}^{\mu}(x)\left\{-Q_{\nu-1}^{-\mu}(x)\cos(\nu-\mu-1)\pi - \frac{1}{2}\pi P_{\nu-1}^{-\mu}(x)\sin(\nu-\mu-1)\pi\right\}$$

$$+ (\nu+\mu)\,P_{\nu-1}^{\mu}(x)\left\{-Q_{\nu}^{-\mu}(x)\cos(\nu-\mu)\pi - \frac{1}{2}\pi P_{\nu}^{-\mu}(x)\sin(\nu-\mu)\pi\right\}$$

$$= (\nu - \mu) P_\nu^\mu (x) \cos (\nu - \mu) \pi \frac{\Gamma(\nu - \mu)}{\Gamma(\nu + \mu)}$$

$$\times \left\{ Q_{\nu-1}^\mu (x) \cos \mu\pi + \frac{1}{2} \pi P_{\nu-1}^\mu (x) \sin \mu\pi \right\}$$

$$+ \frac{1}{2}\pi (\nu - \mu) P_\nu^\mu (x) \sin (\nu - \mu) \pi \frac{\Gamma(\nu - \mu)}{\Gamma(\nu + \mu)} \left\{ P_{\nu-1}^\mu (x) \cos \mu\pi \right.$$

$$\left. - (2/\pi) Q_{\nu-1}^\mu (x) \sin \mu\pi \right\} - (\nu + \mu) P_{\nu-1}^\mu (x) \cos (\nu - \mu) \pi \frac{\Gamma(\nu - \mu + 1)}{\Gamma(\nu + \mu + 1)}$$

$$\times \left\{ Q_\nu^\mu (x) \cos \mu\pi + \frac{1}{2} \pi P_\nu^\mu (x) \sin \mu\pi \right\} - \frac{1}{2} \pi (\nu + \mu) P_{\nu-1}^\mu (x) \sin (\nu - \mu) \pi$$

$$\times \frac{\Gamma(\nu - \mu + 1)}{\Gamma(\nu + \mu + 1)} \left\{ P_\nu^\mu (x) \cos \mu\pi - (2/\pi) Q_\nu^\mu (x) \sin \mu\pi \right\} = \frac{\Gamma(\nu - \mu + 1)}{\Gamma(\nu + \mu)}$$

$$\times [\{P_\nu^\mu (x) Q_{\nu-1}^\mu (x) - P_{\nu-1}^\mu (x) Q_\nu^\mu (x)\} \{\cos (\nu - \mu) \pi \cos \mu\pi$$

$$- \sin (\nu - \mu) \pi \sin \mu\pi\} + \frac{1}{2} \pi \{P_\nu^\mu (x) \; P_{\nu-1}^\mu (x) - P_{\nu-1}^\mu (x) P_\nu^\mu (x)\}$$

$$\times \{\cos (\nu - \mu)\pi \sin \mu\pi + \sin (\nu - \mu) \pi \cos \mu\pi\}]$$

$$= \frac{\Gamma(\nu - \mu + 1)}{\Gamma(\nu + \mu)} \{P_\nu^\mu (x) Q_{\nu-1}^\mu (x) - P_{\nu-1}^\mu (x) Q_\nu^\mu (x)\} \cos \nu\pi. \tag{12}$$

By [2: 3.4 (13) on p. 144] and [2: 3.8 (19) on p. 161],

$$\frac{2}{\pi} \sin \mu\pi . (1 - x^2) \frac{d}{dx} Q_\nu^\mu (x)$$

$$= (1 - x^2) \left\{ \cos \mu\pi \frac{d}{dx} P_\nu^\mu (x) - \frac{\Gamma(\nu + \mu + 1)}{\Gamma(\nu - \mu + 1)} \frac{d}{dx} P_\nu^{-\mu} (x) \right\}$$

$$= \cos \mu\pi \{- \nu x P_\nu^\mu (x) + (\nu + \mu) P_{\nu-1}^\mu (x)\} - \frac{\Gamma(\nu + \mu + 1)}{\Gamma(\nu - \mu + 1)}$$

$$\times \{-\nu x P_\nu^{-\mu} (x) + (\nu - \mu) P_{\nu-1}^{-\mu} (x)\}$$

$$= - \nu x \frac{2}{\pi} \sin \mu\pi Q_\nu^\mu (x) + (\nu + \mu) \left\{ \cos \mu\pi P_{\nu-1}^\mu (x) - \frac{\Gamma(\nu + \mu)}{\Gamma(\nu - \mu)} P_{\nu-1}^{-\mu} (x) \right\}$$

$$= \frac{2}{\pi} \sin \mu\pi \{- \nu x Q_\nu^\mu (x) + (\nu + \mu) Q_{\nu-1}^\mu (x)\};$$

thus if μ is not an integer, and by continuity if it is,

$$(1-x^2)\frac{d}{dx}Q_\nu^\mu(x) = -\nu x\, Q_\nu^\mu(x) + (\nu+\mu)\, Q_{\nu-1}^\mu(x). \tag{13}$$

By (13) and [2: 3.8 (19) on p. 161],

$$P_\nu^\mu(x)\, Q_{\nu-1}^\mu(x) - P_{\nu-1}^\mu(x)\, Q_\nu^\mu(x)$$

$$= \frac{1}{\nu+\mu}\left[p_\nu^\mu(x)\left\{(1-x^2)\frac{d}{dx}Q_\nu^\mu(x) + \nu x\, Q_\nu^\mu(x)\right\}\right.$$

$$\left. - Q_\nu^\mu(x)\left\{(1-x^2)\frac{d}{dx}P_\nu^\mu(x) + \nu x\, P_\nu^\mu(x)\right\}\right]$$

$$= \frac{1-x^2}{\nu+\mu}\left[P_\nu^\mu(x)\frac{d}{dx}Q_\nu^\mu(x) - Q_\nu^\mu(x)\frac{d}{dx}P_\nu^\mu(x)\right]$$

$$= \frac{2^{2\mu}}{\nu+\mu}\,\frac{\Gamma\left(1+\frac{1}{2}\nu+\frac{1}{2}\mu\right)\Gamma\left(\frac{1}{2}+\frac{1}{2}\nu+\frac{1}{2}\mu\right)}{\Gamma\left(1+\frac{1}{2}\nu-\frac{1}{2}\mu\right)\Gamma\left(\frac{1}{2}+\frac{1}{2}\nu-\frac{1}{2}\mu\right)} \tag{14}$$

$$= \frac{2^{2\mu}}{\nu+\mu}\,\frac{\Gamma(1/2)\Gamma(1+\nu+\mu)}{2^{\nu+\mu}}\,\frac{2^{\nu-\mu}}{\Gamma(1/2)\Gamma(1+\nu-\mu)}; \tag{15}$$

here (14) follows from [2: 3.4 (25) on p. 146], and (15) from Legendre's duplication formula [2: 1.2 (15) on p. 5]. We thus obtain

$$P_\nu^\mu(x)\, Q_{\nu-1}^\mu(x) - P_{\nu-1}^\mu(x)\, Q_\nu^\mu(x) = \frac{\Gamma(\nu+\mu)}{\Gamma(1+\nu-\mu)}. \tag{16}$$

This shows that (12) is equal to $\cos \nu\pi$, and so proves Lemma 11.

Remark Besides (16) and Lemma 11 there are several similar identities; for instance

$$(\nu-\mu)\, P_\nu^\mu(x)\, Q_{\nu-1}^{-\mu}(x) - (\nu+\mu)\, P_{\nu-1}^\mu(x)\, Q_\nu^{-\mu}(x) = \cos \mu\pi$$

and

$$(\nu-\mu)\, P_\nu^\mu(x)\, P_{\nu-1}^{-\mu}(-x) + (\nu+\mu)\, P_{\nu-1}^\mu(x)\, P_\nu^{-\mu}(-x) = \frac{2}{\pi}\sin \nu\pi;$$

but they are not needed in this paper.

Lemma 12 If $(1-t^2)^{-1/4} f(t) \in L(-1, 1)$, f is Dini at a certain $x \in (-1, 1)$, μ and ν are fixed and $|\, re\mu \,| < 1/2$, then as $n \to \infty$

$$(-1)^{n-1} \, {}^{*}\!\int_{-1}^{1} \frac{f(t)}{x-t} D(\nu+n, t)\, dt \to f(x) \cos \nu\pi.$$

Proof Replacing μ in Lemma 10 by $-\mu$,

$$\frac{1}{2} \, {}^{*}\!\int_{-1}^{1} \frac{f(t)}{x-t} P_{\nu+n}^{-\mu}(-t)\, dt = -f(x)\, Q_{\nu+n}^{-\mu}(-x) + o(n^{-re\mu-1/2}). \tag{17}$$

Let $\nu_n = \nu + n$ (this differs from the ν_n used in [5]; compare (2.3) therein). Thus $\nu_{n-1} = \nu_n - 1$. By (4).

$$D(\nu_n, t) = \frac{1}{2} \{(\nu_n - \mu) P_{\nu_n}^{\mu}(x) P_{\nu_n - 1}^{-\mu}(-t) + (\nu_n + \mu) P_{\nu_n - 1}^{\mu}(x) P_{\nu_n}^{-\mu}(-t)\}.$$

By Theorem 2, $f(t) P_{\nu_n}^{-\mu}(-t) \in L(-1, 1)$; and by Lemma 7, $f(t) P_{\nu_n}^{-\mu}(-t)$ is Dini at $t = x$. So by Lemma 5 the singular integrals below exist. Thus we have

$$\begin{aligned}
&{}^{*}\!\int_{-1}^{1} \frac{f(t)}{x-t} D(\nu_n, t)\, dt \\
&= (\nu_n - \mu) P_{\nu_n}^{\mu}(x) \frac{1}{2} \, {}^{*}\!\int_{-1}^{1} \frac{f(t)}{x-t} P_{\nu_n - 1}^{-\mu}(-t)\, dt \\
&\quad + (\nu_n + \mu) P_{\nu_n - 1}^{\mu}(x) \frac{1}{2} \, {}^{*}\!\int_{-1}^{1} \frac{f(t)}{x-t} P_{\nu_n}^{-\mu}(-t)\, dt \\
&= (\nu_n - \mu) P_{\nu_n}^{\mu}(x) \{-f(x) Q_{\nu_n - 1}^{-\mu}(-x) + o(n^{-re\mu - \frac{1}{2}})\} \\
&\quad + (\nu_n + \mu) P_{\nu_n - 1}^{\mu}(x) \{-f(x) Q_{\nu_n}^{-\mu}(-x) + o(n^{-re\mu - \frac{1}{2}})\} \\
&= -f(x) \{(\nu_n - \mu) P_{\nu_n}^{\mu}(x) Q_{\nu_n - 1}^{-\mu}(-x) + (\nu_n + \mu) P_{\nu_n - 1}^{\mu}(x) Q_{\nu_n}^{-\mu}(-x)\} \\
&\quad + O(n)\, O(n^{re\mu - \frac{1}{2}})\, o(n^{-re\mu - \frac{1}{2}}),
\end{aligned}$$

using (17) and Lemma 3. So by Lemma 11,

$$\begin{aligned}
{}^{*}\!\int_{-1}^{1} \frac{f(t)}{x-t} D(\nu+n, t)\, dt &= -f(x) \cos(\nu+n)\pi + o(1) \\
&= (-1)^{n+1} f(x) \cos \nu\pi + o(1),
\end{aligned}$$

and this is equivalent to the stated result.

Lemma 13. If $(1-t^2)^{-1/4} f(t) \in L(-1, 1)$, f is Dini at a certain $x \in (-1, 1)$, $|re\mu| < 1/2$, ν is not half an odd integer, and for integers n

$$a_n = (-1)^n \frac{\nu + n + \frac{1}{2}}{2 \cos \nu\pi} \int_{-1}^{1} f(t)\, P_{\nu+n}^{-\mu}(-t)\, dt,$$

then $\sum_{n=0}^{\infty} a_n P_{\nu+n}^{\mu}(x) = \frac{1}{2} f(x) + \frac{1}{2} \sec \nu\pi \; {}^{*}\!\int_{-1}^{1} \frac{f(t)}{x-t} D(\nu, t)\, dt.$

Proof Using Theorem 6 and Lemma 12.

$$\sum_{n=0}^{\infty} a_n P_{\nu+n}^{\mu}(x) = \frac{1}{2} \sec \nu\pi \sum_{n=0}^{\infty} b_n P_{\nu+n}^{\mu}(x)$$

$$= \frac{1}{2} \sec \nu\pi . \lim_{n\to\infty} \{(-1)^{n-1} I_n + I_o\}$$

$$= \frac{1}{2} \sec \nu\pi \left\{ \lim_{n\to\infty} (-1)^{n-1} \; {}^{*}\!\int_{-1}^{1} \frac{f(t)}{x-t} D(\nu+n, t) dt \right.$$

$$\left. + \; {}^{*}\!\int_{-1}^{1} \frac{f(t)}{x-t} D(\nu, t) dt \right\}$$

$$= \frac{1}{2} \sec \nu\pi \left\{ f(x) \cos \nu\pi + \; {}^{*}\!\int_{-1}^{1} \frac{f(t)}{x-t} D(\nu, t) dt \right\},$$

and this gives the stated equation.

Lemma 14 Under the hypotheses of Lemma 13.

$$\sum_{n=1}^{\infty} a_{-n} P_{\nu-n}^{\mu}(x) = \frac{1}{2} f(x) - \frac{1}{2} \sec \nu\pi \; {}^{*}\!\int_{-1}^{1} \frac{f(t)}{x-t} D(\nu, t)\, dt.$$

Proof Using the symmetry property $P_{\nu}^{\mu}(x) = P_{-\nu-1}^{\mu}(x)$ [**2:** 3.4(7) on p. 144], the left side is formally equal to

$$\sum_{n=0}^{\infty} a_{-n-1} P_{\nu-n-1}^{\mu}(x) = \sum_{n=0}^{\infty} a_{-n-1} P_{n-\nu}^{\mu}(x),$$

and

$$a_{-n-1} = (-1)^{-n-1} \frac{\nu - n - 1/2}{2 \cos \nu\pi} \int_{-1}^{1} f(t)\, P_{\nu-n-1}^{-\mu}(-t) dt$$

$$= (-1)^n \frac{-\nu + n + 1/2}{2 \cos (-\nu\pi)} \int_{-1}^{1} f(t)\, P_{-\nu+n}^{-\mu}(-t)\, dt.$$

The last expression is a_n with ν replaced by $-\nu$. Further

$$D(-\nu,t) = \frac{1}{2}\{(-\nu-\mu)\,P^{\mu}_{-\nu}(x)\,P^{-\mu}_{-\nu-1}(-t) + (-\nu+\mu)\,P^{\mu}_{-\nu-1}(x)\,P^{-\mu}_{-\nu}(-t)\}$$

$$= -\frac{1}{2}\{(\nu+\mu)\,P^{\mu}_{\nu-1}(x)\,P^{-\mu}_{\nu}(-t) + (\nu-\mu)\,P^{\mu}_{\nu}(x)\,P^{-\mu}_{\nu-1}(-t)\} \quad (18)$$

$$= -D(\nu,t).$$

So by Lemma 13 with ν replaced by $-\nu$,

$$\sum_{n=0}^{\infty} a_{-n-1}\,P^{\mu}_{\nu-n-1}(x) = \sum_{n=0}^{\infty} a_{-n-1}\,P^{\mu}_{-\nu+n}(x)$$

$$= \frac{1}{2}f(x) - \frac{1}{2}\sec\nu\pi\ {}^{*}\!\!\int_{-1}^{1}\frac{f(t)}{x-t}\,D(\nu,t)\,dt$$

and the series on the left is convergent. This justifies the formal operations; and replacing n by $n-1$ on the left gives the stated result.

Theorem 8 If $(1-t^2)^{-1/4}f(t) \in L(-1,1)$, f is Dini at a certain $x \in (-1,1)$, $|\,re\mu\,| < (1/2)$, ν is not half an odd integer, and for integers n

$$a_n = (-1)^n\,\frac{\nu+n+(1/2)}{2\cos\nu\pi}\int_{-1}^{1} f(t)\,P^{-\mu}_{\nu+n}(-t)\,dt,$$

then $$(x) = \sum_{n=-\infty}^{\infty} a_n\,P^{\mu}_{\nu+n}(x)$$

and the two "halves" of the series are separately convergent.

Proof It is immediate from Lemmas 13 and 14.

Corollary If $(1-t^2)^{-1/4}f(t) \in L(-1,1)$, f is Dini at a certain $x \in (-1,1)$, $|\,re\mu\,| < 1/2$ and for integers n

$$c_n = (-1)^n\left(n+\frac{1}{2}\right)\int_{-1}^{1} f(t)\,P^{-\mu}_{n}(-t)\,dt,$$

then $$f(x) = \sum_{n=0}^{\infty} c_n\,P^{\mu}_{n}(x).$$

Proof By (18), $D(0,t) = 0$. So Lemma 13 with $\nu = 0$ gives

$$\sum_{n=0}^{\infty}(-1)^n\,\frac{n+\frac{1}{2}}{2}\int_{-1}^{1} f(t)\,P^{-\mu}_{n}(-t)\,dt\cdot P^{\mu}_{n}(x) = \frac{1}{2}f(x),$$

from which the stated equation follows.

Remark This corollary generalizes the classical Laplace's expansion in Legendre polynomials $P_n(x)$. For, taking $\mu = 0$, the corollary gives

$$f(x) = \sum_{n=0}^{\infty} c_n P_n(x)$$

where

$$c_n = \left(n + \frac{1}{2}\right) \int_{-1}^{1} f(t)\,(-1)^n P_n(-t)\,dt = \left(n + \frac{1}{2}\right) \int_{-1}^{1} f(t)\, P_n(t)\,dt.$$

An extension The Dini condition permits f to be Hölder-continuous of any order in (0, 1], at x; but it does not permit f to be discontinuous there. However, by a totally different method I have proved the following theorem, which does permit f to have ordinary discontinuity at x:

If $(1 - t^2)^{-1/4} f(t) \in L(-1, 1)$, f has bounded variation on a neighbourhood of a certain $x \in (-1, 1)$, $|\,re\mu\,| < 1/2$ and ν is not half an odd integer, then

$$\lim_{N\to\infty} \sum_{n=-N}^{N} a_n P^{\mu}_{\nu+n}(x) = \frac{1}{2}\,\{f(x+0) + f(x-0)\}.$$

But the two "halves" of the series may not be separately convergent; this is shown by the example

$$f(t) = (1 - t^2)^{\frac{1}{2}\mu} \quad \text{for } t < x, \quad f(t) = 0 \quad \text{for } t > x.$$

This extension is fully described in [4].

Another extension The Legendre functions in all the preceding work should, strictly, be called "modified Legendre functions", or "Legendre functions on the cut;" they are defined in [2: 3.4(1) and (2), on p. 143] in terms of the "Unmodified" ones, which are defined in [2: 3.2(3) and (5), on p. 122]. These latter are holomorphic functions of z in the complex plane cut along the real axis from $-\infty$ to 1; they are mostly discontinuous across the cut.

For these "unmodified" functions $P^{\mu}_{\nu}(z)$ and $Q^{\mu}_{\nu}(z)$ we have proved the following theorem (which has not yet appeared in print):

If $D(\mu, \nu; w, z) = \frac{1}{2}\,\{(\nu - \mu)\, P^{\mu}_{\nu}(w)\, Q^{-\mu}_{\nu-1}(z) - (\nu + \mu)\, P^{\mu}_{\nu-1}(w)\, Q^{-\mu}_{\nu}(z)\}$,

E is an ellipse in the z-plane with foci at $z = \pm 1$, $f(z)$ is bounded on E, $D(\mu, \nu; w, z)\, f(z)$ is a holomorphic function of z inside and on E, $|re\mu| < 1/2$, and for integers n

$$c_n = \frac{\nu + n + \frac{1}{2}}{\pi i}\, e^{\mu\pi i} \int_E f(z)\, Q^{-\mu}_{\nu+n}(z)\,dz,$$

then

$$f(w) = \sum_{n=0}^{\infty} c_n P^{\mu}_{\nu+n}(w)$$

for all w inside E except those on the cut $w \leq 1$; and the series is uniformly convergent on every closed subset.

The special case $\nu = 0$ generalizes a classical expansion given by Karl Neumann [**1:** 11.52 on p. 296]; that special case is:

If $\left(\frac{z-1}{z+1}\right)^{\frac{1}{2}\mu} f(z)$ is holomorphic inside and on the ellipse E, $|\,re\mu\,| < 1/2$, and for integers n

$$c_n = \frac{n+\frac{1}{2}}{\pi i} e^{\mu\pi i} \int_E f(z)\, Q_n^{-\mu}(z)\, dz,$$

then the conclusions stated immediately above hold.

Neumann's expansion is the case $\mu = 0$ of this special case.

References

1. E.T. Copson, Functions of a Complex Variable (Oxford, 1935).
2. A. Erdelyi, W. Magnus, F. Oberhettinger and F.G. Tricomi, Higher Transcendental Functions, Vol. 1 (Bateman Manuscript Project, McGraw-Hill, New York, 1953).
3. E.R. Love, Franz Neumann's Integral of 1848, Proc. Cambridge Philos. Soc. 61 (1965) 445–456.
4. E.R. Love, Abel summability of certain series of Legendre functions, Proc. London Math. Soc. (3) 69 (1994) 629–672.
5. E.R. Love and M.N. Hunter, Expansions in series of Legendre functions, Proc. London Math. Soc. (3) 64 (1992) 579–601.

Function Spaces and Applications
D.E. Edmunds et al (Eds)

15. Overdetermined Weighted Hardy Inequalities on Semiaxis

Maria Nasyrova*

Department of Applied Mathematics, Khabarovsk State University,
Tichookeanskaya, 136, Khabarovsk, 680035, Russia

Introduction

We consider weighted Hardy inequality of the form

$$\| Fu \|_q \leq C \| F^{(k)}v \|_p \tag{1}$$

on semiaxis $(0, \infty)$ with indices $1 < p, q < \infty$ for differentiable functions with finite right hand side of (1) and vanishing at the endpoints with their derivatives up to the order $k - 1$.

The estimates of the norm of functions via the norm of their derivatives appear in various branches of analysis, in particular, in the calculus of variations and the well known problem on the optimal shape of column [2], where the best possible constant estimates the least eigenvalue of the related boundary value problem.

Investigation of the inequality (1) on a finite interval, when a function satisfies to zero boundary conditions in a number of greater then it is necessary for its reconstruction by the k-th derivative, was initiated by A. Kufner, however the first result for $k = 1$ and conditions $F(0) = F(1) = 0$ was obtained by P. Gurka [3]. Such a restriction leads to overdetermined inequalities. In the further development [4-9, 12] on a finite interval the Pólya condition was found [1], which is necessary and sufficient for an overdetermined inequality makes sense and also a number of problems were solved. However, the "most overdetermined" case, when a function vanishes at the both ends with all derivatives up to $(k - 1)$ order, is so far uncertain, except the case, when the weights u and v obey some additional restriction [6].

We consider the inequality (1) on semiaxis $(0, \infty)$. This case is different to some extent from the case of a finite interval and the difference might be expressed by "heuristic principle", when the only one zero condition on

---

[1]1991 *Mathematics Subject Classification:* Primary 26D10; Secondary 34B05, 46N20.

*The research work of the author was partially supported by the Russian Fund of Basic Researches grant 97-01-00604 and the grant 10.98GR of the Ministry of Education of the Russian Federation.

infinity for the least derivative is important. It allows to give a complete solution for $k = 2$ and some cases with $k > 2$.

Our main idea is to turn the problem for the inequality (1) into the equivalent problem for the Riemann-Liouville operator on the narrowed domain. Then, using various kind of decompositions, we restrict the operator on intervals, where it is positive and apply Stepanov's criteria for its boundedness [13, 14]. The case $k = 2$, $p = q = 2$ was studied in [10]. In the present paper we give the precise extension of this result for $1 < p, q < \infty$ (Sections 1 and 2). The main new results are contained in Section 3 and devoted to the case $k > 2$.

1. Preliminaries

Let $(a, b) \subseteq (-\infty, \infty)$, $a < b$, $k \geq 1$, $1 < p, q < \infty$ and $1/r = 1/q - 1/p$. Let u, v be weight functions such that $|u|^q, |v|^p, |v|^{-p'}$ are locally integrable on (a, b). Then the weighted Lebesgue space is given by

$$L_{p,v} = L_{p,v,(a,b)} = \left\{ f : \| fv \|_p^p = \int_a^b |fv|^p < \infty \right\}.$$

Put $\alpha = (\alpha_0, \alpha_1, \ldots, \alpha_{k-1})$, $\alpha_j = 0, 1$; $j = 0, 1, \ldots, k-1$; $|\alpha| = \sum_{0 \leq j \leq k-1} \alpha_j$. Denote

$$F^{(\alpha)}(0) = 0 \Leftrightarrow F^{(j)}(0) = 0 \text{ for all } j, \text{ such that } \alpha_j = 1,$$

$$F^{(\beta)}(\infty) = 0 \Leftrightarrow F^{(i)}(\infty) = 0 \text{ for all } i, \text{ such that } \beta_i = 1.$$

We consider the inequality (1) for functions belonging to the space

$$AC_p^k(\alpha, \beta) = \{F : \| F \|_{AC_p^k(\alpha,\beta)} \equiv \| F^{(k)} v \|_p < \infty,\ F^{(\alpha)}(0) = F^{(\beta)}(\infty) = 0\}.$$

The following notation will be helpful:

$$I_k f(x) = I_{k,(a,b)} f(x) = \frac{1}{\Gamma(k)} \int_a^x (x - y)^{k-1} f(y)\, dy,\ x \in (a, b);$$

$$J_k g(x) = J_{k,(a,b)} g(x) = \frac{1}{\Gamma(k)} \int_x^b (y - x)^{k-1} g(y)\, dy,\ x \in (a, b);$$

$$A_{k,0} = A_{k,0;(a,b),u,v}$$

$$= \begin{cases} \sup\limits_{a<t<b} \left(\int_t^b (x-t)^{q(k-1)} |u(x)|^q\, dx \right)^{1/q} \left(\int_a^t |v|^{-p'} \right)^{1/p'}, & \text{if } p \leq q \\ \left(\int_a^b \left(\int_t^b (x-t)^{q(k-1)} |u(x)|^q\, dx \right)^{r/q} \left(\int_a^t |v|^{-p'} \right)^{r/q'} |v(t)|^{-p'}\, dt \right)^{1/r}, & \text{if } p > q; \end{cases}$$

$$A_{k,1} = A_{k,1;(a,b),u,v}$$

$$= \begin{cases} \sup\limits_{a<t<b} \left(\int_t^b |u|^q \right)^{1/q} \left(\int_a^t (t-x)^{p'(k-1)} |v(x)|^{-p'} dx \right)^{1/p'}, \text{ if } p \le q \\ \left(\int_a^b \left(\int_t^b |u|^q \right)^{r/p} \left(\int_a^t (t-x)^{p'(k-1)} |v(x)|^{-p'} dx \right)^{r/p'} |u(t)|^q dt \right)^{1/r}, \\ \qquad \text{if } p > q; \end{cases}$$

$$B_{k,0} = B_{k,0;(a,b),u,v}$$

$$= \begin{cases} \sup\limits_{a<t<b} \left(\int_a^t (x-t)^{q(k-1)} |u(x)|^q dx \right)^{1/q} \left(\int_t^b |v|^{-p'} \right)^{1/p'}, \text{ if } p \le q \\ \left(\int_a^b \left(\int_a^t (x-t)^{q(k-1)} |u(x)|^q dx \right)^{r/q} \left(\int_t^b |v|^{-p'} \right)^{r/q'} |v(t)|^{-p'} dt \right)^{1/r}, \\ \qquad \text{if } p > q; \end{cases}$$

$$B_{k,1} = B_{k,1;(a,b),u,v}$$

$$= \begin{cases} \sup\limits_{a<t<b} \left(\int_a^t |u|^q \right)^{1/q} \left(\int_t^b (t-x)^{p'(k-1)} |v(x)|^{-p'} dx \right)^{1/p'}, \text{ if } p \le q \\ \left(\int_a^b \left(\int_a^t |u|^q \right)^{r/p} \left(\int_t^b (t-x)^{p'(k-1)} |v(x)|^{-p'} dx \right)^{r/p'} |u(t)|^q dt \right)^{1/r}, \\ \qquad \text{if } p > q; \end{cases}$$

$$A_k = A_{k;(a,b),u,v} = \max\,(A_{k,0}, A_{k,1}),$$

$$B_k = B_{k;(a,b),u,v} = \max\,(B_{k,0}, B_{k,1}).$$

Throughout the paper, expressions of the form $0 \cdot \infty$, ∞/∞, $0/0$ are taken equal to zero, and the inequality $A << B$ means $A \le cB$ with a constant $c > 0$ independent of weight functions, however the relationship $A \approx B$ is interpreted as $A << B << A$ or $A = cB$. χ_E denotes the characteristic function of a set E. Also we need the following.

Theorem 1.1 [13, 14]. *The necessary and sufficient conditions for the inequality* (1) *to be valid on interval* (a, b) *are:*

(a) $A_k < \infty$, if $F(a) = F'(a) = \ldots = F^{(k-1)}(a) = 0$ *and then* $C \approx A_k$,
(b) $B_k < \infty$, if $F(b) = F'(b) = \ldots = F^{(k-1)}(b) = 0$ *and then* $C \approx B_k$.

The following result gives rise to "heuristic principle" (see [10]).

Theorem 1.2 *Let* $-\infty < a < \infty$. *The necessary and sufficient condition for the inequality* (1) *to be true on the interval* (a, ∞), *when* $F(\infty) = 0$, *is* $B_{k,(a,\infty),u,v} < \infty$ *and then* $C \approx B_{k,(a,\infty),u,v}$.

Proof The necessity follows from Theorem 1.1(b).
To prove the sufficiency we suppose that $\| F^{(k)}v\|_p < \infty$, $F(\infty) = 0$ and $B_{k,(a,\infty),u,v} < \infty$. Let $F^{(k)} = f$ and put

$$\tilde{F}(x) = J_k f(x) = \frac{1}{\Gamma(k)} \int_x^\infty (y-x)^{k-1} f(y)\, dy, x > a.$$

Since $B_{k;(a,\infty),u,v} < \infty$, then

$$|\tilde{F}(x)| \le \frac{1}{\Gamma(k)} \| fv \|_p \left(\int_x^\infty (y-x)^{p'(k-1)} |v(y)|^{-p'} dy \right)^{1/p'} \to 0,$$

if $x \to \infty$ and, consequently, $F = \tilde{F}$. Once again applying Theorem 1.1(b), we obtain the sufficiency.

2. The Case $k = 2$

The full list of nontrivial cases of the inequality

$$\| Fu \|_q \le C \| F''v \|_p \tag{3}$$

on the semiaxis $(0, \infty)$ is the following.

$\vert\alpha\vert + \vert\beta\vert = 1.$	(1.1) $F(\infty) = 0.$
$\vert\alpha\vert + \vert\beta\vert = 2.$	(2.1) $F(0) = F(\infty) = 0,$
	(2.2) $F(0) = F'(0) = 0,$
	(2.3) $F(\infty) = F'(\infty) = 0,$
	(2.4) $F(0) = F'(\infty) = 0,$
	(2.5) $F'(0) = F(\infty) = 0.$
$\vert\alpha\vert + \vert\beta\vert = 3.$	(3.1) $F(0) = F'(0) = F(\infty) = 0,$
	(3.2) $F(0) = F'(0) = F'(\infty) = 0,$
	(3.3) $F(0) = F(\infty) = F'(\infty) = 0,$
	(3.4) $F'(0) = F(\infty) = F'(\infty) = 0.$
$\vert\alpha\vert + \vert\beta\vert = 4.$	(4.1) $F(0) = F'(0) = F(\infty) = F'(\infty) = 0.$

The cases (2.2) and (2.3) follow from Theorem 1.1. Applying Theorem 1.2, we obtain (1.1). The case (2.4) was solved in [4]. Observe, that the "heuristic principle" (see [10]) hints that

$$(1.1) \Leftrightarrow (2.3), (2.1) \Leftrightarrow (3.3), (2.5) \Leftrightarrow (3.4), (3.1) \Leftrightarrow (4.1)$$

and we prove this equivalences as well as the cases, when $|\alpha| + |\beta| = 3$.

Case (3.2).

$$\| Fu \|_q \le C \| F''v \|_p, F(0) = F'(0) = F'(\infty) = 0$$

Setting $F'' = f$ we obtain the equivalent statement of the initial problem of the form

$$\| (I_2 f)\, u \|_q \le C \| f v \|_p, \quad \int_0^\infty f = 0. \tag{4}$$

Theorem 2.1 *Let $\tau \in (0, \infty)$ be defined by*

$$\int_0^\tau | v |^{-p'} = \int_\tau^\infty | v |^{-p'} \tag{5}$$

and let

$$D_\tau = \left(\int_\tau^\infty | u |^q \right)^{1/q} \left(\int_0^\tau (\tau - x)^{p'} | v(x) |^{-p'}\, dx \right)^{1/p'}$$

Then the two-sided estimate of the best constant C in (4) is

$$C \approx A_{2;(0,\tau),u,v} + D_\tau + A_{1;(\tau,\infty),u,(x-\tau)^{-1}v(x)} + B_{1;(\tau,\infty),(x-\tau)u(x),v} \tag{6}$$

If one or both of the integrals of (5) is infinite, then

$$C \approx \inf_{\tau \ge 0} \left(A_{2;(0,\tau),u,v} + D_\tau + A_{1;(\tau,\infty),u,(x-\tau)^{-1}v(x)} + B_{1;(\tau,\infty),(x-\tau)u(x),v} \right) \tag{6'}$$

Proof I. Using $\int_0^\infty f = 0$ we find for $x > \tau$

$$I_2 f(x) = \int_0^x \left(\int_0^s f \right) ds = \int_0^\tau \left(\int_0^s f \right) ds + \int_\tau^x \left(\int_0^s f \right) ds$$

$$= \int_0^\tau (\tau - y) f(y)\, dy - \int_\tau^x \left(\int_s^\infty f \right) ds$$

$$= \int_0^\tau (\tau - y) f(y)\, dy - (x - \tau) \int_x^\infty f - \int_\tau^x (y - \tau) f(y)\, dy. \tag{7}$$

Then we have the upper bound

$$\| (I_2 f)\, u \|_q \le \| \chi_{[0,\tau]} (I_2 f)\, u \|_q + \| \chi_{[\tau,\infty]} (I_2 f)\, u \|_q$$

$$\le \| \chi_{[0,\tau]} (I_2 f)\, u \|_q + \| \chi_{[\tau,\infty]}(x) u(x) \int_0^\tau (\tau - y) f(y)\, dy \|_q$$

$$+ \| \chi_{[\tau,\infty]}(x) (x - \tau)\, u(x) \int_x^\infty f \|_q$$

$$+ \| \chi_{[\tau,\infty]}(x)u(x) \int_\tau^x (y-\tau)f(y)\,dy \|_q$$

$$= Q_1 + Q_2 + Q_3 + Q_4.$$

Applying Theorem 1.1 to the above operators we get

$$Q_1 \le A_{2;(0,\tau),u,v} \| fv \|_p,$$

$$Q_4 \le A_{1;(\tau,\infty),u,(x-\tau)^{-1}v(x)} \| fv \|_p,$$

$$Q_3 \le B_{1;(\tau,\infty),(x-\tau)u(x),v} \| fv \|_p.$$

For Q_2 we use the Hölder inequality

$$Q_2 = \| \chi_{[\tau,\infty]} u \|_q \left| \int_0^\tau (\tau - y)\, f(y)\, dy \right|$$

$$\le \| \chi_{[\tau,\infty]}\, u \|_q \, \| \chi_{[0,\tau]}(\tau - y)v^{-1}(y) \|_{p'} \, \| fv \|_p = D_\tau \| fv \|_p.$$

Thus,

$$C \le A_{2;(0,\tau),u,v} + D_\tau + A_{1;(\tau,\infty),u,(x-\tau)^{-1}v(x)} + B_{1;(\tau,\infty),(x-\tau)u(x),u(x),v}$$

and since $\tau \ge 0$ is arbitrary the upper bounds (6) and (6′) follow.

II. Let us consider two cones

$$\mathcal{L}_1 = \{ f \in L_{p,v} : f \ge 0, \operatorname{supp} f \subseteq [0, \tau] \}$$

$$\mathcal{L}_2 = \{ f \in L_{p,v} : f \le 0, \operatorname{supp} f \subseteq [\tau, \infty] \}$$

and such a one-to-one isometrical map ω_τ between the cones, that

$$\int_0^\infty (f + \omega_\tau f) = 0, f \in \mathcal{L}_1. \tag{8}$$

We construct ω_τ using an idea of ([11], Section 1.8). Let the map $\rho : [0, \tau] \to [\tau, \infty)$ be given by

$$\int_0^s | v |^{-p'} = \int_{\rho(s)}^\infty | v |^{-p'}, s \in [0, \tau].$$

Then (5) and the equality

$$\| fv\chi_{[0,\tau]} \|_p = \| (\omega_\tau f)v\chi_{[\tau,\infty)} \|_p,$$

imply

$$(\omega_\tau f)(x) = -\frac{f(\rho^{-1}(x)) \,| v\, (\rho^{-1}(x)) |^{p'}}{| v(x) |^{p'}}, x \ge \tau. \tag{9}$$

Now, suppose that (4) is true and the right hand side of (5) is finite. For arbitrary function of the form $f + \omega_\tau f, f \in \mathcal{L}_1$ or $\omega_\tau^{-1} f + f, f \in \mathcal{L}_2$ the right hand side of (7) is nonnegative, therefore

$$\| \chi_{[0,\tau]}(I_2 f)u \|_q \leq \| (I_2(f + \omega_\tau f)) \, u \|_q \leq C \, \| (f + \omega_\tau f)v \|_p$$
$$\leq 2C \, \| \chi_{[0,\tau]} fv \|_p, \, f \in \mathscr{L}_1$$

and, similarly,

$$\| \chi_{[\tau,\infty)}(I_2 f) \, u \|_q \leq 2C \, \| \chi_{[\tau,\infty)} \, fv \|_p, \, f \in \mathscr{L}_2.$$

Using representation formulae (7), the reverse Hölder inequality and Theorem 1.1 we obtain the required lower bound. If one or both integrals in (5) are infinite we replace the weight $v(x)$ by $\varepsilon\chi_{[0,1]}(x) + \varepsilon x\chi_{[1,\infty]}(x) + | \, v(x) \, |$ in the above argument we prove the lower bound (6′) letting $\varepsilon \to 0$. Besides, it implies that if inequality (4) is true with a finite constant C then D_τ is finite for all $\tau > 0$.

Theorem 2.1 is proved.

Remark 1 Theorem 2.1 remains true under restriction on an interval (a, b) with corresponding amendment at the right hand side of (6) and (6′).

Case (3.4). Now we consider the problem

$$\| Fu \|_q \leq C \, \| F''v \|_p, \, F'(0) = F(\infty) = F'(\infty) = 0, \tag{10}$$

which is equivalent to

$$\| (J_2 g) \, u \|_q \leq C \, \| gv \|_p, \, \int_0^\infty g = 0,$$

and dual to (4).

Theorem 2.2 *The least constant C in* (10) *is estimated by*

$$C \approx B_{2;(\tau,\infty),u,v} + D_{1,\tau} + A_{1;(0,\,\tau),(\tau-x)u(x),v} + B_{1;(0,\tau),u,(\tau-x)^{-1}v(x)}, \tag{11}$$

where τ is defined by (5) *and*

$$D_{1,\tau} = \left(\int_0^\tau | \, u \, |^q \right)^{1/q} \left(\int_\tau^\infty (x - \tau)^{p'} \, | \, v(x) \, |^{-p'} \, dx \right)^{1/p'}.$$

If one or both of the integrals of (5) *is infinite, then*

$$C \approx \inf_{\tau \geq 0} (B_{2;(\tau,\infty),u,v} + D_{1,\tau} + A_{1;(0,\,\tau),(\tau-x)u(x),v} + B_{1;(0,\tau),u,(\tau-x)^{-1}v(x)}). \tag{11′}$$

Proof For $0 < x < \tau$ we have the formulae

$$J_2 g(x) = \int_\tau^\infty (y - \tau) g(y) \, dy - (\tau - x) \int_0^x g - \int_x^\tau (\tau - y) g(y) \, dy,$$

using which similar to the proof of Theorem 2.1 we find the required estimate.

Case (2.5). We have

$$\| Fu \|_q \leq C \| F''v \|_p, \, F'(0) = F(\infty) = 0. \tag{12}$$

Theorem 2.3 *The estimate of the least constant C in* (12) *is given by* (11′).

Proof Since

$$\{F : F'(0) = F(\infty) = 0\} \supset \{F : F'(0) = F(\infty) = F'(\infty) = 0\},$$

then (12) implies (10). Therefore by Theorem 2.2 we find the lower bound

$$C >> B_{2;(\tau,\infty),u,v} + D_{1,\tau} + A_{1;(0,\tau),(\tau-x)u(x),v} + B_{1;(0,\tau),u,(\tau-x)^{-1}v(x)}.$$

Now suppose that

$$B_{2;(\tau,\infty),u,v} + D_{1,\tau} + A_{1;(0,\tau),(\tau-x)u(x),v} + B_{1;(0,\tau),u,(\tau-x)^{-1}v(x)} < \infty.$$

Then, in particular,

$$\int_s^\infty (x-s)^{p'} \mid v(x) \mid^{-p'} dx < \infty, \; s > \tau.$$

Consequently, the function

$$\tilde{F}(s) = \int_s^\infty (x-s)\, F''(x)\, dx$$

is well defined for all $s > 0$, moreover $\tilde{F}(\infty) = 0$ and $\tilde{F}'' = F''$. Hence, $\tilde{F} = F$ and Theorem 2.2 gives the upper bound.

Case (3.3).

$$\| Fu \|_q \le C \| F''v \|_p, \; F(0) = F(\infty) = F'(\infty) = 0.$$

The equivalent problem is

$$\| (J_2 g)\, u \|_q \le C \| gv \|_p, \; \int_0^\infty y g(y)\, dy = 0. \tag{13}$$

We reduce this to the problems considered above by the substitution $yg(y) \to g(y)$. Thus we obtain the next equivalence in the form

$$\| (\overline{J_2} g)\, u \|_q \le C \| g v_1 \|_p, \; \int_0^\infty g = 0,$$

where $v_1(x) = v(x)/x$ and

$$\overline{J_2} g(x) = \int_x^\infty \left(1 - \frac{x}{y}\right) g(y)\, dy.$$

For the norm of the operator we find

$$\| J_2 \|_{L_{p,v,(\tau,\infty)} \to L_{q,u,(\tau,\infty)}} = \sup_{f \neq 0} \frac{\| (J_2 f) u \|_q}{\| f v \|_p}$$

$$= \sup_{f = g(x)/x \neq 0} \frac{\| J_2(g(x)/x) u \|_q}{\| g(x) v(x)/x \|_p}$$

$$= \sup_{g \neq 0} \frac{\| (\bar{J}_2 g) u \|_q}{\| g v_1 \|_p} = \| \bar{J}_2 \|_{L_{p,v_1,(\tau,\infty)} \to L_{q,u,(\tau,\infty)}}.$$

Using the decomposition

$$\bar{J}_2 g(x) = \frac{x}{\tau} \int_\tau^\infty (y - \tau) g(y) \frac{dy}{y} - \frac{\tau - x}{\tau} \int_0^x g - \frac{x}{\tau} \int_x^\tau (\tau - y) g(y) \frac{dy}{y}$$

for $0 < x < \tau$ and arguing similar to the proof of Theorem 2.1 we obtain the following.

Theorem 2.4 *The least constant C in* (13) *is estimated by*

$$C \approx \inf_{\tau \geq 0} \left(B_{2;(\tau,\infty),u,v} + \tau^{-1} \left(D_{2,\tau} + A_{1;(0,\tau),(\tau - x)u(x), x^{-1} v(x)} \right.\right.$$

$$\left.\left. + B_{1;(0,\tau), x u(x), (\tau - x)^{-1} v(x)} \right)\right) \tag{14}$$

where

$$D_{2,\tau} = \left(\int_0^\tau x^q \,|\, u(x) \,|^q \, dx \right)^{1/q} \left(\int_\tau^\infty (x - \tau)^{p'} \,|\, v(x) \,|^{-p'} \, dx \right)^{1/p'}$$

Case (2.1). The problem is to characterize the following inequality

$$\| F u \|_q \leq C \,\| F'' v \|_p, \; F(0) = F(\infty) = 0. \tag{15}$$

Theorem 2.5 *The best constant C in* (15) *is estimated by* (14).
The proof is analogous to the case (2.5).

Case (3.1). We have

$$\| F u \|_q \leq C \,\| F'' v \|_p, \; F(0) = F'(0) = F(\infty) = 0$$

with the equivalent form

$$\| (I_2 f) \, u \|_q \leq C \, \| f v \|_p, \; \int_0^\infty \left(\int_0^y f \right) dy = 0. \tag{16}$$

For characterization of the inequality we need the following assertion from [9].

Lemma *A locally integrable function f satisfies the condition*

$$\int_0^\infty \left(\int_0^y f \right) dy = 0 \tag{17}$$

if and only if there exists $\lambda \in (0, \infty)$, such that

$$\int_0^\lambda f = 0 \text{ and } \int_0^\lambda \left(\int_y^\lambda f \right) dy = \int_\lambda^\infty \left(\int_\lambda^y f \right) dy. \tag{18}$$

Let τ_λ for any fixed $\lambda \in (0, \infty)$ be defined by

$$\int_0^{\tau_\lambda} | v |^{-p'} = \int_{\tau_\lambda}^\lambda | v |^{-p'} \tag{19}$$

Theorem 2.6 *In the inequality* (16) *the least constant C has the following estimation*

$$C \approx A_{2;(0,\infty),u,v} + \sup_{\lambda>0} (A_{1;(\tau_\lambda,\lambda),u,(x-\tau_\lambda)^{-1}v(x)} + B_{1;(\tau_\lambda,\lambda),(x-\tau_\lambda)u(x),v}). \tag{20}$$

Proof The upper bound. Let $f \in L_{p,v}$ and (16) is true. Then by Lemma there exists $\lambda \in (0, \infty)$, such that (18) is fulfilled. Hence, for $x > \lambda$ the following decomposition is valid

$$I_2 f(x) = \int_0^\lambda (\lambda - y) f(y)\, dy + \int_\lambda^x (x - y) f(y)\, dy.$$

Applying Theorems 1.1 and 2.1 and taking into account Remark 1, we obtain

$$\| (I_2 f)\, u \|_q \le \| \chi_{[0,\lambda]}(I_2 f)\, u \|_q + \| \chi_{[\lambda,\infty]}(I_2 f)\, u \|_q$$

$$<< \sup_{\lambda>0} \left(A_{2;(0,\tau_\lambda),u,v} + \left(\int_{\tau_\lambda}^\lambda | u |^q \right)^{1/q} \left(\int_0^{\tau_\lambda} (\tau_\lambda - x)^{p'} | v(x) |^{-p'} dx \right)^{1/p'} \right.$$

$$+ A_{1;(\tau_\lambda,\lambda),u,(x-\tau_\lambda)^{-1}v(x)} + B_{1;(\tau_\lambda,\lambda),(x-\tau_\lambda)u(x),v}$$

$$\left. + \left(\int_\lambda^\infty | u |^q \right)^{1/q} \left(\int_0^\lambda (\lambda - x)^{p'} | v(x) |^{-p'} dx \right)^{1/p'} + A_{2;(\lambda,\infty),u,v} \right) \| f v \|_p .$$

For $p \le q$ we find

$$\sup_{\lambda>0} \left[\left(\int_{\tau_\lambda}^\lambda | u |^q \right)^{1/q} \left(\int_0^{\tau_\lambda} (\tau_\lambda - x)^{p'} | v(x) |^{-p'} dx \right)^{1/p'} \right]$$

$$\le \sup_{0<t<\infty} \left(\int_t^\infty | u |^q \right)^{1/q} \left(\int_0^t (t - x)^{p'} | v(x) |^{-p'} dx \right)^{1/p'}$$

$$= A_{2,1;(0,\infty),u,v} \le A_{2;(0,\infty),u,v}.$$

For the case $p > q$ we need the following notation

$$V(t) = \int_0^t (t-x)^{p'} \mid v(x) \mid^{-p'} dx, \; U(t) = \int_t^\infty \mid u \mid^q$$

and write

$$\left(\int_{\tau_\lambda}^{\lambda} \mid u \mid^q\right)^{1/q} \left(\int_0^{\tau_\lambda} (\tau_\lambda - x)^{p'} \mid v(x) \mid^{-p'} dx\right)^{1/p'} \leq U^{1/q}(\tau_\lambda) V^{1/p'}(\tau_\lambda)$$

$$= \left[U^{r/q}(\tau_\lambda) V^{r/p'}(\tau_\lambda)\right]^{1/r} \leq \left[V^{r/p'}(\tau_\lambda) \int_{\tau_\lambda}^{\infty} dU^{r/q}(t)\right]^{1/r}$$

$$\leq \left[\int_{\tau_\lambda}^{\infty} V^{r/p'}(t) dU^{r/q}(t)\right]^{1/r} = \left[\frac{r}{q} \int_{\tau_\lambda}^{\infty} V^{r/p'}(t) U^{r/p}(t) \mid u(t) \mid^q dt\right]^{1/r}$$

$$\leq \left[\frac{r}{q} \int_0^{\infty} V^{r/p'}(t) U^{r/p}(t) \mid u(t) \mid^q dt\right]^{1/r} = \frac{r}{q} A_{2,1;(0,\infty),u,v} \leq \frac{r}{q} A_{2;(0,\infty),u,v}.$$

Analogously we find, that

$$\left(\int_{\lambda}^{\infty} \mid u \mid^q\right)^{1/q} \left(\int_0^{\lambda} (\lambda - x)^{p'} \mid v(x) \mid^{-p'} dx\right)^{1/p'}$$

$$= U^{1/q}(\lambda) V^{1/p'}(\lambda) \leq \frac{r}{q} A_{2,1;(0,\infty),u,v} \leq \frac{r}{q} A_{2;(0\,\infty),u,v}.$$

Finally we get

$$\| (I_2 f)\, u \|_q \leq \left(\frac{r}{q} A_{2;(0,\infty),u,v} + \sup_{\lambda>0} \left(A_{1;(\tau_\lambda,\lambda),u,(x-\tau_\lambda)^{-1} v(x)}\right.\right.$$
$$\left.\left. + B_{1;(\tau_\lambda,\lambda),(x-\tau_\lambda)u(x),v}\right)\right) \|fv\|_p.$$

The upper bound is established.

The lower bound. Suppose $\lambda \in (0, \infty)$ and τ_λ is chosen in accordance with (19). We define the function $\rho_1 : [0, \tau_\lambda] \to [\tau_\lambda, \lambda]$ by

$$\int_0^s \mid v \mid^{-p'} = \int_{\rho_1(s)}^{\lambda} \mid v \mid^{-p'}, \; s \in [0, \tau_\lambda].$$

Now we take a function $f \in L_{p,v}$ such that $\operatorname{supp} f \subseteq [0, \tau_\lambda]$, $f \geq 0$. Let the extension $f_1(x)$ of $f(x)$ on the interval $[\tau_\lambda, \lambda]$ be chosen so that the equalities

$$\| \chi_{[0,\tau_\lambda]} f v \|_p = \| \chi_{[\tau_\lambda,\lambda]} f_1 v \|_p,$$

$$\int_0^{\tau_\lambda} f + \int_{\tau_\lambda}^{\lambda} f_1 = 0 \tag{21}$$

are fulfilled. It follows from these conditions that the extension has the form

$$f_1(x) = \begin{cases} -\dfrac{f(\rho_1^{-1}(x))\,|\,v(\rho_1^{-1}(x))\,|^{p'}}{|\,v(x)\,|^{p'}}, x \in (\tau_\lambda, \lambda), \\ 0, x \notin (\tau_\lambda, \lambda). \end{cases}$$

Put

$$f_0 = f + f_1.$$

Note, that $\operatorname{supp} f_0 \subseteq [0, \lambda]$. Given $\lambda \in (0, \infty)$ we find $\mu = \mu(\lambda) \in (0, \infty)$ from

$$\int_0^\lambda x^{p'}\,|\,v(x)\,|^{-p'}\,dx = \int_\lambda^\mu (\mu - x)^{p'}\,|\,v(x)\,|^{-p'}\,dx$$

and define the function $\rho_2 : [0, \lambda] \to [\lambda, \mu]$ by

$$\int_0^s x^{p'}\,|\,v(x)\,|^{-p'}\,dx = \int_{\rho_2(s)}^\mu (\mu - x)^{p'}\,|\,v(x)\,|^{-p'}\,dx.$$

The extension $f_2(x)$ of $f_0(x)$ on $[\lambda, \mu]$ we choose so that the conditions

$$\|\,\chi_{[0,\lambda]} f_0 v\,\|_p = \|\,\chi_{[\lambda,\mu]} f_2 v\,\|_p,$$

$$\int_0^\lambda \left(\int_y^\lambda f_0 \right) dy = \int_\lambda^\mu \left(\int_\lambda^y f_2 \right) dy \tag{22}$$

are fulfilled. Then the function $f_2(x)$ is given by

$$f_2(x) = \begin{cases} \dfrac{(\mu - x)^{p'-1}\, f_0(\rho_2^{-1}(x))\,|\,v(\rho_2^{-1}(x))\,|^{p'}}{(\rho_2^{-1}(x))^{p'-1}\,|\,v(x)\,|^{p'}}, x \in (\lambda, \mu), \\ 0, x \notin (\lambda, \mu). \end{cases}$$

Set $\tilde{f} = f_0 + f_2$. If follows from (21) and (22) that the function $\tilde{f}$ satisfies

$$\|\,\tilde{f} v\,\|_p = 4^{1/p}\,\|\,f v\,\|_p,$$

$$\int_0^\lambda \tilde{f} = 0,$$

and

$$\int_0^\lambda \left(\int_y^\lambda \tilde{f} \right) dy = \int_\lambda^\infty \left(\int_\lambda^y \tilde{f} \right) dy.$$

Hence, $\tilde{f}$ is admissible for (16) on the strength of Lemma. Since the functions

f_0 form the cone used in the proof of lower bound of Theorem 2.1, then we obtain

$$C >> \sup_{\lambda>0} (A_{2;(0,\tau_\lambda),u,v} + D_{\tau_\lambda} + A_{1;(\tau_\lambda,\lambda),u,(x-\tau_\lambda)^{-1}v(x)} + B_{1;(\tau_\lambda,\lambda),(x-\tau_\lambda)u(x),v})$$

$$>> \sup_{\lambda>0} (A_{1;(\tau_\lambda,\lambda),u,(x-\tau_\lambda)^{-1}v(x)} + B_{1;(\tau_\lambda,\lambda),(x-\tau_\lambda)u(x),v}).$$

Now suppose $0 < \lambda < \sigma < \infty$ and take $\mu \in (\lambda, \sigma)$ such that

$$\int_\lambda^\mu (\sigma - x)^{p'} \mid v(x) \mid^{-p'} dx = \int_\mu^\sigma (\sigma - x)^{p'} \mid v(x) \mid^{-p'} dx.$$

Obviously, $\mu \to \infty$ if $\sigma \to \infty$. Let $f \in L_{p,v}$, $\operatorname{supp} f \subseteq [\lambda, \mu]$, $f(x) \geq 0$. We define the function $\rho : [\lambda, \mu] \to [\mu, \sigma]$ by

$$\int_\lambda^s (\sigma - x)^{p'} \mid v(x) \mid^{-p'} dx = \int_{\rho(s)}^\sigma (\sigma - x)^{p'} \mid v(x) \mid^{-p'} dx$$

and arrange the extension $f_1(x)$ of $f(x)$ on $[\mu, \sigma]$ so, that

$$\| \chi_{[\lambda,\mu]} f v \|_p = \| \chi_{[\mu,\sigma]} f_1 v \|_p,$$

$$\int_\lambda^\mu (\sigma - x) f(x)\, dx = \int_\mu^\sigma (\sigma - x) f_1(x)\, dx.$$

Then the extension has the form

$$f_1(x) = \begin{cases} \dfrac{(\sigma - x)^{p'-1} f(\rho^{-1}(x)) \mid v(\rho^{-1}(x)) \mid^{p'}}{(\sigma - \rho^{-1}(x))^{p'-1} \mid v(x) \mid^{p'}}, x \in (\mu, \sigma), \\ 0, x \notin (\mu, \sigma). \end{cases}$$

For the test function

$$\tilde{f} = f + f_1$$

we have
$$\| \tilde{f} v \|_p = 2^{1/p} \| f v \|_p$$

and

$$\int_0^\infty \left(\int_0^y \tilde{f} \right) dy = \int_\lambda^\sigma \left(\int_\lambda^y \tilde{f} \right) dy = \int_\lambda^\sigma (\sigma - x) \tilde{f}(x)\, dx = 0.$$

Applying Theorem 1.1 we find

$$C >> A_{2;(\lambda,\mu),u,v},\ 0 < \lambda < \mu < \infty.$$

Letting $\sigma \to \infty$, and then $\lambda \to 0$ we obtain

$$C >> A_{2;(0,\infty),u,v}.$$

Theorem 2.6 is proved.
Case (4.1).

$$\| Fu \|_q \leq C \| F''v \|_p, \ F(0) = F'(0) = F(\infty) = F'(\infty) = 0. \qquad (23)$$

Theorem 2.7 *The least constant C in* (23) *is estimated by* (20).

Proof Using Theorem 2.6 we obtain the upper bound.

Now suppose the inequality (23) is true. Then it is also true for the weight $v_\varepsilon(x) = \varepsilon x^2 + | v(x) |$ instead of $v(x)$. We have

$$\left| \int_x^\infty F'' \right| \leq \left(\int_x^\infty | F''(t)\, v_\varepsilon(t) |^p \, dt \right)^{(1/p)} \left(\int_x^\infty | v_\varepsilon(t) |^{-p'} \right)^{1/p'} < \infty.$$

Hence $\int_x^\infty F'' \to 0$ if $x \to \infty$. Therefore for all $x, y \to \infty$

$$| F'(y) - F'(x) | = \left| \int_x^y F'' \right| \to 0$$

Consequently, there exists a finite limit A of the derivative $F'(x)$ at infinity.

If $A \neq 0$, then the function $F(x)$ behaves as $y = Ax$ when $x \to \infty$, which contradicts to the boundary condition $F(\infty) = 0$. Thus, $A = 0$ and

$$\{F : \| F''v_\varepsilon \|_p < \infty, \ F(0) = F'(0) = F(\infty) = F'(\infty) = 0\}$$
$$= \{F : \| F''v_\varepsilon \|_p < \infty, \ F(0) = F'(0) = F(\infty) = 0\}.$$

Theorem 2.6 yields

$$C >> A_{2;(a,\infty),u,v_\varepsilon} + \sup_{\lambda>0} \left(A_{1;(\tau_\lambda,\lambda),u,(x-\tau_\lambda)^{-1}v_\varepsilon(x)} + B_{1;(\tau_\lambda,\lambda),(x-\tau_\lambda)u(x),v_\varepsilon}\right).$$

Letting ε tend to zero and applying Fatou theorem we establish the required lower bound.

3. Certain Results for $k > 2$

The following notation will be useful for $\tau \in (a, b)$

$$D_{\tau,m,n} = D_{\tau,m,n,(a,b),u,v}$$
$$= \left(\int_\tau^b (x - \tau)^{mq} \, | u(x) |^q \, dx \right)^{1/q} \left(\int_a^\tau (\tau - x)^{np'} \, | v(x) |^{-p'} \, dx \right)^{1/p'}$$

Proposition 1 $D_{\tau,m,n,(a,b),u,v} \leq D_{\tau,m+n,0,(a,b),u,v} + D_{\tau,0,m,+n,(a,b),u,v}$.

Proof Let us put $\alpha = m/(m + n)$ and consider each of the integrals defining $D_{\tau,m,n}$ separately. We write

$$\int_\tau^b (x - \tau)^{mq} \, | u(x) |^q \, dx = \int_\tau^b (x - \tau)^{(m+n)q\alpha} \, | u(x) |^{q\alpha} \, | u(x) |^{q(1-\alpha)} \, dx$$

and applying Hölder's inequality with exponents $1/\alpha$ and $1/(1-\alpha)$ we find

$$\leq \left(\int_\tau^b (x-\tau)^{(m+n)q} \,|u(x)|^q \, dx\right)^\alpha \left(\int_\tau^b |u(x)|^q \, dx\right)^{1-\alpha}$$

Analogously, since $1-\alpha = n/(m+n)$, we have

$$\int_a^\tau (\tau-x)^{np'} \,|v(x)|^{-p'} \, dx = \int_a^\tau |v(x)|^{-p'\alpha} (\tau-x)^{(m+n)p'(1-\alpha)} \,|v(x)|^{-p'(1-\alpha)} \, dx$$

and again using Hölder's inequality we obtain

$$\leq \left(\int_a^\tau |v(x)|^{-p'} \, dx\right)^\alpha \left(\int_a^\tau (\tau-x)^{(m+n)p'} \,|u(x)|^{-p'} \, dx\right)^{1-\alpha}$$

Thus,

$$\begin{aligned} D_{\tau,m,n} &= \left(\int_\tau^b (x-\tau)^{mq} \,|u(x)|^q \, dx\right)^{1/q} \left(\int_a^\tau (\tau-x)^{np'} \,|v(x)|^{-p'} \, dx\right)^{1/p'} \\ &\leq \left(\left(\int_\tau^b (x-\tau)^{(m+n)q} \,|u(x)|^q \, dx\right)^{1/q} \left(\int_a^\tau |v(x)|^{-p'} \, dx\right)^{1/p'}\right)^\alpha \\ &\quad \left(\left(\int_\tau^b |u(x)|^q \, dx\right)^{1/q} \left(\int_a^\tau (\tau-x)^{(m+n)p'} \,|v(x)|^{-p'} \, dx\right)^{1/p'}\right)^{1-\alpha} \\ &= D^\alpha_{\tau,m+n,0} \cdot D^{1-\alpha}_{\tau,0,m+n}. \end{aligned}$$

Since $0 < \alpha < 1$ the required estimate follows.

Proposition 2

$$D_{\tau,0,n,(a,b),u,v} \leq A_{n+1;(a,b),u,v};$$
$$D_{\tau,m,0,(a,b),u,v} \leq A_{m+1;(a,b),u,v}.$$

Proof follows similar to the proof of the upper bound in Theorem 2.6.

Case 1. Here we characterize the inequality

$$\| Fu \|_q \leq C \,\| F^{(k)} v \|_p,$$
$$F(0) = F'(0) = \dots = F^{(k-1)}(0) = F^{(k-1)}(\infty) = 0 \tag{24}$$

for arbitrary $k > 2$.

Theorem 3.1 *Let* $1 < p, q < \infty$. *Then the least constant* C *in* (24) *is estimated by*

$$\begin{aligned} C \approx \inf_{\tau \geq 0} \big(&A_{k;(0,\tau),u,v} + D_{\tau,k-2,1} + D_{\tau,0,k-1} \\ &+ A_{1;(\tau,\infty),(x-\tau)^{k-2} u(x),\,(x-\tau)^{-1} v(x)} + B_{1;(\tau,\infty),(x-\tau)^{k-1} u(x),v}\big). \end{aligned} \tag{25}$$

Proof We denote $F^{(k)} = f$ and pass to the equivalent problem

$$\| (I_k f)\, u \|_q \le C \| fv \|_p, \quad \int_0^\infty f = 0. \tag{26}$$

(1) The upper bound. Let $\tau > 0$. Using that $\int_0^\infty f = 0$ we find for $x > \tau$

$$| I_k f(x) | = | I_{k-1}(I_1 f)(x) | \le I_{k-1} | I_1 f | (x) \ll \int_0^x (x-t)^{k-2} \left| \int_0^t f \right| dt$$

$$= \int_0^\tau (x-t)^{k-2} \left| \int_0^t f \right| dt + \int_\tau^x (x-t)^{k-2} \left| - \int_t^\infty f \right| dt$$

$$\le \int_0^\tau (x-t)^{k-2} \left(\int_0^t |f| \right) dt + \int_\tau^x (x-t)^{k-2} \left(\int_t^\infty |f| \right) dt$$

$$\ll \int_0^\tau [(x-\tau)^{k-2} + (\tau-x)^{k-2}] \left(\int_0^t |f| \right) dt$$

$$+ \int_x^\infty |f(s)| \left(\int_\tau^x (x-t)^{k-2}\, dt \right) ds + \int_\tau^x |f(s)| \left(\int_\tau^s (x-t)^{k-2}\, dt \right) ds$$

$$\ll (x-\tau)^{k-2} \int_0^\tau \left(\int_0^t |f| \right) dt + \int_0^\tau (\tau-t)^{k-2} \left(\int_0^t |f| \right) dt$$

$$+ (x-\tau)^{k-1} \int_x^\infty |f|$$

$$+ \int_\tau^x |f(s)| \, [(x-s)^{k-2}(s-\tau) + (s-\tau)^{k-1}]\, ds$$

$$\ll (x-\tau)^{k-2} \int_0^\tau (\tau - s)\, |f(s)|\, ds + \int_0^\tau (\tau-s)^{k-1} |f(s)|\, ds$$

$$+ (x-\tau)^{k-1} \int_x^\infty |f|$$

$$+ \int_\tau^x (s-\tau)\, [(x-s)^{k-2} + (s-\tau)^{k-2}]\, |f(s)|\, ds$$

$$\ll (x-\tau)^{k-2} \int_0^\tau (\tau-s)\, |f(s)|\, ds + \int_0^\tau (\tau-s)^{k-1} |f(s)|\, ds$$

$$+ (x-\tau)^{k-1} \int_x^\infty |f| + (x-\tau)^{k-2} \int_\tau^x (s-\tau)\, |f(s)|\, ds.$$

Then the following estimate takes place

$$\| (I_k f)\, u \|_q \le \| (I_k f)\, u\chi_{[0,\tau]} \|_q + \| (I_k f)\, u\chi_{[\tau,\infty)} \|_q$$

$$<< (A_{k;(0,\tau),u,v}) + \left(\int_\tau^\infty (x-\tau)^{q(k-2)} \, |u(x)|^q \, dx \right)^{1/q}$$

$$+ \left(\int_0^\tau (\tau - s)^{p'} \, |v(x)|^{-p'} \, dx \right)^{1/p'}$$

$$+ \left(\int_\tau^\infty |u(x)|^q \, dx \right)^{1/q} \left(\int_0^\tau (\tau - s)^{p'(k-1)} \, |v(x)|^{-p'} dx \right)^{1/p'} \| fv \|_p$$

$$+ \| (J_{1,(\tau,\infty)} f)(x)(x-\tau)^{k-1} u(x) \, \chi_{[\tau,\infty)} \|_q$$

$$+ \| (x-\tau)^{k-2} \left(\int_\tau^x (s-\tau) f(s) \, dx \right) u(x) \, \chi_{[\tau,\infty]} \|_q$$

$$\le (A_{k;(0,\tau),u,v} + D_{\tau,k-2,1} + D_{\tau,0,k-1}$$

$$+ A_{1;(\tau,\infty),(x-\tau)^{k-2}u(x),(x-\tau)^{-1}v(x)} + B_{1;(\tau,\infty),(x-\tau)^{k-1}u(x),v}) \, \| fv \|_p .$$

Thus, the upper bound is proved.

II. The lower bound. Similar to the proof of Theorem 2.1 we consider two cones

$$\mathscr{L}_1 = \{ f \in L_{p,v} : f \ge 0, \ \mathrm{supp}\, f \subseteq [0, \tau] \},$$

$$\mathscr{L}_2 = \{ f \in L_{p,v} : f \le 0, \ \mathrm{supp}\, f \subseteq [\tau, \infty] \}$$

and constructed there by the formulae (9) one-to-one correspondence ω_τ between them such that

$$\int_0^\infty (f + \omega_\tau f) = 0, f \in \mathscr{L}_1$$

and

$$\| fv \|_p = \| (\omega_\tau f) \, v \|_p.$$

Every function of the form $f_1 = f + \omega_\tau f$, $f \in \mathscr{L}_1$ or $f_2 = \omega_\tau^{-1} f + f$, $f \in \mathscr{L}_2$ is admissible for the inequality (26). Moreover,

$$2^{1/p} \| fv \|_p = \| f_i v \|_p, f \in \mathscr{L}_i, \ i = 1, 2.$$

Put $f_1 = f + \omega_\tau f$. Since $f_1(x) = f(x)$ for $x < \tau$, we have

$$2^{1/p} C \, \| \chi_{[0,\tau]} fv \|_p = C \, \| f_1 v \|_p \ge \| (I_k f_1) u \|_q \ge \| \chi_{[0,\tau]} (I_k f) u \|_q.$$

Hence,

$$C >> A_{k;(0,\tau),u,v} .$$

For $x > \tau$

$$I_k f_1(x) = \frac{1}{\Gamma(k-1)} \int_0^x (x-t)^{k-2} \left(\int_0^t f_1 \right) dt$$

$$= \frac{1}{\Gamma(k-1)} \int_0^\tau (x-t)^{k-2} \left(\int_0^t f_1 \right) dt$$

$$+ \frac{1}{\Gamma(k-1)} \int_\tau^x (x-t)^{k-2} \left(-\int_t^\infty f_1 \right) dt.$$

Since both the terms on the right hand side are non-negative we find

$$I_k f_1(x) \geq \frac{1}{\Gamma(k-1)} \int_0^\tau (x-t)^{k-2} \left(\int_0^t f \right) dt$$

$$\gg \int_0^\tau [(x-\tau)^{k-2} + (\tau - t)^{k-2}] \left(\int_0^t f \right) dt$$

$$= (x-\tau)^{k-2} \int_0^\tau \left(\int_0^t f \right) dt + \int_0^\tau (\tau - t)^{k-2} \left(\int_0^t f \right) dt$$

$$\gg (x-\tau)^{k-2} \int_0^\tau (\tau - s) f(s)\, ds + \int_0^\tau (\tau - s)^{k-1} f(s)\, ds.$$

Then

$$C \, \| f v \chi_{[0,\tau]} \|_p \gg \| (I_k f_1)\, u \|_q$$

$$\geq \| u(x) \chi_{[\tau,\infty)}(x) \|_q \left| \int_0^\tau (\tau - s)^{k-1} f(s)\, ds \right|$$

and by the reverse Hölder inequality we get

$$C \gg D_{\tau,0,\mathrm{k}-1}.$$

Analogously, the inequality

$$C \, \| f v \chi_{[0,\tau]} \|_p \gg \| (I_k f_1)\, u \|_q$$

$$\geq \| (x-\tau)^{k-2} u(x) \, \chi_{[\tau,\infty)} (x) \|_q \left| \int_0^\tau (\tau - s) f(s)\, ds \right|,$$

implies

$$C \gg D_{\tau,k-2,1}.$$

Now we substitute in (24) the function f_2 and find for $x > \tau$

$$I_k f_2(x) = \frac{1}{\Gamma(k-1)} \int_0^x (x-t)^{k-2} \left(\int_0^t f_2 \right) dt$$

$$= \frac{1}{\Gamma(k-1)} \int_0^\tau (x-t)^{k-2} \left(\int_0^t f_2 \right) dt$$

$$+ \frac{1}{\Gamma(k-1)} \int_\tau^x (x-t)^{k-2} \left(-\int_t^\infty f_2 \right) dt.$$

The function $f_2(x)$ is chosen so that the both last terms are non-negative. Hence, for $f \in \mathscr{L}_2$ we obtain

$$I_k f_2(x) >> \int_\tau^x (x-t)^{k-2} \left(-\int_t^\infty f_2 \right) dt$$

$$= \int_x^\infty (-f(s)) \left(\int_\tau^x (x-t)^{k-2}\, dt \right) ds + \int_\tau^x (-f(s)) \left(\int_\tau^s (x-t)^{k-2}\, dt \right) ds$$

$$>> (x-\tau)^{k-1} \int_x^\infty (-f) + \int_\tau^x (-f(s))[(x-s)^{k-2}(s-\tau) + (s-\tau)^{k-1}]\, ds$$

$$= (x-\tau)^{k-1} \int_x^\infty (-f) + \int_\tau^x (s-\tau)[(x-s)^{k-2} + (s-\tau)^{k-2}]\,(-f(s))\, ds$$

$$>> (x-\tau)^{k-1} \int_x^\infty (-f) + (x-\tau)^{k-2} \int_\tau^x (s-\tau)(-f(s))\, ds.$$

Applying Theorem 1.1 and using the first term, we find

$$C \,\| f v \chi_{[\tau,\infty)} \|_p >> \| (I_k f_2) u \|_q \geq \| u(x) \chi_{[\tau,\infty)}(x)(x-\tau)^{k-1} \int_x^\infty (-f) \|_q$$

and

$$C >> B_{1;(\tau,\infty),(x-\tau)^{k-1} u(x),v}.$$

Using the second term we obtain

$$C \,\| f v \chi_{[\tau,\infty)} \|_p >> \| (I_k f_2) u \|_q$$

$$\geq \| u(x) \chi_{[\tau,\infty)}(x)(x-\tau)^{k-2} \int_\tau^x (s-\tau)\,(-f(s))\, ds \|_q,$$

therefore

$$C >> A_{1;(\tau,\infty),(x-\tau)^{k-2} u(x),(x-\tau)^{-1} v(x)}$$

and the lower bound is proved.

Remark 2 Theorem 3.1 remains valid under restriction on an interval (a, b) with the related variation of the right hand side of (25).

Case 2. Now we characterize the inequality

$$\| Fu \|_q \le C \| F^{(k)} v \|_p \tag{27}$$

with the boundary condition

$$F(0) = F'(0) = \dots = F^{(k-1)}(0) = F(\infty) = 0$$

and arbitrary $k > 2$ on $(0, \infty)$. Passing to the equivalent form, we have

$$\| (I_k f)\, u \|_q \le C \| fv \|_p, \; (I_k f)\,(\infty) = 0.$$

Theorem 3.2 *Let $1 < p, q < \infty$ and $\tau_\lambda \in (0, \lambda)$ be defined by*

$$\int_0^{\tau_\lambda} | v |^{-p'} = \int_{\tau_\lambda}^{\lambda} | v |^{-p'}.$$

Then the least constant C in (27) is estimated by

$$C \approx A_{k;(0,\infty),u,v} + \sup_{\lambda>0} \left(A_{1;(\tau_\lambda,\lambda),(x-\tau_\lambda)^{k-2} u(x),(x-\tau_\lambda)^{-1} v(x)} + B_{1;(\tau_\lambda,\lambda),(x-\tau_\lambda)^{k-1} u(x),v} \right). \tag{28}$$

Proof Since $(I_k f)\,(\infty) = 0$, then there exists the sequence $\lambda_1, \lambda_2, \dots, \lambda_{k-1}$ such that

$$\lambda_{k-1} \in (0, \infty) \text{ and } (I_{k-1} f)\,(\lambda_{k-1}) = 0,$$

$$\lambda_{k-2} \in (0, \lambda_{k-1}) \text{ and } (I_{k-2} f)\,(\lambda_{k-2}) = 0,$$

$$\dots$$

$$\lambda_1 \in (0, \lambda_2) \text{ and } (I_1 f)\,(\lambda_1) = 0.$$

Let $x \in (\lambda_{k-1}, \infty)$. Then

$$(I_k f)\,(x) = \int_0^x (I_{k-1} f)(t)\, dt = \int_0^{\lambda_{k-1}} (I_{k-1} f)(t)\, dt + \int_{\lambda_{k-1}}^x (I_{k-1} f)\,(t)\, dt$$

$$= (I_k f)(\lambda_{k-1}) + \frac{1}{\Gamma(k-1)} \int_{\lambda_{k-1}}^x \left(\int_0^t (t-s)^{k-2} f(s)\, ds \right) dt.$$

Using the condition $\displaystyle\int_x^{\lambda_{k-1}} (\lambda_{k-1} - s)^{k-2} f(s)\, ds = 0$, we transform the integral as follows

$$\int_{\lambda_{k-1}}^x \left(\int_0^t (t-s)^{k-2} f(s)\, ds \right) dt$$

$$= (k-2)\int_{\lambda_{k-1}}^{x}\left(\int_0^t\int_0^s (s-u)^{k-3} f(u)\, duds\right) dt$$

$$= (k-2)\int_{\lambda_{k-1}}^{x}\left(\int_0^{\lambda_{k-1}}\int_0^s (s-u)^{k-3} f(u)\, duds\right.$$

$$\left. + \int_{\lambda_{k-1}}^{t}\int_0^s (s-u)^{k-3} f(u)\, duds\right) dt$$

$$= (k-2)\int_{\lambda_{k-1}}^{x}\left(\int_{\lambda_{k-1}}^{t}\int_0^s (s-u)^{k-3} f(u)\, duds\right) dt$$

$$= (k-2)\int_{\lambda_{k-1}}^{x}\left[\int_0^{\lambda_{k-1}}\left(\int_{\lambda_{k-1}}^{t} (s-u)^{k-3}\, ds\right) f(u)\, du\right.$$

$$\left. + \int_{\lambda_{k-1}}^{t}\left(\int_u^t (s-u)^{k-3}\, ds\right) f(u)\, du\right] dt$$

$$= \int_{\lambda_{k-1}}^{x}\left[\int_0^{\lambda_{k-1}}\left((t-u)^{k-2} - (\lambda_{k-1}-u)^{k-2}\right) f(u)\, du\right.$$

$$\left. + \int_{\lambda_{k-1}}^{t} (t-u)^{k-2} f(u)\, du\right] dt$$

$$= \int_{\lambda_{k-1}}^{x}\left[\int_0^{\lambda_{k-1}} (t-u)^{k-2} f(u)\, du\right] dt + \int_{\lambda_{k-1}}^{x} f(u)\int_u^x (t-u)^{k-2}\, dtdu$$

$$= \int_0^{\lambda_{k-1}} f(u)\left(\int_{\lambda_{k-1}}^{x} (t-u)^{k-2}\, dt\right) du + \frac{1}{k-1}\int_{\lambda_{k-1}}^{x} (x-u)^{k-1} f(u)\, du.$$

Thus, for $x \in (\lambda_{k-1}, \infty)$ we obtain

$$(I_k f)\,(x) = (I_k f)\,(\lambda_{k-1}) + \frac{1}{\Gamma(k)}\int_{\lambda_{k-1}}^{x} (x-u)^{k-1} f(u)\, du$$

$$+ \frac{1}{\Gamma(k)}\int_0^{\lambda_{k-1}} f(u)((x-u)^{k-1} - (\lambda_{k-1}-u)^{k-1})\, du.$$

Estimating the last term from above we obtain

$$\left| \int_0^{\lambda_{k-1}} f(u)\,((x-u)^{k-1} - (\lambda_{k-1}-u)^{k-1})\,du \right|$$

$$= \left| \int_0^{\lambda_{k-1}} f(u)\,(\sum_{j=0}^{k-1} C_{k-1}^j (x-\lambda_{k-1})^{k-1-j}(\lambda_{k-1}-u)^j - (\lambda_{k-1}-u)^{k-1})\,du \right|$$

$$\leq \sum_{j=0}^{k-3} C_{k-1}^j (x-\lambda_{k-1})^{k-1-j} \int_0^{\lambda_{k-1}} (\lambda_{k-1}-u)^j\,|f(u)|\,du,$$

where C_{k-1}^j is a Binomial coefficient. Then

$$|(I_k f)(x)| << |(I_k f)(\lambda_{k-1})| + |(I_{k,(\lambda_{k-1},\infty)} f)(x)|$$

$$+ \sum_{j=0}^{k-3} (x-\lambda_{k-1})^{k-1-j} \int_0^{\lambda_{k-1}} (\lambda_{k-1}-u)^j\,|f(s)|\,ds.$$

Consequently,

$$\|(I_k f) u \chi_{[\lambda_{k-1},\infty]}\|_q$$

$$<< \left(\left(\int_{\lambda_{k-1}}^{\infty} |u(x)|^q\,dx \right)^{1/q} \left(\int_0^{\lambda_{k-1}} (\lambda_{k-1}-s)^{p'(k-1)}\,|v(x)|^{-p'}\,dx \right)^{1/p'} \right.$$

$$+ \sum_{j=0}^{k-3} \left(\int_{\lambda_{k-1}}^{\infty} (x-\lambda_{k-1})^{q(k-1-j)}\,|u(x)|^q\,dx \right)^{1/q}$$

$$\left. \times \left(\int_0^{\lambda_{k-1}} (\lambda_{k-1}-u)^{p'j}\,|v(x)|^{-p'}\,dx \right)^{1/p'} \right) \|fv\chi_{[0,\lambda_{k-1}]}\|_p$$

$$+ A_{k;(\lambda_{k-1},\infty),u,v} \|fv\chi_{[\lambda_{k-1},\infty)}\|_p$$

$$\leq (A_{k;(\lambda_{k-1},\infty),u,v} + D_{\lambda_{k-1},0,k-1} + \sum_{j=0}^{k-3} D_{\lambda_{k-1},k-1-j,j}) \|fv\|_p$$

$$\leq (A_{k;(\lambda_{k-1},\infty),u,v} + D_{\lambda_{k-1},0,k-1} + D_{\lambda_{k-1},k-1,0}) \|fv\|_p.$$

Taking into account that $\int_0^{\lambda_i} (\lambda_i - s)^{i-1} f(s)\,ds = 0$ we find for $x \in (\lambda_i, \lambda_{i+1})$

$$(I_k f)(x) = \int_0^x I_{k-1} f(t)\,dt = \int_0^{\lambda_i} I_{k-1} f(t)\,dt + \int_{\lambda_i}^x I_{k-1} f(t)\,dt$$

$$= I_k f(\lambda_i) + \int_{\lambda_i}^{x}\int_0^t I_{k-2} f(s)\, ds dt = I_k f(\lambda_i)$$

$$+ \frac{1}{\Gamma(k-i-1)} \int_{\lambda_i}^{x}\int_0^t (t-s)^{k-i-2} I_i f(s)\, ds dt$$

$$= I_k f(\lambda_i) + \frac{1}{\Gamma(k-i-1)} \int_{\lambda_i}^{x}\int_0^{\lambda_i} (t-s)^{k-i-2} I_i f(s)\, ds\, dt$$

$$+ \frac{1}{\Gamma(k-i-1)} \int_{\lambda_i}^{x}\int_{\lambda_i}^t (t-s)^{k-i-2} I_i f(s)\, ds\, dt$$

$$= I_k f(\lambda_i) + Q_1 + Q_2.$$

Now

$$Q_1 = \frac{1}{\Gamma(k-i-1)} \int_0^{\lambda_i} I_i f(s) \int_{\lambda_i}^{x} (t-s)^{k-i-2}\ dt\, ds$$

$$= \frac{1}{\Gamma(k-i)} \int_0^{\lambda_i} I_i f(s)\, ((x-s)^{k-i-1} - (\lambda_i - s)^{k-i-1})\, ds.$$

Then for $1 < i < k-1$ we have

$$|\, Q_1 \,| \le \int_0^{\lambda_i} I_i \,|f|\, (s)\, ((x-s)^{k-i-1} - (\lambda_i - s)^{k-i-1})\, ds$$

$$= \int_0^{\lambda_i} I_i \,|f|\, (s) (\sum_{j=0}^{k-i-1} C^j_{k-i-1} (x-\lambda_i)^{k-i-1-j} (\lambda_i - s)^j - (\lambda_i - s)^{k-i-1})\, ds$$

$$= \sum_{j=0}^{k-i-2} C^j_{k-i-1} (x-\lambda_i)^{k-i-1-j} \int_0^{\lambda_i} (\lambda_i - s)^j\, I_i \,|f|\, (s)\, ds$$

$$\ll \sum_{j=0}^{k-i-2} (x-\lambda_i)^{k-i-1-j} \int_0^{\lambda_i} (\lambda_i - s)^{j+i}\, |\, f(s)\,|\, ds.$$

Similarly

$$Q_2 = \frac{1}{\Gamma(k-i-1)} \int_{\lambda_i}^{x}\int_{\lambda_i}^{t} (t-s)^{k-i-2} I_i f(s)\, ds\, dt$$

$$= \frac{1}{\Gamma(k-i)} \int_{\lambda_i}^{x} I_i f(s)\, (x-s)^{k-i-1}\, ds$$

$$= \frac{1}{\Gamma(k-i)} \int_{\lambda_i}^{x} \left[\int_{0}^{\lambda_i} I_{i-1} f(u)\, du + \int_{\lambda_i}^{s} I_{i-1} f(u)\, du \right] (x-s)^{k-i-1}\, ds$$

$$= \frac{1}{\Gamma(k-i)} \int_{\lambda_i}^{x} \left(\int_{\lambda_i}^{s} I_{i-1} f(u)\, du \right) (x-s)^{k-i-1}\, ds$$

$$= \frac{1}{\Gamma(k-i)\Gamma(i-1)} \int_{\lambda_i}^{x} \left(\int_{\lambda_i}^{s} \int_{0}^{u} (u-y)^{i-2} f(y)\, dy du \right) (x-s)^{k-i-1}\, ds$$

$$= \frac{1}{\Gamma(k-i)\Gamma(i-1)} \int_{\lambda_i}^{x} \left(\int_{0}^{\lambda_i} f(y) \int_{\lambda_i}^{s} (u-y)^{i-2}\, du\, dy \right.$$

$$\left. + \int_{\lambda_i}^{s} f(y) \int_{y}^{s} (u-y)^{i-2}\, du\, dy \right) (x-s)^{k-i-1}\, ds$$

$$= \frac{1}{\Gamma(k-i)\Gamma(i)} \int_{\lambda_i}^{x} \left(\int_{0}^{\lambda_i} f(y)[(s-y)^{i-1} - (\lambda_i - y)^{i-1}]\, dy \right) (x-s)^{k-i-1}\, ds$$

$$+ \frac{1}{\Gamma(k-i)\Gamma(i)} \int_{\lambda_i}^{x} \left(\int_{\lambda_i}^{s} f(y)(s-y)^{i-1}\, dy \right) (x-s)^{k-i-1}\, ds$$

Estimating Q_2 under the same conditions as Q_1, we get

$$|Q_2| \le \int_{\lambda_i}^{x} \left(\int_{0}^{\lambda_i} |f(y)|\,[(s-y)^{i-1} - (\lambda_i - y)^{i-1}]\, dy \right) (x-s)^{k-i-1}\, ds$$

$$+ \Gamma(k-i-1)\, I_{k,[\lambda_i,\lambda_{i+1}]} |f|(x)$$

$$<< \int_{\lambda_i}^{x} \left(\int_{0}^{\lambda_i} |f(y)|\,[\sum_{j=0}^{i-2} C_{i-1}^{j} (s-\lambda_i)^{i-1-j} (\lambda_i - y)^{j}]\, dy \right) (x-s)^{k-i-1}\, ds$$

$$+ I_{k,[\lambda_i,\lambda_{i+1}]} |f|(x)$$

$$= \sum_{j=0}^{i-2} C_{i-1}^{j} \left(\int_{0}^{\lambda_i} (\lambda_i - y)^{j} |f(y)|\, dy \right) \int_{\lambda_i}^{x} (x-s)^{k-i-1} (s-\lambda_i)^{i-1-j}\, ds$$

$$+ I_{k,[\lambda_i,\lambda_{i+1}]} |f|(x)$$

$$<< \sum_{j=0}^{i-2} (x-\lambda_i)^{k-1-j} \int_{0}^{\lambda_i} (\lambda_i - y)^{j} |f(y)|\, dy + I_{k,[\lambda_i,\lambda_{i+1}]} |f|(x).$$

Thus, if $x \in (\lambda_i, \lambda_{i+1})$, then

$$| (I_k f)(x) | << (I_k |f|)(\lambda_i) + \sum_{j=0}^{k-i-2} (x-\lambda_i)^{k-i-1-j} \int_0^{\lambda_i} (\lambda_i - s)^{j+i} |f(s)| \, ds + \sum_{j=0}^{i-2} (x-\lambda_i)^{k-1-j} \int_0^{\lambda_i} (\lambda_i - y)^j |f(y)| \, dy + (I_{k[\lambda_i,\lambda_{i+1}]} |f|)(x).$$

For all $i = 1, 2, \ldots, k-2$ we obtain

$$\begin{aligned}
&\| (I_k f) u \chi_{[\lambda_i,\lambda_{i+1}]} \|_q \\
&\le \Bigg[\left(\int_{\lambda_i}^{\lambda_{i+1}} |u(x)|^q \, dx \right)^{1/q} \left(\int_0^{\lambda_i} (\lambda_i - x)^{p'(k-1)} |v(x)|^{-p'} \, dx \right)^{1/p'} \\
&\quad + \sum_{j=0}^{k-i-2} \left(\int_{\lambda_i}^{\lambda_{i+1}} (x-\lambda_i)^{q(k-i-1-j)} |u(x)|^q \, dx \right)^{1/q} \\
&\quad \times \left(\int_0^{\lambda_i} (\lambda_i - x)^{p'(j+i)} |v(x)|^{-p'} \, dx \right)^{1/p'} \\
&\quad + \sum_{j=0}^{i-2} \left(\int_{\lambda_i}^{\lambda_{i+1}} (x-\lambda_i)^{q(k-1-j)} |u(x)|^q \, dx \right)^{1/q} \\
&\quad \times \left(\int_0^{\lambda_i} (\lambda_i - x)^{p'j} |v(x)|^{-p'} \, dx \right)^{1/p'} \Bigg] \\
&\quad \times \| f v \chi_{[0,\lambda_i]} \|_p + A_{k;(\lambda_i,\lambda_{i+1}),u,v} \| f v \chi_{[\lambda_i,\lambda_{i+1}]} \|_p \\
&\le \Big(A_{k;(\lambda_i,\lambda_{i+1}),u,v} + D_{\lambda_i,0,k-1} + \sum_{j=0}^{k-i-2} D_{\lambda_i,k-i-1-j,j+i} + \sum_{j=0}^{i-2} D_{\lambda_i,k-1-j,j} \Big) \| f v \|_p \\
&<< \left(A_{k;(\lambda_i,\lambda_{i+1}),u,v} + D_{\lambda_i,0,k-1} + D_{\lambda_i,k-1,0} \right) \| f v \|_p .
\end{aligned}$$

On the last interval $(0, \lambda_1)$ the condition $\int_0^{\lambda_1} f = 0$ is true, therefore by Theorem 3.1 and Remark 2 we write for $\tau_\lambda \in (0, \lambda_1)$

$$\begin{aligned}
| (I_k f)(x) | &\le (x-\tau_\lambda)^{k-2} \int_0^{\tau_\lambda} (\tau_\lambda - s) |f(s)| \, ds \\
&\quad + \int_0^{\tau_\lambda} (\tau_\lambda - s)^{k-1} |f(s)| \, ds - (x-\tau_\lambda)^{k-1} \int_x^{\lambda_1} |f| \\
&\quad - (x-\tau_\lambda)^{k-2} \int_{\tau_\lambda}^{x} (s-\tau_\lambda) |f(s)| \, ds.
\end{aligned}$$

For the norm on this interval we have

$$\| (I_k f)u\, \chi_{[0,\lambda_1]} \|_q \le (A_{k;(0,\tau_\lambda),u,v}$$

$$+ \left(\int_{\tau_\lambda}^{\lambda_1} (x - \tau_\lambda)^{q(k-2)} \,|\, u(x) \,|^q \, dx \right)^{1/q} \left(\int_0^{\tau_\lambda} (\tau_\lambda - x)^{p'} \,|\, v(x) \,|^{-p'} \, dx \right)^{1/p'}$$

$$+ \left(\int_{\tau_\lambda}^{\lambda_1} |\, u(x) \,|^q \, dx \right)^{1/q} \left(\int_0^{\tau_\lambda} (\tau_\lambda - x)^{p'(k-1)} \,|\, v(x) \,|^{-p'} \, dx)^{1/p'} \right) \| fv \|_p$$

$$+ \| (J_{1,(\tau_\lambda,\lambda_1)} f)(x)(x - \tau_\lambda)^{k-1} u(x)\, \chi_{[\tau_\lambda,\lambda_1)} \|_q$$

$$+ \| (x - \tau_\lambda)^{k-2} \left(\int_{\tau_\lambda}^{x} (s - \tau_\lambda) f(s)\, ds \right) u(x) \chi_{[\tau_\lambda,\lambda_1)} \|_q$$

$$\le (A_{k;(0,\tau_\lambda),u,v} + D_{\tau_\lambda,k-2,1} + D_{\tau_\lambda,0,k-1}$$

$$+ A_{1;(\tau_\lambda,\lambda_1),(x-\tau_\lambda)^{k-2}u(x),(x-\tau_\lambda)^{-1}v(x)} + B_{1;(\tau_\lambda,\lambda_1),(x-\tau_\lambda)^{k-1}u(x),v}) \| fv \|_p .$$

Combining the upper bounds on the intervals, we obtain

$$\| (I_k f)u \|_q$$

$$\le \| (I_k f)u\, \chi_{[0,\lambda_1]} \|_q + \sum_{i=1}^{k-2} \| (I_k f)u\, \chi_{[\lambda_i,\lambda_{i+1}]} \|_q + \| (I_k f)u\, \chi_{[\lambda_{k-1},\infty]} \|_q$$

$$<< (A_{k;(0,\infty),u,v} + \sup_{\lambda>0} (D_{\lambda,k-1,0} + D_{\lambda,k-1,0} + D_{\lambda,1,k-2}$$

$$+ A_{1;(\tau_\lambda,\lambda),(x-\tau_\lambda)^{k-2}u(x),(x-\tau_\lambda)^{-1}v(x)} + B_{1;(\tau_\lambda,\lambda),(x-\tau_\lambda)^{k-1}u(x),v})) \| fv \|_p$$

$$<< (A_{k;(0,\infty),u,v} + \sup_{\lambda>0} (A_{1;(\tau_\lambda,\lambda),(x-\tau_\lambda)^{k-2}u(x),(x-\tau_\lambda)^{-1}v(x)}$$

$$+ B_{1;(\tau_\lambda,\lambda),(x-\tau_\lambda)^{k-1}u(x),v})) \| fv \|_p .$$

Thus, the total upper bound for C is proved.

2. The lower bound. Let the inequality be true. Fix a number λ_1 and choose $\lambda_k \in (\lambda_1, \infty)$ from the condition

$$\int_0^{\lambda_1} (\lambda_k - x)^{p'(k-1)} \,|\, v\ (x) \,|^{-p'} \, dx = \int_{\lambda_1}^{\lambda_k} (\lambda_k - x)^{p'(k-1)} \,|\, v\ (x) \,|^{-p'} \, dx$$

and $\tau_\lambda \in (0, \lambda_1)$ from

$$\int_0^{\tau_\lambda} |\, v \,|^{-p'} = \int_{\tau_\lambda}^{\lambda_1} |\, v \,|^{-p'} .$$

Let $f \in L_{p,v}$ be any function such that $f \geq 0$ and $\operatorname{supp} f \subseteq [0, \tau_\lambda]$. Defining the function $\rho_1 : [0, \tau_\lambda] \to [\tau_\lambda, \lambda_1]$ from the relation

$$\int_0^s |v|^{-p'} = \int_{\rho_1(s)}^{\lambda_1} |v|^{-p'}, s \in [0, \tau_\lambda],$$

we construct extension f^* of f with the properties $f^* \leq 0$, $\operatorname{supp} f^* \subseteq [\tau_\lambda, \lambda_1]$,

$$\int_0^{\tau_\lambda} f(x)\,dx + \int_{\tau_\lambda}^{\lambda_1} f^*(x)\,dx = 0$$

and with the equality of the norms

$$\| fv \|_p = \| f^*v \|_p.$$

The extension is

$$f^*(x) = \begin{cases} -\dfrac{f(\rho_1^{-1}(x))\,|\,v(\rho_1^{-1}(x))\,|^{p'}}{|\,v(x)\,|^{p'}}, x \in (\tau_\lambda, \lambda_1), \\ 0, x \notin (\tau_\lambda, \lambda_1). \end{cases}$$

Now, for the function $f_1 = f + f^*$ we again construct the extension f_1^*, satisfying supp $f_1^* \subseteq [\lambda_1, \lambda_k]$,

$$\int_0^{\lambda_1} (\lambda_k - x)^{k-1} f_1(x)\,dx + \int_{\lambda_1}^{\lambda_k} (\lambda_k - x)^{k-1} f_1^*(x)\,dx = 0$$

and

$$\| f_1^* v \|_p = \| f_1 v \|_p.$$

To this end we define the function $\rho_2 : [0, \lambda_1] \to [\lambda_1, \lambda_k]$ by

$$\int_0^s (\lambda_k - x)^{p'(k-1)} |\,v\,|^{-p'}(x)\,dx$$

$$= \int_{\rho_2(s)}^{\lambda_k} (\lambda_k - x)^{p'(k-1)} |\,v\,|^{-p'}(x)\,dx,\ s \in [0, \lambda_1]$$

and obtain

$$f_1^*(x) = \begin{cases} \dfrac{(\lambda_k - x)^{(p'-1)(k-1)} f_1(\rho_2^{-1}(x))\,|\,v(\rho_2^{-1}(x))\,|^{p'}}{(\lambda_k - \rho_2^{-1}(x))^{(p'-1)(k-1)}|\,v(x)\,|^{p'}}, x \in (\lambda_1, \lambda_k), \\ 0, x \notin (\lambda_1, \lambda_k). \end{cases}$$

The test function $f_2 = f_1 + f_1^*$ is admissible for (27) and

$$\| f_2 v \|_p = 2^{1/p} \| f_1 v \|_p.$$

Since $I_k f_2(\infty) = I_k f_2(\lambda_k) = 0$ and $\int_0^{\lambda_1} f_2 = 0$, then for the test function on the strength of definition of f_1 on the interval $[0, \lambda_1]$ we obtain

$$2^{1/p} C \| f_1 v \chi_{[0,\lambda_1]} \|_p = C \| f_2 v \|_p \geq \| (I_k f_2) u \|_q \geq \| \chi_{[0,\lambda_1]} (I_k f_1) u \|_q$$

and by Theorem 3.1 and Remark 2 we find

$$C >> \sup_{\lambda_1 > 0} (A_{1;(\tau_\lambda,\lambda_1),(x-\tau_\lambda)^{k-2} u(x),(x-\tau_\lambda)^{-1} v(x)} + B_{1;(\tau_\lambda,\lambda_1),(x-\tau_\lambda)^{k-1} u(x),v}).$$

Now, for any λ_k we choose $\lambda_{i+1} \in (0, \lambda_k)$ such, that

$$\int_0^{\lambda_{i+1}} (\lambda_k - x)^{p'(k-1)} | v(x) |^{-p'} dx = \int_{\lambda_{i+1}}^{\lambda_k} (\lambda_k - x)^{p'(k-1)} | v(x) |^{-p'} dx.$$

Let λ_i be an arbitrary number from $(0, \lambda_{i+1})$ and $\mu \in (\lambda_i, \lambda_{i+1})$ is determined from the equality

$$\int_{\lambda_i}^{\mu} (\lambda_{i+1} - x)^{p'i} | v(x) |^{-p'} dx = \int_{\mu}^{\lambda_{i+1}} (\lambda_{i+1} - x)^{p'i} | v(x) |^{-p'} dx.$$

Let $f \in L_{p,v}$, $\operatorname{supp} f \subseteq [\lambda_i, \mu]$, $f(x) \geq 0$. We define the extension $f^{**}(x)$ of $f(x)$ on $[\mu, \lambda_{i+1}]$ so, that the following conditions

$$\| \chi_{[\lambda_i,\mu]} f v \|_p = \| \chi_{[\mu,\lambda_{i+1}]} f^{**} v \|_p$$

and

$$\int_{\lambda_i}^{\mu} (\lambda_{i+1} - x)^i f(x)\, dx + \int_{\mu}^{\lambda_{i+1}} (\lambda_{i+1} - x)^i f^{**}(x)\, dx = 0$$

are fulfilled. Taking $f_3 = f + f^{**}$ we construct the extension f_3^{**} satisfying $\operatorname{supp} f_3^{**} \subseteq [\lambda_{i+1}, \lambda_k]$ and

$$\int_0^{\lambda_{i+1}} (\lambda_k - x)^{k-1} f_3(x)\, dx + \int_{\lambda_{i+1}}^{\lambda_k} (\lambda_k - x)^{k-1} f_3^{**}(x)\, dx = 0,$$

$$\| \chi_{[\lambda_i,\lambda_{i+1}]} f_3 v \|_p = \| \chi_{[\lambda_{i+1},\lambda_k]} f_3^{**} v \|_p.$$

Then the function $f_4 = f_3 + f_3^{**}$ obeys $I_k f_4(\infty) = I_k f_4(\lambda_k) = 0$, $I_{i+1} f_4(\lambda_{i+1}) = 0$, $I_i f_4(\lambda_i) = 0$ and

$$\| f_4 v \|_p = 4^{1/p} \| fv \|_p.$$

Consequently, for the test function f_4 on $[\lambda_i, \mu]$ we get

$$4^{1/p} C \| fv\chi_{[\lambda_i,\mu]} \|_p = C \| f_4 v \|_p \geq \| (I_k f_4)u \|_q \geq \| \chi_{[\lambda_i,\mu]}(I_k f)u \|_q$$

and Theorem 1.1 yields

$$C >> A_{k;(\lambda_i,\mu),u,v}.$$

Since $\lambda_k \to \infty$ implies $\mu \to \infty$, then letting $\lambda_k \to \infty$, and then $\lambda_i \to 0$ we find

$$C >> A_{k;(0,\infty),u,v}.$$

Thus, the lower bound for the constant C is proved.

Case 3. Now we study the inequality

$$\| Fu \|_q \leq C \| F^{(k)} v \|_p$$

under the boundary conditions

$$F(0) = F'(0) = \ldots = F^{(k-1)}(0) = F(\infty) = \ldots = F^{(k-1)}(\infty) = 0$$

and arbitrary $k > 2$ on the interval $(0, \infty)$. The equivalent form is the characterization problem for

$$\| (I_k f)\, u \|_q \leq C \| fv \|_p$$

with

$$(I_k f)\,(\infty) = (I_{k-1} f)\,(\infty) = \ldots = (I_1 f)\,(\infty) = 0. \tag{29}$$

Theorem 3.3 *Let the hypothesis of Theorem* 3.2 *the estimate of the least constant C in* (29) *is given by* (28).

Proof The upper bound follows from Theorem 3.2.
Now suppose that the inequality (29) is true. Then it is also true for the weight $v_\varepsilon(x) = \varepsilon x^{k+1} + | v(x) |$ instead of $v(x)$. Then let us show that for any function $F(x)$ the conditions $F(\infty) = 0$ and $F^{(k)} \in L_{p,v_\varepsilon}$ imply $F'(\infty) = \ldots = F^{(k-1)}(\infty) = 0$. We prove it inductively.

Step 1. By the Hölder inequality we write

$$\left| \int_x^\infty F^{(k)} \right| \leq \left(\int_x^\infty | F^{(k)}(t) v_\varepsilon(t) |^p \, dt \right)^{1/p} \left(\int_x^\infty | v_\varepsilon(t) |^{-p'} \right)^{1/p'} < \infty.$$

Hence, $\int_x^\infty F^{(k)} \to 0$ when $x \to \infty$. Therefore for any $x, y \to \infty$ we have

$$| F^{(k-1)}(y) - F^{(k-1)}(x) | = \left| \int_x^y F^{(k)} \right| \to 0.$$

Consequently, there exists a finite limit A_{k-1} of $F^{(k-1)}(x)$ at infinity.

If $A_{k-1} \neq 0$, then the function behaves as $y = A_{k-1}x^{k-1}$ for $x \to \infty$ and can not tend to zero, which contradicts with $F(\infty) = 0$. Thus, $A_{k-1} = 0$.

Step 2. Let $F^{(n)}(\infty) = \ldots = F^{(k-1)}(\infty) = 0$ for $1 < n \leq k-1$. Then

$$F^{(n)}(t) = \frac{1}{\Gamma(k-n)} \int_t^{\infty} F^{(k)}(s)(s-t)^{k-n-1}\, ds;$$

$$\int_x^{\infty} F^{(n)} = \frac{1}{\Gamma(k-n)} \left(\int_x^{\infty} \int_t^{\infty} F^{(k)}(s)(s-t)^{k-n-1}\, ds \right) dt$$

$$= \frac{1}{\Gamma(k-n+1)} \int_x^{\infty} F^{(k)}(s)(s-x)^{k-n}\, ds;$$

$$\left| \int_x^{\infty} F^{(n)} \right| \leq \left| \int_x^{\infty} F^{(k)}(s)(s-x)^{k-n}\, ds \right|$$

$$\leq \left(\int_x^{\infty} | F^{(k)}(s)\, v_\varepsilon(s) |^p\, ds \right)^{1/p} \left(\int_x^{\infty} (s-x)^{p'(k-n)}\, | v_\varepsilon(s) |^{-p'}\, ds \right)^{1/p'}$$

$$< \infty.$$

Hence, $\int_x^{\infty} F^{(n)} \to 0$ when $x \to \infty$. Therefore, similar to the previous case, for any x, $y \to \infty$ we have

$$| F^{(n-1)}(y) - F^{(n-1)}(x) | = \left| \int_x^{y} F^{(n)} \right| \to 0$$

and a finite limit A_{n-1} of $F^{(n-1)}(x)$ at infinity exists. Clearly $A_{n-1} = 0$ otherwise the function asymptotically coincides with $y = A_{n-1}x^{n-1}$ which contradicts with $F(\infty) = 0$. Thus,

$$\{F : \| F^{(k)} v_\varepsilon \|_p < \infty,$$
$$F(0) = \ldots = F^{(k-1)}(0) = F(\infty) = \ldots = F^{(k-1)}(\infty) = 0\}$$
$$= \{F : \| F^{(k)} v_\varepsilon \|_p < \infty,\ F(0) = \ldots = F^{(k-1)}(0) = F(\infty) = 0\}.$$

Consequently, by Theorem 3.2, we conclude

$$C >> A_{k;(0,\infty),u,v_\varepsilon}$$
$$+ \sup_{\lambda > 0} \left(A_{1;(\tau_\lambda,\lambda),(x-\tau_\lambda)^{k-2}u(x),(x-\tau_\lambda)^{-1} v_\varepsilon(x)} + B_{1;(\tau_\lambda,\lambda),(x-\tau_\lambda)^{k-1}u(x),v_\varepsilon} \right).$$

Letting $\varepsilon \to 0$ the required lower bound follows by Fatou theorem.

References

1. P. Drábek and A. Kufner, The Hardy inequalities and Birkhoff interpolation, *Bayreuther Math. Schriften* **47** (1994), 99–104.
2. Yu.V. Egorov and V.A. Kondratiev, The estimates of the first eigenvalue in some Sturm-Liouville problems, *Uspekhy Mat. Nauk* **51** (1996), No. 3(309), 73–144. (Russian)
3. P. Gurka, Generalized Hardy inequalities for functions vanishing on both ends of the interval, 1987, Preprint.
4. H.P. Heing and A. Kufner, Hardy's inequality for higher order derivatives, *Proc. Steklov Inst. Math.* **192**, Issue 3(1993).
5. A. Kufner, Higher order Hardy inequalities, *Bayreuther Math. Schriften* **44** (1993), 105–146.
6. A. Kufner and H. Leinfelder, On overdetermined Hardy inequalities, 1998, Preprint.
7. A. Kufner and C.G. Simader, Hardy ineqalities for overdetermined classes of functions, *J. for Anal. and its Appl.* **16** (1997), No. 2, 387–403.
8. A. Kufner and G. Sinnamon, Overdetermined Hardy inequalities, *J. Math. Anal. Appl.* **213** (1997), 468–486.
9. A. Kufner and A. Wannebo, Some remarks on the Hardy inequality for higher order derivatives, General Inequalities VI, Internat. Series of Numer. Math. Vol. 103, Birkhäuser Verlag, Basel, 1992, 33–48.
10. M. Nasyrova and V.D. Stepanov, On weighted Hardy inequalities on semiaxis for functions vanishing at the endpoints, *J. of Inequal. & Appl.* **1** (1997), 223–238.
11. B. Opic and A. Kufner, Hardy-type Inequalities, Pitman Research Notes in Math., Series 219, Longman Sci&Tech., Harlow, 1990.
12. G. Sinnamon, Kufner's conjecture for higher order Hardy inequalities, *Real Analysis Exchange* **21** (1995/96), 590–603.
13. V.D. Stepanov, Weighted inequalities for a class of Volterra convolution operators, *J. London Math. Soc.* (2) **45** (1992), 232–242.
14. V.D. Stepanov, Weighted norm inequalities of Hardy type for a class of integral operators, *J. London Math. Soc.* (2) **50** (1994), 105–120.

Function Spaces and Applications
D.E. Edmunds et al (Eds)

16. Hankel Convolution on Some Ultra-Differentiable Function Spaces

R.S. Pathak and K.K. Shrestha

Department of Mathematics, Banaras Hindu University, Varanasi - 221 005, India

1. Introduction

Zemanian [7] introduced the function space H_μ consisting of all complex valued infinitely differentiable functions ϕ defined on $I = (0, \infty)$ satisfying

$$\gamma^{\mu}_{m,k}(\phi) = \sup_{x\in I}\left| x^m \left(x^{-1}\frac{d}{dx}\right)^k (x^{-\mu-1/2}\,\phi(x))\right| < \infty, \ \forall\, m, k \in \mathbf{N}_0 \quad (1.1)$$

for extending the Hankel transformation

$$(h_\mu\phi)(y) = \int_0^\infty \phi(x)\sqrt{xy}\,J_\mu(xy)dx,\ \mu \geq -\frac{1}{2}, \quad (1.2)$$

to generalized functions belonging to H'_μ, the dual of H_μ. Following the techniques of Gel'fand and Shilov [2], spaces $H_{\mu,\,\alpha,\,A}$, $H_\mu^{\beta,B}$ and $H^{\beta,B}_{\mu,\alpha,A}$ were defined by Lee [3]. Pathak and Pandey [4] introduced spaces $H_{\mu,a_k,A}$, $H_{\mu,}^{b_q,B}$ and $H^{b_q,B}_{\mu,a_k,A}$ of ultradifferentiable functions generalizing the aforesaid Lee spaces and proved that the conventional Hankel transformation h_μ is a continuous linear mapping from each of these ultradifferentiable function spaces into certain other spaces of the same type. The purpose of our present work is to study the Hankel translation and the Hankel convolution transformation on these ultradifferentiable function spaces and ultradistribution spaces. Some other generalizations are also obtained.

The following definitions and results will be needed in the sequel. The space L^1_μ ($\mu \geq -1/2$) is the set of all measurable functions ϕ on $I = (0, \infty)$ such that

$$\|\phi\|_\mu = \int_0^\infty |\phi(x)|\, x^{\mu+1/2}\,dx < \infty.$$

The Hankel translation of $\phi \in L^1_\mu(I)$ is defined by

$$(\tau_x\phi)(y) = \int_0^\infty \phi(z)D_\mu\ (x, y, z)\, dz,\ (x, y \in I) \tag{1.3}$$

where
$$D_\mu(x, y, z) = \int_0^\infty t^{-\mu-1/2}\, j_\mu\,(xt)\, j_\mu\,(yt)\, j_\mu\,(zt)\, dt \tag{1.4}$$

and
$$j_\mu(xt) = \sqrt{xt}\, J_\mu(xt). \tag{1.5}$$

The Hankel convolution transform of two functions ϕ and ψ belonging to $L^1_\mu(I)$ is defined by

$$(\phi \# \psi)(x) = \int_0^\infty \phi(y)\ (\tau_x\ \psi)\ (y)\ dy \ \ (a.e.\ x \in I) \tag{1.6}$$

We shall also make use of the following results [1, p. 285].

$$h_\mu(\tau_x\phi)(t) = t^{-\mu-1/2}\, j_\mu(tx)(h_\mu\phi)\ (t)\ (x, t \in I) \tag{1.7}$$

and
$$h_\mu(\phi \# \psi)(t) = t^{-\mu-1/2}\ (h_\mu\phi)(t)(h_\mu\ \psi)(t), \quad (t \in I) \tag{1.8}$$

The following formulae are given in [7, pp. 129, 134] and [5, pp. 242, 240]

$$\left(x^{-1}\frac{d}{dx}\right)^k (x^{-\mu-1/2}\,\theta\phi) \tag{1.9}$$

$$= \sum_{v=0}^{k} \binom{k}{v} \left(x^{-1}\frac{d}{dx}\right)^v \theta \left(x^{-1}\frac{d}{dx}\right)^{k-v} (x^{-\mu-1/2}\phi)$$

$$\left(x^{-1}\frac{d}{dx}\right)^k (x^{-\mu}\, J_\mu\,(x)) = (-1)^k\, x^{-(\mu+k)}\, J_{\mu+k}\,(x) \tag{1.10}$$

$$\left(x^{-1}\frac{d}{dx}\right)^k (x^{\mu}\, J_\mu\,(x)) = x^{\mu-k}\, J_{\mu-k}\,(x). \tag{1.11}$$

2. The Spaces $H_{\mu,a_K,A}$, $H_\mu^{b_q,B}$ and $H_{\mu,a_k,A}^{b_q,B}$

This section recalls the definitions of function spaces due to Pathak and Pandey [4], see also [5].

Let $\{a_k\}_{k\in N_0}$ and $\{b_q\}_{q\in N_0}$ be arbitrary sequences of positive numbers. We shall impose some of the following constraints on these sequences.

$$a_k^2 \le a_{k-1}\, a_{k+1}\,,\ \forall\ k \ge 1 \tag{2.1}$$

$$b_q^2 \le b_{q-1}\, b_{q+1}\,,\ \forall\ q \ge 1 \tag{2.2}$$

Immediate consequences of these inequalities are

$$a_p\, a_k \le a_o\, a_{p+k}, \qquad p, k = 0, 1, 2, \dots \tag{2.3}$$

$$b_p\, b_q \le b_o\, b_{p+q}, \qquad p, q = 0, 1, 2, \dots \tag{2.4}$$

Non-quasi-analyticity

$$\sum_{q=1}^{\infty} \frac{a_{q-1}}{a_q} < \infty; \tag{2.5}$$

Stability Under Multiplication by *x*

There are constants c, c_1, h, h_1 such that $\forall\ k \ge 0$, $\forall\ q \ge 0$,

$$a_{k+1} \le ch^k\, a_k \tag{2.6}$$

$$b_{q+1} \le ch_1^q\, b_q; \tag{2.7}$$

Stability Under Hankel Transformations

Conditions (2.6) and (2.7) may be replaced by the following stronger conditions whenever necessary

$$a_{r+k} \le LR^{r+k}\, a_r\, a_k,\ \forall\ r, k \ge 0; \tag{2.8}$$

$$b_{r+q} \le L_1 R_1^{r+q}\, b_r\, b_q,\ \forall\ r, q \ge 0; \tag{2.9}$$

where L, R, L_1 and R_1 are positive constants.

Let ϕ be infinitely differentiable function on I. Then $\phi \in H_{\mu,a_k,A}$ if and only if

$$|\, x^k (x^{-1}D)^q\, x^{-\mu-1/2}\, \phi(x)\,| \le C_q^\mu (A + \delta)^k\, a_k,\ \forall\ k, q \in \mathbf{N}_0 \tag{2.10}$$

where the constants A, C_q^μ depend on ϕ and $\delta > 0$ is an arbitrary constant. The space $H_\mu^{b_q,B}$ is defined to be the set of all C^∞-functions ϕ on I such that

$$|\, x^k (x^{-1}D)^q\, x^{-\mu-1/2} \phi(x)\,| \le C_k^\mu (B + \rho)^q b_q,\ \forall\ k, q \in \mathbf{N}_0 \tag{2.11}$$

where B, C_k^μ depend on ϕ and $\rho > 0$ is an arbitrary constant.
$\hat{H}_\mu^{b_q,B}$ is defined to be the space of all ϕ in $H_\mu^{b_q,B}$ satisfying the condition

$$\sup_k C_{k+2q}^\mu = C_q^{*\mu} \tag{2.12}$$

where C_q^* are constants restraining the ϕ's in $H_\mu^{b_q,B}$.

Next $\phi \in H_{\mu,a_k,A}^{b_q,B}$ if and only if

$$|\, x^k (x^{-1}D)^q\, x^{-\mu-1/2}\, \phi(x)\,| \le C^\mu\, (A + \delta)^k\, (B + \rho)^q a_k b_q,\quad k, q \in \mathbf{N}_0 \tag{2.13}$$

where C^μ, A and B certain positive constants and δ and ρ are as above.

3. Hankel Convolution on Spaces $H_{\mu,a_k,A}$, $H_{\mu}^{b_q,B}$ and $H_{\mu,a_k,A}^{b_q,B}$

This section investigates the Hankel translation τ and the Hankel convolution transform in the spaces $H_{\mu,a_k,A}$, $H_{\mu}^{b_q,B}$ and $H_{\mu,a_k,A}^{b_q,B}$.

Theorem 3.1 For each fixed x, $0 < x < x_0$ and $\mu \geq -1/2$, the mapping $\phi \to \tau_x\, \phi$ is linear and continuous from

$$\hat{H}_{\mu}^{b_q,B} \text{ into } H_{\mu}^{b_q^2,B_2}, \text{ where } B_2 = B^2(R^*)^6,\ R^* = \max\,(1, R). \tag{a}$$

Moreover if a_k satisfies (2.1) and (2.8) then it is linear and continuous from

$$H_{\mu,a_k,A,} \text{ into } \hat{H}_{\mu,a_k^2,A_2,} \tag{b}$$

where $A_2 = R^2[B_1 + (x_0 a_0/a_1)^2 + \delta]$, $B_1 = A^2(R^*)^6$, and from

$$\hat{H}_{\mu,a_k,A}^{b_q,B} \text{ into } H_{\mu,a_k^3 b_k,A_3}^{a_q^2 b_q^2,B_4}, \tag{c}$$

where $A_3 = A_1 B_3\,(R^{\otimes})^2$, $B_3 = R^2[B_1 + (x_0 a_0/a_1)^2 + \rho]$, $A_1 = AB(R^*)^2$, $B_4 = A_1^2\,(R^{\otimes})^6$, $R^{\otimes} = \max\,(1, RR_1)$ and R, R_1 are determined by (2.8) and (2.9).

Proof. Here we prove (c). The other two parts can be proved in a similar way.

Let $\phi \in H_{\mu,a_k,A}^{b_q,B}$. Then applying Leibnitz type formula (1.9) and using (1.7), (2.8) and (2.1) we obtain

$$\gamma_{k,q}^{\mu}[(h_\mu\, \tau_x \phi)\,(t)] = \sup_t \mid t^k\,(t^{-1}D)^q\, t^{-\mu-1/2}\, j_\mu\,(tx)\,(h_\mu \phi)(t) \mid$$

$$\leq \sum_{r=0}^{q} \binom{q}{r} \sup_t \mid t^k (t^{-1}D)^{q-r}\, t^{-\mu-1/2}\,(h_\mu \phi)(t) \mid$$

$$\times \sup_t \mid \sqrt{x}\,(t^{-1}\,D)^r\, t^{-\mu}\, J_\mu\,(tx) \mid$$

$$\leq \sum_{r=0}^{q} \binom{q}{r} C(A_1 + \delta)^k\, a_k b_k\,(B_1 + \rho)^{q-r}\, a_{q-r}^2\, x^{\mu+2r+1/2}$$

$$\times \sup_{t,x} \mid (tx)^{-(\mu+r)}\, J_{\mu+r}\,(tx) \mid$$

$$\leq \sum_{r=0}^{q} \binom{q}{r} C(A_1 + \delta)^k\, a_k b_k\,(B_1 + \rho)^{q-r}\, a_{q-r}^2\, x^{\mu+2r+1/2}\, A_{\mu,r}$$

$$\leq A_\mu C x^{\mu+1/2} \sum_{r=0}^{q} \binom{q}{r} x^{2r}\,(A_1 + \delta)^k\, a_k b_k\,(B_1 + \rho)^{q-r}\, a_0\, a_{2q-2r}$$

$$(A_\mu = \max_{0 \leq r \leq q} A_{\mu,r})$$

$$\leq A_\mu \, Ca_0 \, x^{\mu+1/2} \sum_{r=0}^{q} \binom{q}{r} x^{2r} (A_1 + \delta)^k a_k b_k \, (B_1 + \rho)^{q-r} \left(\frac{a_0}{a_1}\right)^{2r} a_{2q}$$

$$\leq A_\mu \, Ca_0 \, x^{\mu+1/2} \sum_{r=0}^{q} \binom{q}{r} x^{2r} (A_1 + \delta)^k a_k b_k \, (B_1 + \rho)^{q-r}$$

$$\times \left(\frac{a_0}{a_1}\right)^{2r} LR^{2q} \, a_q^2$$

$$\leq A_\mu \, Ca_o L x_o^{\mu+1/2} \, R^{2q} \left[B_1 + \left(\frac{a_o x_o}{a_1}\right)^2 + \rho \right]^q a_q^2 (A_1 + \delta)^k a_k b_k$$

$$\leq C'(B_3 + \rho')^q \, a_q^2 (A_1 + \delta)^k \, a_k b_k$$

where $B_3 = R^2 \left[B_1 + \left(\frac{a_o x_o}{a_1}\right)^2 + \rho \right]$, $C' = A_\mu C a_0 \, L x_0^{\mu+1/2}$;

so that $$(h_\mu (\tau_x \phi))(t) \in H_{\mu, a_k, b_k \cdot A_1}^{a_q^2, B_3}$$

Again, let $(h_\mu (\tau_x \phi))(t) = \Phi_x(t)$. Then

$$\tau_x \phi = h_\mu^{-1} \, \Phi_x(t) = h_\mu \, \Phi_x(t).$$

Now, by integration by parts, we have

$$| z^k (z^{-1} D)^q \, z^{-\mu-1/2} \, (h_\mu \, \Phi_x(t)(z) \, |$$

$$= \left| \int_0^\infty t^{2\mu+k+2q+1} \, \{(t^{-1} D)^k \, t^{-\mu-1/2} \, \Phi_x(t)\} \, (tz)^{-(\mu+q)} \, J_{\mu+q+k}(tz) \, dt \right|.$$

Assume that v is a positive integer such that $v \geq 2\mu + 1$. Set $n = v + k + 2q$, and use the fact that $| z^{-(\mu+q)} J_{\mu+k+q}(z) | \leq B_\mu$, where B_μ is independent of k and q [5, pp. 309–310]. Then, for $\Phi_x(t) \in H_{\mu, a_k, b_k, A_1}^{a_q^2, B_3}$ using (2.8) and (2.9) we obtain

$$\gamma_{k,q}^\mu (\tau_x \phi) \leq \int_0^1 t^{2\mu+k+2q+1} \, | (t^{-1} D)^k \, t^{-\mu-1/2} \, \Phi_x(t) \, | \, \operatorname{Sup}_z | (tz)^{-(\mu+q)}$$

$$\times J_{\mu+q+k}(tz) \, | \, dt + \int_1^\infty t^{2\mu+k+2q+3} \, | (t^{-1} D)^k \, t^{-\mu-1/2} \, \Phi_x(t) \, |$$

$$\times \sup_z | (tz)^{-(\mu+q)} \, J_{\mu+q+k}(tz) \, | \, t^{-2} \, dt \leq B_\mu C^\mu (A_1 + \delta)^{k+2q} \, a_{k+2q} \, b_{k+2q}$$

$$\times (B_3 + \rho)^k \, a_k^2 + B_\mu C^\mu \, (A_1 + \delta)^{n+2} \, a_{n+2} \, b_{n+2} \, (B_3 + \rho)^k \, a_k^2$$

$$\leq B_\mu C^\mu (A_1 + \delta)^{k+2q} a_{k+2q} b_{k+2q} (B_3 + \rho)^k a_k^2 [1 + (A_1 + \delta)^{\upsilon+2}$$

$$\times LR^{n+2} a_{\upsilon+2} R_1^{n+2} \cdot L_1 b_{\upsilon+2}] \leq B_\mu C^\mu (A_1 + \delta)^{k+2q} LR^{k+2q} a_k a_{2q}$$

$$\times L_1 R_1^{k+2q} b_k b_{2q} (B_3 + \rho)^k a_k^2 [1 + LL_1 (RR_1)^{k+2q+\upsilon+2} (A_1 + \delta)^{\upsilon+2}$$

$$\times a_{\upsilon+2} b_{\upsilon+2}] \leq B_\mu C^\mu (LL_1)^2 [RR_1(A_1 + \delta)(B_3 + \rho)]^k$$

$$\times a_k^3 b_k [(RR_1)^4 (A_1 + \delta)^2]^q a_q^2 b_q^2 [1 + (RR_1)^{k+2q} LL_1 (RR_1)^{\upsilon+2}$$

$$\times (A_1 + \delta)^{\upsilon+2} a_{\upsilon+2} b_{\upsilon+2}].$$

Let $R^{\otimes} = \max(1, RR_1)$. Then

$$\gamma_{k,q}^\mu(\tau_x\phi) \leq C_1^\mu [(R^{\otimes})^2 (A_1 + \delta)(B_3 + \rho)]^k a_k^3 b_k [(R^{\otimes})^6$$

$$\times (A_1 + \delta)^2]^q a_q^2 b_q^2 \leq C_1^\mu (A_3 + \delta')^k a_k^3 b_k (B_4 + \rho')^q a_q^2 b_q^2 \qquad (3.1)$$

where $\qquad A_3 = (R^{\otimes})^2 A_1 B_3,\ B_4 = (R^{\otimes})^6 A_1^2,$

so that

$$\tau_x \phi \in H_{\mu, a_k^3 b_k . A_3}^{a_q^2 b_q^2, B_4}$$

From (3.1) it follows that τ_x maps a bounded set in $H_{\mu,a_k,A}^{b_q.B}$ into a bounded set in $H_{\mu,a_k^3 b_k . A_3}^{a_q^2 b_q^2, B_4}$ and hence is continuous. Linearity is obvious.

Theorem 3.2 For $\mu \geq -1/2$, the mapping $(\phi, \psi) \to \phi \# \psi$ from

$\hat{H}_\mu^{b_q, B} \times \hat{H}_\mu^{b_q, B}$ into $H_\mu^{b_q^2 . B_2}$, where $B_2 = B^2 (R^*)^6$, $R^* = \max(1, \mathrm{R})$

is linear and continuous.

Moreover, if a_k satisfies (2.1) and (2.8) then the convolution mappings from

(b) $H_{\mu,a_k,A} \times H_{\mu,a_k,A}$ into $\hat{H}_{\mu,a_k^2,B_1}$, where $B_1 = A^2 (R^*)^6$

and

(c) $H_{\mu,a_k,A}^{b_q,B} \times H_{\mu,a_k,A}^{b_q,B}$ into $H_{\mu,a_k^3 b_k, A_4}^{a_q^2, b_q^2, B_5}$

where $A_4 = (R^{\otimes})^2 A_1 B_1$, $B_5 = (R^{\otimes})^6 A_1^2$, $R^{\otimes} = \max(1, RR_1)$, are linear and continuous.

Proof Here also we prove (c). The other parts can be proved similarly.

Let $\phi, \psi \in H_{\mu,a_k,A}^{b_q,B}$. Then by [5, p. 294] $h_\mu \phi$, $h_\mu \psi$ all belong to $H_{\mu,a_k b_k, A_1}^{a_q^2, B_1}$, where $A_1 = AB\,(R^*)^2$, $B_1 = A^2(R^*)^6$, $R^* = \max\,(1, R)$. Using 1.8 and applying Leibnitz type formula (1.9) we obtain

$$\gamma^{\mu}_{k,q}[h_\mu(\phi\#\,\psi)(t)] = \sup_t \mid t^k(t^{-1}D)^q\; t^{-\mu-1/2}\; t^{-\mu-1/2}(h_\mu\,\phi)(t)\;(h_\mu\,\psi)(t)\mid$$

$$\leq \sum_{r=0}^{q}\binom{q}{r}\sup_t \mid t^k\,(t^{-1}D)^{q-r}\; t^{-\mu-1/2}\,(h_\mu\phi)(t)$$

$$\times \mid \sup_t \mid (t^{-1}D)^r\; t^{-\mu-1/2}\,(h_\mu\psi)(t)\mid$$

$$\leq \sum_{r=0}^{q}\binom{q}{r}\; C^\mu\;(A_1+\delta)^k\; a_k b_k\,(B_1+\rho)^{q-r}\; a^2_{q-r}$$

$$\times\; C^\mu a_0 b_0\,(B_1+\rho)^r\; a^2_r$$

$$\leq C^\mu_1\,(A_1+\delta)^k\; a_k b_k\,(B_1+\rho)^q\; a^2_q\,,$$

so that

$$h_\mu\,(\phi\;\#\;\psi)\in H^{a_q^2,B_1}_{\mu,a_k b_k,A_1}.$$

Now, following the method of the proof of Theorem 3.1 it can be shown that $\phi\;\#\;\psi\in H^{a_q^2 b_q^2,B_5}_{\mu,a_k^3 b_k,A_4}$, where $A_4 = A_1\,B_1\,(R^{\otimes})^2$, $B_5 = A_1^2\,(R^{\otimes})^6$, $R^{\otimes} = \max\,(1, R_1)$, and the Hankel convolution is a continuous and linear operator from

$$H^{b_q,B}_{\mu,a_k,A}\times H^{b_q,B}_{\mu,a_k,A}\text{ into }H^{a_q^2 b_q^2,B_5}_{\mu,a_k^3 b_k.A_4}\,.$$

4. The Generalized Hankel Convolution Operator

Definition 4.1 The generalized Hankel translation operator τ'_x is defined as the adjoint of τ_x through the relation

$$\langle \tau'_x\,f,\,\phi\rangle = \langle f,\,\tau_x\,\phi\rangle,$$

where ϕ belongs to $H_{\mu,a_k,A}$, $\hat{H}^{b_q,B}_{\mu}$ or $H^{b_q,B}_{\mu,a_k,A}$ and f belongs to $(\hat{H}_{\mu,a_k^2,A_2})'$, $(H^{b_q^2,B_2}_{\mu})'$ or $(H^{a_q^2 b_q^2,B_4}_{\mu,a_k^3 b_k,A_3})'$ respectively. Here A, B, A_2, A_3, B_4 are the same constants as used in Theorems 3,1 and 3.2.

Theorem 4.2 If $f\in(H^{a_q^2 b_q^2,B_4}_{\mu,a_k^3 b_k,A_3})'$, then $\tau'_x\,f\in(H^{b_q,B}_{\mu,a_k,A})'$.

Proof. Let $\{\phi_\upsilon\}_{\upsilon=1}^{\infty}$ be a sequence of functions in $H^{b_q,B}_{\mu,a_k,A}$ that converges to zero in $H^{b_q,B}_{\mu,a_k,A}$. Then by Definition 4.1 we have

$$\langle \tau'_x f, \phi_\upsilon \rangle = \langle f, \tau_x \phi_\upsilon \rangle.$$

By theorem 3.1 (c), we have

$$\tau_x \phi_\upsilon \in H^{a_q^2 b_q^2, B_4}_{\mu, a_k^3 b_k, A_3}$$

and hence,

$$\sup_{x \in I} | x^k (x^{-1} D)^q x^{-\mu-1/2} \tau_x \phi_\upsilon | \le C^\mu (A_3 + \delta)^k \ (B_4 + \rho)^q \ a_k^3 b_k a_q^2 b_q^2 .$$

Since $\phi_\upsilon \to 0$ in $H^{b_q, B}_{\mu, a_k, A}$, therefore $\tau_x \ \phi_\upsilon \to 0$ in $H^{a_q^2 b_q^2, B_4}_{\mu, a_k^3 b_k, A_3}$; so that $\langle f, \tau_x \phi_\upsilon \rangle \to 0$ as $\upsilon \to \infty$.

Thus $\tau'_x f$ is continuous on $H^{b_q, B}_{\mu, a_k, A}$. Linearity is obvious.

In a similar way, we can show that

(i) if $f \in (\hat{H}_{\mu, a_k^2, A_2})'$ then $\tau'_x f \in (\hat{H}_{\mu, a_k, A})'$

(ii) if $f \in (H^{b_q^2, B_2}_\mu)'$, then $\tau'_x f \in (\hat{H}^{b_q, B}_\mu)'$.

The above results yield the following:

Theorem 4.3 For $\mu \ge -{}^1/_2$, the generalized Hankel translation operator τ'_x is a continuous linear mapping from $(H^{b_q^2, B_2}_\mu)'$ into $(\hat{H}^{b_q, B}_\mu)'$. Moreover, if a_k satisfies (2.1) and (2.8) then it is a continuous linear mapping from $(\hat{H}_{\mu, a_k^2, A_2})'$ and $(H^{a_q^2 b_q^2, B_4}_{\mu, a_k^3 b_k, A_3})'$ into $(H_{\mu, a_k, A})'$ and $(H^{b_q, B}_{\mu, a_k, A})'$ respectively.

Definition 4.4 The generalized Hankel convolution operator #′ is defined by the relation:

$$\langle f \#' \phi, \psi \rangle = \langle f, \phi \# \psi \rangle$$

where ϕ, ψ belong to $H_{\mu, a_k, A}$, $\hat{H}^{b_q, B}_\mu$ or $H^{b_q, B}_{\mu, a_k . A}$ and f belongs to

$$(\hat{H}_{\mu, a_k^2, B_1})', \ (H^{b_q^2, B_2}_\mu)' \text{ or } (H^{a_q^2 b_q^2, B_5}_{\mu, a_k^3 b_k, A_4})' \text{ respectively.}$$

Theorem 4.5 If a_k satisfies eqs. (2.1) and (2.8), $f \in (H^{a_q^2 b_q^2, B_5}_{\mu, a_k^3 b_k, A_4})'$ and $\phi \in H^{b_q, B}_{\mu, a_k, A}$, then $f \#' \phi \in (H^{b_q, B}_{\mu, a_k, A})'$.

Proof. Let $\{\psi_\upsilon\}$ be a sequence of functions that converges to zero in $H^{b_q, B}_{\mu, a_k, A}$. By Definition 4.4,

$$\langle f \#' \phi, \psi_\upsilon \rangle = \langle f, \phi \# \psi_\upsilon \rangle.$$

Now $\phi \# \psi_\upsilon \in H^{a_q^2 b_q^2, B_5}_{\mu, a_k^3 b_k, A_4}$; hence

$$\sup_{x \in I} | x^k (x^{-1} D)^q \; x^{-\mu-1/2} (\phi \# \psi_\upsilon) | \leq C^\mu (A_4 + \delta')^k (B_5 + \rho')^q a_k^3 b_k a_q^2 b_q^2$$

Since $\psi_\upsilon \to 0$ in $H^{b_q, B}_{\mu, a_k, A}$, $\phi \# \psi_\upsilon \to 0$ in $H^{a_q^2 b_q^2, B_5}_{\mu, a_k^3 b_k, A_4}$; so that $\langle f, \phi \# \psi_\upsilon \rangle \to 0$ as $\upsilon \to \infty$.

Linearity is obvious. Thus $f \#' \phi \in (H^{b_q, B}_{\mu, a_k, A})'$.

Similarly, we can show that

(i) for $\phi \in H_{\mu, a_k, A}$ and $f \in (\hat{H}_{\mu, a_k^2, B_1})'$, $f \#' \phi \in (H_{\mu, a_k, A})'$ if a_k satisfies (2.1) and (2.8)

(ii) for $\phi \in \hat{H}^{b_q, B}_\mu$ and $f \in (H^{b_q, B}_\mu)'$, $f \#' \phi \in (\hat{H}^{b_q, B}_\mu)'$.

The above results yield the following:

Theorem 4.6 For $\mu \geq -1/2$, the generalized Hankel convolution operator $\#'$ is a continuous linear mapping from $(H^{b_q^2, B_2}_\mu)'$ into $(\hat{H}^{b_q, B}_\mu)'$. Moreover, if a_k satisfies (2.1) and (2.8), then it is a continuous linear mapping from $(\hat{H}_{\mu, a_k^2, B_1})'$ and $(H^{a_q^2 b_q^2, B_5}_{\mu, a_k^3 b_k, A_4})'$ into $(H_{\mu, a_k, A})'$ and $(H^{b_q, B}_{\mu, a_k, A})'$ respectively.

Remark 4.7 In theorem 3.2 one may define $\phi \# \psi$ for $\phi \in H_{\mu, a_k, A}$ and $\psi \in \hat{H}^{b_q, B}_\mu$ or $\psi \in H^{b_q, B}_{\mu, a_k, A}$ and study its properties. The corresponding results of Theorem 4.6 are also accordingly modified.

5. Some Generalizations

In this section a generalization of the space $H^{b_q, B}_{\mu, a_k A}$ is obtained. In what follows we consider a sequence $\{m_{k,q}\}_{k,q \in N_0}$ of positive numbers satisfying one or both of the following conditions:

(i) For any $\varepsilon > 0$.

$$m_{k+2q+2,k} / m_{m+2q,k} \leq \mu_\varepsilon (1 + \varepsilon)^{2(k+q)} \;\forall\; k, q \in \mathbf{N}_0 \tag{5.1}$$

(ii) $m_{r,k} \, m_{p,q} \leq m_{0,0} \, m_{r+p, k+q}$; $r, p, k, q \in \mathbf{N}_0$. (5.2)

Theorem 5.1 If an infinitely differentiable function ϕ on $(0, \infty)$ satisfies the inequality

$$| x^k (x^{-1} D)^q \; x^{-\mu - \frac{1}{2}} \phi(x) | \leq C A^k B^q m_{kq} \; (k, q \in \mathbf{N}_0) \tag{5.3}$$

and the numbers m_{kq} are such that condition (5.1) is satisfied, then the Hankel transform $\psi(y)$ of the function $\phi(x)$ satisfies the inequality

$$| y^k(y^{-1}D)^q \, y^{-\mu-1/2} \, \psi(y) \, | \leq C_1 \, A_1^q \, B_1^k \, m_{k+2q,k}, \tag{5.4}$$

where $C_1 = B_\mu C \, [1 + A^{2n} \, \mu_\varepsilon^n \, (1+\varepsilon)^{n(n-1)}], A_1 = A^2 \, (1+\varepsilon)^{2n}, B_1 = AB \, (1+\varepsilon)^{2n}$.

Proof Using (1.2) and (1.10) we get

$$| y^k(y^{-1} \, D)^q \, y^{-\mu-1/2} \, \psi(y) \, | = | \, y^k(y^{-1}D)^q \, y^{-\mu-1/2} \int_0^\infty \phi(x)\sqrt{xy} \; J_\mu \, (xy)dx \, |$$

$$= \left| \int_0^\infty \sqrt{x}\phi(x) \, y^k \, (y^{-1} \, D)^q \; (y^{-\mu} J_\mu (xy)) \, dx \right|$$

$$= \left| \int_0^\infty \sqrt{x}\phi(x) \, y^k \, (-1)^q \; x^q \, y^{-(\mu+q)} \; J_{\mu+q} \, (xy) \; dx \right|$$

$$= \left| \int_0^\infty x^{-\mu+1/2} \; \phi(x) \, y^{-(m+q)} \, [y^k \; x^{\mu+q} \; J_{\mu+q} (xy)] \; dx \right|$$

$$= \left| \int_0^\infty x^{-\mu+1/2} \; \phi(x) \, y^{-(\mu+q)} \, (x^{-1} \, D)^k \; [x^{\mu+k+q} J_{\mu+k+q} \, (xy)] \, dx \right| .$$

Integrating by parts we obtain

$$| \, y^k(y^{-1}D)^q \, y^{-\mu-1/2} \, \psi(y) \, |$$

$$= | \, (-1)^k \int_0^\infty y^{-(\mu+q)} \, x^{\mu+q+k+1} \; J_{\mu+q+k}(xy) \; (x^{-1}D)^k \; [x^{-\mu-1/2} \; \phi(x)] \; dx \, |$$

$$\leq \int_0^\infty | \, x^{2\mu+2q+k+1} \; (xy)^{-(\mu+q)} \; J_{\mu+q+k} \; (xy) \; (x^{-1}D)^k \; [x^{-\mu-1/2} \; \phi(x)] \, | \; dx.$$

Let $n \in \mathbf{N}_0$ be such that $2n \geq 2\mu + 3$. Then using $| \, z^{-(\mu+q)} \, J_{\mu+k+q}(z) \, | \leq B_\mu$, where B_μ is independent of k and q [5, pp 309–310] and applying (5.1) and (5.3) we obtain

$$| \, y^k(y^{-1}D)^q \, y^{-\mu-1/2} \, \psi(y) \, | \leq B_\mu \int_0^1 | \, x^{2\mu+2q+k+1} \; (x^{-1} \; D)^k \; (x^{-\mu-1/2} \; \phi(x) \, | \; dx$$

$$+ \int_1^\infty | \, x^{k+2q+2n} \; (x^{-1}D)^k \; [x^{-\mu-1/2} \; \phi(x)] \, | \; x^{-2} \; dx$$

$$\leq B_\mu \; [CA^{k+2q} \; B^k m_{k+2q,k} + CA^{k+2q+2n} \; B^k m_{k+2q+2n,k}]$$

$$\leq B_\mu C A^{k+2q} \; B^k m_{k+2q,k} \left[1 + A^{2n} \; \frac{m_{k+2q+2n,k}}{m_{k+2q,k}} \right].$$

Now,

$$\frac{m_{k+2q+2n,k}}{m_{k+2q,k}} = \frac{m_{k+2q+2n,k}}{m_{k+2q+2n-2,k}} \cdot \frac{m_{k+2q+2n-2,k}}{m_{k+2q+2n,k}} \cdots \frac{m_{k+2q+2k}}{m_{k+2q,k}}$$

$$\leq \mu_\varepsilon^n \; (1+\varepsilon)^{(2k+2q+2n-2)+(2k+2q+2n-4)+(2k+2q+2n-6)+\ldots+(2k+2q)}$$

$$= \mu_\varepsilon^n \; [(1+\varepsilon)^{2n}]^{k+q} \; (1+\varepsilon)^{n(n-1)};$$

so that

$$| \, y^k (y^{-1}D)^q \, y^{-\mu-1/2} \; \psi(y) \, | \leq B_\mu \, CA^{k+2q} B^k . \; m_{k+2q,k}$$

$$[1 + A^{2n} \; \mu_\varepsilon^n \; (1+\varepsilon)^{n(n-1)} \; \{(1+\varepsilon)^{2n}\}^{k+q}]$$

$$\leq B\mu \; CA^{k+2q} \; B^k \cdot \{(1+\varepsilon)^{2n}\}^{k+q} \; [1 + A^{2n} \cdot \; \mu_\varepsilon^n (1+\varepsilon)^{n(n-1)}] \; m_{k+2q,k}$$

$$= B_\mu C[AB(1+\varepsilon)^{2n}]^k \; [A^2(1+\varepsilon)^{2n}]^q, \; [1 + A^{2n}. \; \mu_\varepsilon^n (1+\varepsilon)^{n(n-1)}] m_{k+2q,k}$$

$$= C_1 A_1^q B_1^k \, m_{k+2q,k}$$

where $A_1 = A^2(1+\varepsilon)^{2n}$, $B_1 = AB(1+\varepsilon)^{2n}$ and $C_1 = B_\mu C[1 + A^{2n} \; \mu_\varepsilon^n (1+\varepsilon)^{n(n-1)}]$.

Remark 5.2 Let $m_{kq} = a_k \; b_q$ where a_k satisfies (2.8). Then the space of functions $\phi(x)$ satisfying (5.3) reduces to the space $H_{\mu,a_k,A}^{b_4,B}$ if c is replaced by C^μ, A by $A + \delta$, B by $B + \rho$; and the Hankel transform of $\phi(x)$ belongs to the space $H_{\mu,a_k b_k, A_1}^{a_q^2, B_1}$, where $A_1 = ABR$, $B_1 = A^2 R^4$.

Theorem 5.3 If the infinitely differentiable functions ϕ and ψ satisfy the inequality (5.4) of the theorem 5.1 and the numbers m_{kq} are such that conditions (5.1) and (5.2) are satisfied, then the Hankel transform of the convolution $\phi \# \psi$ satisfies the inequality.

$$| \, y^k (y^{-1}D)^q \; y^{-\mu-\frac{1}{2}} \; (h_\mu \, (\phi \# \psi)) \; (y) \, | \leq C_1 \; A_1^q \; B_1^q \, m_{k+2q,k.}$$

Proof. Using (1.8) and applying Leibnitz type formula (1.9) we obtain

$$| \, y^k (y^{-1}D)^q \; y^{-\mu-\frac{1}{2}} \; (h_\mu(\phi \# \psi)) \; (y) \, |$$

$$= | \, y^k \; (y^{-1}D)^q \; y^{-\mu-\frac{1}{2}} \; y^{-\mu-\frac{1}{2}} \times (h_\mu \phi) \; (y) \; (h_\mu \psi) \; (y) \, |$$

$$\leq \sum_{r=0}^{q} \binom{q}{r} \, | \, y^k \; (y^{-1}D)^r \; y^{-\mu-\frac{1}{2}} \; (h_\mu \, \phi) \; (y) \, | \, | \, (y^{-1}D)^{q-r} \; y^{-\mu-\frac{1}{2}} \; (h_\mu \psi) \; (y) \, |$$

$$\leq \sum_{r=0}^{q} \binom{q}{r} \; C_1 \, A_1^r \, B_1^k \, C_1 \, A_1^{q-r} \; B_1^0 \; m_{k+2r,k} \; m_{2(q-r),0}$$

$$\leq \sum_{r=0}^{q} \binom{q}{r} \; C_1^2 \, A_1^q \; B_1^k \; m_{k+2r,k} \; m_{2(q-r),0}$$

$$\leq \sum_{r=0}^{q} \binom{q}{r} C_1^2 A_1^q B_1^k m_{0,0} m_{k+2q,k}$$

$$\leq CA_1^q B_1^k m_{k+2q,k}.$$

This completes the proof of the theorem.

References

1. Betancor, J.J. and Marrero, I., "The Hankel convolution and the Zemanian spaces β_μ and β'_μ", Math. Nachr. 160 (1993), 277–298.
2. Gel'fand, I.M. and Shilov, G.E. "Generalized Functions" Vol 2 (Academic Press, New York, 1968).
3. Lee, W.Y.K. "On spaces of type H_μ and their Hankel transformations" SIAM J. Math. Anal. 5 (2), 336–348 (1974).
4. Pathak, R.S. and Pandey, A.B., "On Hankel transforms of ultradistributions", Applicable Analysis 20 (1985), 245–268.
5. Pathak, R.S. "Integral Transforms of Generalized Functions and Their Applications", Gordon and Breach Science Publishers, Amsterdam (1997).
6. Pinto, J. de Sousa "A Generalized Hankel Convolution", SIAM, J. Math. Anal 16 (1985), 1335–1346.
7. Zemanian, A.H. "Generalized Integral Transformations" Interscience Publishers, New York, 1968.

Function Spaces and Applications
D.E. Edmunds et al (Eds)

17. Embedding Theorems in Functional Analysis

M.A. Sofi

Postgraduate Department of Mathematics and Statistics, University of Kashmir, Srinagar-190006, India

1. Introduction

Generally speaking, an embedding theorem is concerned with the possibility of realizing a mathematical object in a given category as included in a larger object of the same category possessing a simpler and richer structure. Thus, for instance, one knows that every n-dimensional real manifold can be embedded as a submanifold of R^{2n+1} (Whitneys embedding theorem) whereas according to a celebrated theorem of Banach and Mazur [12], every separable Banach space can be realised as a (closed) subspace of C[0, 1], the Banach space of continuous functions on [0, 1] which, apart from possessing a whole lot of nice properties, also possesses what is called a Schauder base (See below for definition). It is also possible to prove (via Hahn-Banach theorem) that every separable Banach space can also be (isometrically) embedded as a subspace of l_∞-the Banach space of all bounded sequences. As a consequence of these embeddings, it is possible for example to show the optimality of '2' in Grothendieck's theorem on the square summability of eigenvalues of nuclear operators on a Banach space. As another useful consequence, it is also possible to renorm a given separable Banach space in a strictly convex manner. In what follows we shall see that certain variations on the Banach Mazur theme described above can be used to answer (at least partially) some outstanding open problems in the structure theory of Banach and nuclear Frêchet spaces. One of these is the so-called.

Pelczynski's Problem

(P). Whether complemented subspaces of nuclear Frêchet spaces with basis have a basis?

Before we put this problem in the general perspective of embedding theorems, we fix notations and include some of the more important definitions. We shall follow [20] for Banach space theory and [16], [26] for the necessary background material on (nuclear) locally convex spaces. Throughout this article, E, F, ... will denote Banach spaces whereas X, Y, ... will be reserved for Frêchet spaces. By a nuclear Frêchet space (nFS, for short), we shall mean a Frêchet space with its (locally convex) topology generated by a sequence of seminorms such that the linking maps connecting the associated Banach spaces are nuclear. Besides the Frêchet space(s) of all numerical sequences ω or all rapidly decreasing sequences, s, the class of all *nFS* includes the all-important spaces of entire functions, infinitely differentiable functions on R, test functions on an open subset Ω of R^n or the space of distributions on Ω. By a *complemented subspace* of a Frêchet space X, we shall mean a closed subspace of X which can be 'complemented' by a subspace of X which is itself closed. It is well-known that whereas finite-dimensional subspaces are always complemented, the (closed) space C_o (of all null sequences) is *not* complemented in l_∞ [36]—a fact noticed for the first time by R.S. Philips (see also [21] for a simple proof). Moreover, Hilbert spaces are the only Banach spaces in which every closed subspace is complemented [19]. The last statement is the celebrated 'complemented subspaces' theorem of Lindenstrauss and Tzafriri. Besides the well-known L_p-spaces, all the spaces encountered above can be shown to possess what is called a (Schauder) basis. A system $\{x_n; f_n\}$ is called a *Schauder basis* for X if $\{x_n\} \subset X$, $\{f_n\} \subset X^*$ and each $x \in X$ can be uniquely written as

$$x = \sum_{n=1}^{\infty} f_n(x) x_n$$

where convergence takes place in the (original) topology of X. Such spaces are easily seen to be separable. It was an old question of Banach whether every sparable Banach space had a basis. In 1974, P. Enflo [8] came up with a counter example. In fact, he was able to locate a closed subspace of C_o which lacked even the so-called approximation property (A.P.)—a property weaker than that of possessing a Schauder basis. It also turns out that A.P. behaves well under complemented subspaces. In view of this, it follows that $C[0, 1]$ does not include all separable Banach spaces as complemented subspaces. In other words $C[0, 1]$ is *not complementably universal.*

2. Complemented Embeddings

We start with the following question which is motivated by the above considerations.

Question 1 Does there exist a Banach space which is complementably universal for all separable Banach spaces?

The fact that the (universal) spaces l_∞ and $C[0, 1]$ are not complementably universal has already been noted above. However, it needs a modicum of solid effort to show that no such Banach spaces can exist that are complementably universal for all separable Banach spaces! Making use of the fact that there even exist closed subspaces of $l_p(2 \neq p \geq 1)$ lacking the (much weaker) compact approximation property, Johnson and Szankowski [6] were able to prove the non-existence of such spaces. However, it seems to be unknown whether there do not even exist Frêchet spaces which are complementably universal for all separable Banach spaces. Moreover, it turns out that the absence of (a suitable variant of) the A.P. is the main impediment in effecting a complemented embedding. This was demonstrated in the following theorem of Pelczynski:

Theorem 2.1 [23] Any separable Banach space with the B.A.P. can be embedded as a complemented subspace of a Banach space with a basis. Such a space is unique up to isomorphism.

The *B.A.P. (bounded approximation property)* mentioned above refers to the possibility of approximating pointwise the identity operator on a Banach space by a sequence of finite rank operators. In case the sequence of finite rank operators in the above definition is replaced by a series of finite rank projections, one gets what is called a *finite dimensional decomposition* (FDD). This leads to the following chain of implications:

$$\text{Schauder base} \Rightarrow \text{FDD} \Rightarrow \text{BAP} \Rightarrow \text{AP} \Rightarrow \text{CAP}.$$

However, in general the implications cannot be reversed. This has already been noted above in the case of A.P. versus Schauder base (Enflo) and the fact that A.P. is weaker than BAP was observed by Figiel and Johnson in [9]. However, the status of the most important of these implications involving BAP and the existence of a Schauder base remained unknown until 1989 when S.J. Szarek [32] was able to prove the existence of a separable Banach space with B.A.P. but without a Schuader base! Combining this counterexample with Pelczynski's theorem 2.1, we get a negative solution to Pelczynski Problem (P) in the setting of Banach spaces. However, for nuclear Frêchet spaces, the problem continues to remain open. Let us see why? Treading carefully the same path as in the case of Banach spaces, one is naturally led to look for a Frêchet space analogue of Pelczynski's theorem 2.1. In search of this analogue, Pelczynski and Wojtaszczyk [24] were able to prove the following embedding theorem for Frêchet spaces:

Theorem 2.2 [24] A Frêchet space X has BAP if and only if it can be embedded as a complemented subspace of a Frêchet space X_0 admitting an FDD.

In order to be able to tackle problem (P) in the case of nuclear Frêchet spaces, the above theorem leads, in turn to the following very natural questions:

Question 2 In the above theorem, is it possible to choose for nuclear space X an X_0 which is also nuclear?

Question 3 Can we choose the space X_0 in such a way where FDD gives way to a Schauder basis?

Happily, the answer to Question 2 turns out to be in the affirmative. This was proved by Benndorff [4] where he was also able to prove a Frêchet-Schwartz analogue of this theorem (see also [6]). However, regarding Question 3, it is interesting to note that in his work on the approximation of analytic mappings by polynomials in Frêchet spaces (Stud. Math. Vol. 60 (3), 1977, pp. 223–238), C. Matysczyk was able to give a simple and elegant proof of Theorem 2.2 where he showed that X_0 could be shown to even have a Schauder basis! This settles Question 3 in the affirmative but without any assumptions involving nuclearity. In order to settle Question 3 in its full generality, it is essential to be able to choose X_0 to be nuclear for nuclear X in Matysczyk's theorem. To this end, we recall the famous Djakov-Mityagin example of a nuclear Frêchet space X_* without a Schauder basis (*nfs*-analogue of Enflo's famous counterexample, see [7]), which has the additional feature that for no nuclear Frêchet Space X does $X_* \times X$ have a Schauder basis. This means that X_* cannot be embeded as a complemented subspace of a nuclear Frêchet space with a Schauder basis. In the setting of nuclearity, Question 3 is thus resolved in the negative! In any case, it is still conceivable that X_* could be realized as a complemented subspace of a Frêchet Schwartz space having a Schauder basis. That this indeed is so was shown by J. Taskinen in 1993, when he proved the following theorem. This settles problem (P) in the negative at least in the setting of Frêchet-Schwartz spaces.

Theorem 2.3 [33] The *nFS* X_* can be embedded as a complemented subspace of a *Frêchet-Schwartz* space having a Schauder basis.
This result suggests the following:

Conjecture (Taskinen): Every *nFS* (and not merely X_*) with BAP can be embedded as a complemented subspace of a Frêchet-Schwartz space having a Schauder basis.

3. Nuclear Spaces

In the following we shall see to what extent the Banach-Mazur theorem on the universality of $C[0, 1]$ for separable Banach spaces can be extended to the case of Frêchet spaces. It is folklore that amongst all *nFS*, the space s of rapidly decreasing scalar sequences defined by

$$s = \{\bar{x} = (x_n); p_k(\bar{x}) = \sup_{n \geq 1} n^k \mid x_n \mid < \infty, \text{for } k \geq 1\}$$

and equipped with the nuclear Frêchet topology determined by the sequence $\{p_k;\ k \geq 1\}$ of (semi)-norms enjoys a privileged position in view of the following famous theorem of T. Komura and Y. Komura.

Theorem 3.1 [26, 16] Every nuclear space can be embedded as a subspace of a suitable product s^I of s.

Remark The index set I in the above theorem can be chosen to depend upon the density character of the space in question. As a consequence, it follows that every nuclear Frêchet space embeds isomorphically as a subspace of s^N—the countable product of s—which has the same place in the category of *nFS* as $C[0, 1]$ does in the category of separable Banach spaces. Again, as in the case of $C[0, 1]$, the space s^N has a Schauder basis which is even absolute (and hence unconditional) in view of the Dynin-Mityagin theorem. This suggests the question whether the same holds even for Frêchet-Schwartz (FS)-spaces. More precisely, we ask.

Question 4 Whether every (FS)-space can be embedded as a subspace of an (FS)-space having an unconditional basis?

An affirmative answer to this question was provided by S. Bellenot [3] in 1981 who in an earlier work [14] had shown that there is a limit to this nice behavior beyond the class of (FS)-spaces. More precisely, he showd that there exist *Frêchet Montel* spaces which do not imbed as subspaces of Frêchet-Montel spaces admitting an unconditional basis!

In the case of nuclear Frêchet spaces, it turns out that spaces admitting a continuous norm possess a much richer structure than those admitting none. The space ω (the countable product of the scalar field) is a typical (and even ubiquitous) example of a *nFS* lacking a continuous norm. The space s^N encountered above is yet another example exhibiting this phenomenon. The question whether every *nFS* can be embedded in a *nFS* with basis and admitting a continuous norm assumes significance in view of its far-reaching implications in the structure theory of nuclear Frêchet spaces. It turns out that it is not always possible to have a continuous norm unless the given space is what has come to be called countably normed.

Definition 3.2 A locally convex space X is said to be *countably normed* (CN) if there exists a generating family of seminorms on X such that the linking maps are injective.

A useful characterization of such spaces was achieved by L. Holmstrom in 1983.

Theorem 3.3 [14] A *nFS* X embeds as a subspace of a *nFS* having a basis and a continuous norm precisely when X is CN.

At this point it is worthwhile to mention that unlike the Banach-Mazur theorem (for separable Banach spaces) or the Komura-Komura theorem (for nuclear Frêchet spaces) which guarantee the existence of a single (universal) space for the category in question, it is not possible to have a single universal space for the class of all countably normed spaces. This follows from another result of Holmstrom [13] which says that the 'universal class' for the stated category has to be necessarily uncountable. Such a class may be obtained by considering, for example, all quotients of s with a basis. A remarkable unification of the aforesaid results of Bellenot and Holmstrom was achieved in the following theorem of Vogt and Waldorff [35].

Theorem 3.4 [35] Every countably normed Frêchet-Schwartz space can be embedded as a subspace of a Frêchet-Schwartz-Kothe space admitting a continuous norm.

Remark Contrary to the class of separable Banach spaces and of nuclear Frêchet spaces which have been seen to admit universal spaces, it is still unknown whether the same is true for the class of all Frêchet spaces or of all nuclear spaces. The case of Schwartz spaces is treated in the next section.

4. Schwartz Spaces

It is a well known fact due to Pelczynski that there exists no single (FS)-space which is universal for all (FS)-spaces. This also follows from the results of Moscatelli [22] in which he was able to characterize the existence of Frêchet universal generators for certain classes of locally convex spaces. However, in spite of the guaranteed existence of a universal generator for the class of Schwartz spaces, no such example was known prior to 1972 when, almost simultaneously, H. Jarchow [15] and D. Randtke [27] came up with almost similar examples of universal generators for this class. The term (universal) generator is used in the sense that every member of the class in question is included in a suitable power of the generator. In the theorem below, we describe these examples.

Theorem 4.1 [15, 27] Let E_0 be the space C_0 (or l_∞) equipped with the topology of uniform convergence on null sequences in E_0^* the normed dual of E_0. Then E_0 is a universal generator for the class of all Schwartz space.

As a first consequence it follows, in particular, that every Schwartz space embeds into a suitable product of C_0 and of l_∞. This result may be compared with a result of S.A. Saxon [30] (see also [34]) which says that every nuclear space embeds as a subspace of a sufficiently high power of an arbitrarily given infinite-dimensional Banach space. Conversely, it turns out that a locally convex space embeddable into a sufficiently high power of every ∞-dimensional Banach space is necessarily Schwartz. However, there are examples of Schwartz spaces which do not embed into any product of l_p, $p \geq 1$, a fact discovered by Bellenot [1].

This motivates the problem of classifying Banach spaces E such that every Schwartz space embeds as a subspace of a suitable product of E. Such a classification was achieved in the following theorem of D. Randtke.

Theorem 4.2 [28] For a Banach space E, the following statements are equivalent:

(i) Every Schwartz space embeds into a sufficiently high power of E.

(ii) The space E equipped with the topology of uniform convergence on null sequences in E^* is a universal generator for the class of Schwartz spaces.

(iii) Every compact operator between Banach spaces factors over a subspace of E.

(iv) E finitely contains l_∞.

A word of explanation is in order. The statement in (iv) means that E contains isometric copies of n-dimensional spaces l_∞^n for all $n > 1$. It turns out that, apart from C_0 and l_∞, such spaces also include $C(K)$, K, a compact Hausdorff space, besides many more.

A unification of the above results of Jarchow, Valdivia, Bellenot and Randtke has recently been achieved by the author [31] in terms of the factorisability of certain operator ideals generalizing the ideals of operators encountered above.

Remark It is interesting to remark that $C[0, 1]^N$, the countable product of $C[0, 1]$, is large enough to include all separable Banach spaces as well as all nuclear Frêchet spaces as subapaces. This follows by combining the Banach-Mazur theorem with the theorem of Saxon-Valdivia quoted in the paragraph preceding Theorem 4.2.

Now we get back to the Banach-Mazur embedding theorem quoted in the beginning and one which has motivated a whole lot of questions discussed in the preceding sections. Here we address ourselves to yet another natural question that arises in this circle of ideas and one which has attracted a great deal of attention in recent years.

Question 5 Is it possible to embed a given separable Banach spaces-a'la Banach-Mazur theorem -as a subspace of $C[0, 1]$ consisting of only *differentiable* or of only *nowhere differentiable* functions?

It turns out that the first part of this question is not difficult to settle. Indeed, as is well known, the space of all continuously differentiable functions on $[0, 1]$ is not closed in $C[0, 1]$. It also turns out that a closed subspace of $C[0, 1]$ consisting exclusively of differentiable functions is necessarily finite-dimensional [11]. In other words, it is not possible to embed an infinite dimensional separable Banach space as a (closed) subspace of $C[0, 1]$ consisting entirely of differntiable functions. Regarding the other half of this question, we note the following facts:

(a) A closed subspace E of $C[0, 1]$ is necessarily finite dimensional if every function in E is of bounded variation [18].

(b) Contrary to demanding that every function in a closed subspace E of $C[0, 1]$ is differentiable on $[0, 1]$, in which case it is forced to be finite-dimensional, if differentiability is assumed merely on $(0, 1]$, then E necessarily contains an isomorphic copy of C_0 [11].

(c) In the case of embedding l_1 into $C[0, 1]$, it is always possible to find functions in the image of the embedding which are non-differentiable at every point of a perfect subset of $[0, 1]$ (See [25]).

(d) The subset of $C[0, 1]$ consisting of everywhere-differentiable functions on $[0, 1]$ is considerably smaller in the sense of Baire's category-they constitute a set of first category! In view of (d) and in spite of (a), (b) and (c), it is still conceivable that the Banach-Mazur embedding can be chosen in such a way that the (non-zero) members of l_1 (or C_0) act as continuous functions on $[0, 1]$ which are nowhere differentiable. That this indeed is so has been established in a recent work of L. Rodriguez Piazza [29] where he proves the same conclusion for all separable Banach spaces.

Theorem 5.1 [29] Every separable Banach space can be embedded as a closed subspace of $C[0, 1]$ consisting of the zero-function and nowhere-differentiable functions only.

The method of proof of this remarkable result is highly constructive and consists in the construction of a closed subset K of $[0, 1]$ homeomorphic with the Cantour set and an isometric extension operator $T: C(K) \rightarrow C[0, 1]$ whose range consists of nowhere-differentiable functions only bedsides, of course, the zero function.

We conclude with a brief discussion of embedding theorems in the setting of finite-dimensional spaces. Since all norms on a finite dimensional space are equivalent it follows, in particular, that R (resp. $\mathbb{C}$) is universal (in the isometric sense) for all 1-dimensional real (resp. complex) Banach spaces. For 2-dimensional spaces, however, the situation turns out to be much more difficult. It was way back in 1958 that Grunbaum [10] was able to show that there exists no 3-dimensional space which is universal (in the isometric sense) for all 2-dimensional Banach spaces. The decisive step which consisted in proving the non-existence of even an n-dimensional space ($n > 2$) universal for all 2-dimensional Banach spaces, was taken by C. Bessaga [5] immediately thereafter. In other words a universal space for 2-dimensional space is necessarily infinite-dimensional! This leads to the following very natural question:

Question 6 Let E be a Banach space universal for all 2-dimensional spaces. Is E universal for all separable Banachspaces?

In an attempt to answer this question, Szankowski (An example of a universal Banach space-Israel. Jour. Math., vol 11, 1972 pp. 292–296) was able to construct an example of an infinite dimensional separable reflexive Banach space X_s which contains a linear isometric copy of every finite dimensional Banach space in such a way that the range of the isometric embedding is complemented in X_s by a norm-one projection. In particular, X_s is 2-universal which *cannot* be universal for all separable Banach spaces. This situation may be compared with the well-known fact that $L_1[0, 1]$ is universal for all 2-dimensional Banach spaces which, however, cannot be universal for all separable Banach spaces. (For a simple and elementary proof of this fact see [37]).

Remark Contrary to isometric embeddings of finite-dimensional spaces as considered in the preceding paragraphs, if one insists on 'uniformly isomorphic' embeddings into larger spaces, the situation turns out to be no less interesting. It is both natural and desirable to be able to embed all finite-dimensional spaces into a (necessarily infinite-dimensional) Banach space E_0 in such a manner that the Banach-Mazur distances are (uniformly) bounded. In other words, there exists $C > 0$ such that for each finite dimensional space F there exists an isomorphism T of F onto a subspace E of E_0 such that $\| T \| \| T^{-1} \| \leq C$, i.e., $d(F, E) \leq C$. It is known that whereas there are examples of such spaces E_0 which are reflexive, no such spaces can be uniformly convex. Also it is not difficult to show that such (universal) spaces E_0 are precisely those which have no finite cotype, i.e., which contain l_∞^n's almost isometrically—a property similar to that already encountered as condition (iv) of Theorem 4.2 where such spaces are characterized as universal generators for the class of all Schwartz spaces when endowed with the topology of uniform convergence on null sequences in E_0^*. This provides a nice instance of a situation where certain Banach space properties are characterized in terms of suitable linear topological properties of certain locally convex topologies associated with the space in question. These considerations motivate the following question involving the isomorphic analogue of the Grunbaum-Bessaga Theorem.

Question 7 Let E_0 be a Banach space which contains for each 2-dimensional space F a subspace E with $d(F, E) \leq C$ for some $C > 0$. Does E_0 have to be ∞-dimensional?

We conjecture the answer is in the affirmative. However, in view of Szankowski's negative answer to Question 6 as described above, the isomorphic analogue of Question 6 makes no sense.

Further, in a recent path-breaking work (subspaces of L_p isometric to subspaces of l_p; positivity 2 : 339–367 (1998)), Delbaen, Jarchow and Pelczynski relate the type and cotype of a given infinite-dimensional Banach space to the embeddability of its 2-dimensional subspaces into a finite-dimensional universal space. As a consequence and as a far-reaching generalisation of the Grünbaum-Bessaga theorem, it follows that as long as p is not a positive even number, no finite-dimensional Banach space can be (isometrically) universal for all 2-dimensional subspaces of L_p (or l_p). Not only that, every infinite-dimensional Banach space can be renormed in such a way that a universal space for all of its 2-dimensional subspaces (in the new norm) has to be necessarily infinite-dimensional! In other words, a finite-dimensional universal space fails to exist in a very strong (and a ubiquitous) sense.

In this article which is essentially of survey nature, we have confined ourselves exclusively to the Banach space and locally convex space settings. We would like to point out that on the other side of the spectrum, there is an enormous literature available on embedding theorems in the non-locally-convex setting. Suffice it to say that whereas a number of positive results are already available on isomorphic embeddings of (separable) non-locally convex F-spaces, the isometric theory is in marked contrast with the Banach space setting. The last statement may be justified by the fact there are no F-spaces which are (isometrically) universal even for the class of all 1-dimensional F-spaces!

References

1. S.F. Bellenot, Factorable bounded operators and Schwartz spaces; Proc. Amer. Math. Soc. A2 (1974), pp. 551–554.
2. S.F. Bellenot. Basic sequences in non-Schwartz Frêchet spaces; Trans. Amer. Math. Soc. 258 (1980), pp. 199–216.

3. S.F. Bellenot. Each Schwartz Frêchet space is a subspace of a Schwartz Frêchet space with an unconditional basis; Composition Mathematica, Vol. 42 (1981), pp. 273–278.
4. A. Benndorf. On the relation of the bounded approximation property and a finite dimensional decomposition in nuclear Frêchet spaces; Stud. Math. 75 (1983), pp. 103–119.
5. C. Bessaga. A. Note on universal Banach space of a finite dimension; Bull. Acad. Pol. Sci. 6 (1958), pp. 97–101.
6. J.M.F. Castillo. On the BAP in Frêchet Schwartz spaces and their duals; Monat. Math.; 105 (1988), pp. 43–46.
7. P.V. Djakov and B.S. Mityagin. Modified construction of a nuclear Frêchet space without a basis; Jour. Func. Anal. 23 (1976), pp. 415–423.
8. P. Enflo. A counter example to the approximation problem in Banach spaces; Acta Math. 130. (1973), pp. 309–317.
9. T. Figiel and W.B. Johnson. The approximation property does not imply the bounded approximation property; Proc. Amer. Math. Soc. 41 (1973), pp. 197–200.
10. B. Grunbaum. On a problem of Mazur; Bull. Res. Council of Israel 7F (1958), pp. 133–135.
11. V.I. Guraii. Subspaces of differentiabl functions in the space of continuous functions; Teor. Funktsii Funktsional Anal i Prilozhen. 4 (1967), pp. 116–121.
12. Holmes. Geometric Functional Analysis and its Applications; GTM (24), Springer Verlag.
13. L. Holmstrom. Universal classes of nuclear Kothe spaces with a continuous norm; Jor. Func. Anal. 48 (1), (1982), pp. 12–19.
14. L. Holmstrom. A note on countably normed nuclear spaces; Proc. Amer. Math. Soc. 89, (1983), pp. 453–456.
15. H. Jarchow. Die Universalitat des Raumes Co fur die Klasse der Schwartz-Raume; Math. Ann. 203 (1973), pp. 211–214.
16. H. Jarchow. Locally Convex Spaces; Teubner Text Stuttgart (1981).
17. W.B. Johnson and A. Szankowski. Complementably Universal Banach Spaces; Stud. Math. 58. (1976), pp. 91–97.
18. B. Levine and D.P. Milman. On linear sets in space C_0 consisting of functions of bounded variation; Comm. Inst. Sci. Math. Mec. Jinv. Kharkov 16 (1940). pp. 102–105.
19. L. Lindenstrauss and L. Tzafriri. On complemented subspaces problem; Israel Jour. Math. 9 (1971) pp. 263–269.
20. L. Lindenstrauss. Classical Banach Spaces I, Sequence Spaces: Spriger-Verlag (1977).
21. V. Mascioni. Topics in the theory of complemented subspaces in Banach spaces; Expositiones Mathematicae 7 (1989), pp. 3–47.
22. V.B. Moscatelli. On the existence of universal λ-nuclear Frêchet spaces; Jour. fur die reine und angew. Math. 301 (1978), pp. 1–26.
23. A. Pelczynski. any separable banach space with the bounded approximation property is a complemented subspace of a banach space with a basis; Stud. Math. 40 (1971), pp. 239–242.
24. A Pelcznskii and P. Wojtaszczyk. Banach spaces with finite dimensional expansions of identity and universal basis of finite dimensional subspaces; Stud. math. 40 (1971), pp. 91–108.

25. P.P. Petrushev and S.L. Troyanski. On the Banach-Mazur theorem on the universality of *C*[0, 1]; C.R. Acad. Bulgare Sci. 37 (1984), pp. 283–285.
26. A. Pietsch. Nuclear Locally Convex Spaces; Springer Verlag (1972).
27. D. Randtke. A simple example of a universal Schwartz space; Proc. Amer. math. Soc. 37 (1973). pp. 185–188.
28. D. Randtke. On the embedding of Schwartz spaces into product spaces; Proc. Amer. Math. Soc. 55 (1976), pp. 87–92.
29. L. Rodriguez Piazza. Every separable Banach space is isometric to a subspace of continuous nowhere differentiable functions; Proc. Amer. Math. Soc. 123 (12), 1995, pp. 3649–54.
30. S.A. Saxon. Embedding nuclear spaces in products of an arbitrary Banach space; Proc. Amer. Math. Soc. 34 (1972), pp. 138–140.
31. M.A. Sofi. Embedding A spaces into product spaces (Communicated).
32. S.J. Szarek. A Banach space without a basis which has the bounded approximation property; Acta Math. 159 (1987), pp. 81–98.
33. J. Taskinen. A Frêchet Schwarz space with basis having a complemented subspce without basis; Proc. Amer. Math. Soc. 113 (1), 1991, pp. 151–155.
34. M. Valdivia. Nuclearity and Banach spaces; Proc. Edin. Math. Soc. 20 (1977), pp. 205–209.
35. D. Vogt and V. Walldrof. Two results in Frêchet Schwartz spaces; Arch. der Math. 61 (1993), pp. 459–464.
36. R. Whitley. Projecting m on to Co.; Amer. Math. Monthly, 73 (1966), pp. 285–286.
37. D. Yost. L_1 contains every two-dimensional normed spaces; Ann. Polon. Math. vol. 49 (1988), pp. 17–19.

Function Spaces and Applications
D.E. Edmunds et al (Eds)

18. Four Questions Related to Hardy's Inequality

Gord Sinnamon*

Department of Mathematics, The University of Western Ontario, London, Ontario, N6A 5B7, Canada

Introduction: The Weighted Hardy Inequality

All of the questions in this paper depend on or are inspired by the weight characterization for the Hardy inequality. It is appropriate to begin with that well-known result.

Proposition 0.1 Suppose that $1 < p < \infty$, $0 < q < \infty$ and μ and ν are non-negative, regular measures on the interval (a, b) with $-\infty \le a < b \le \infty$. Then there exists a constant C such that

$$\left(\int_a^b \left| \int_a^x f(t)\, d\nu(t) \right|^q d\mu(x)\right)^{1/q} \le C\left(\int_a^b |f(t)|^p\, d\nu(t)\right)^{1/p}$$

holds for all $f \in L^p_\nu[a, b]$ if and only if either
$p \le q$ and

$$\sup_{a\le y\le b} \left(\int_a^y d\nu\right)^{1/p'} \left(\int_y^b d\mu\right)^{1/q} < \infty, \text{ or}$$

$q < p$, $1/r = 1/q - 1/p$, and

$$\left(\int_a^b \left(\int_a^y d\nu\right)^{r/p'} \left(\int_y^b d\mu\right)^{r/p} d\mu(y)\right)^{1/r} < \infty.$$

Various proofs of Proposition 0.1 in the case that μ and ν are absolutely continuous with respect to Lebesgue measure (weight functions) may be found in [5] and the references therein. The extension to measures may be

---

1991 *Mathematics Subject Classification.* Primary 26D15; Secondary 26A33.
*A portion of the research presented here was done during the author's visit to the Mathematical Institute of the Czech Academy of Sciences in Prague. The hospitality and support of the Institute is much appreciated. Support from the Natural Sciences and Engineering Research Council of Canada is gratefully acknowledged.

found in [7] and [8]. It is important to point out that the usual form of the weighted Hardy inequality,

$$\left(\int_a^b \left|\int_a^x f(t)\,dt\right|^q u(x)\,dx\right)^{1/q} \le C\left(\int_a^b |f(t)|^p v(t)\,dt\right)^{1/p} \tag{0.1}$$

for non-negative weight functions u and v, can be cast in the form of Proposition 0.1 by replacing f by $fv^{1-p'}$ in (0.1) and taking $d\mu(x) = u(x)\,dx$ and $d\nu(t) = v(t)^{1-p'}\,dt$.

Although we have introduced the Hardy inequality on the interval $[a, b]$ we will work on $[0, 1]$ for simplicity except in Section 3 where it is simpler to work on $[0, \infty)$. Generally speaking, results on one interval translate readily to any other. As usual we denote the harmonic conjugate of p by p' so that $1/p + 1/p' = 1$. Integrals are taken to include their endpoints so $\int_a^b d\mu = \int_{[a,b]} d\mu$ and χ_S denotes the function with value 1 on S and 0 otherwise.

1. Hardy's Inequality on Hyperplanes

Suppose that $1 < p < \infty$ and $0 < q < \infty$, let u and v be weights, and set $w = v^{1-p'}$ Fix a function $m \in L_w^{p'}(0, 1) \equiv L_w^{p'}$ and set

$$H_m = \{h \in L_v^p : \int_0^1 hm = 0\}.$$

Question 1.1 What conditions on p, q, u, and v are necessary and sufficient for there to exist a constant C such that

$$\left(\int_0^1 \left|\int_0^x f(t)\,dt\right|^q u(x)\,dx\right)^{1/p} \le C\left(\int_0^1 \left|f(t)\right|^p v(t)\,dt\right)^{1/p} \tag{1.1}$$

for all $f \in H_m$?

The case $m = 0$ is the weighted Hardy inequality of Proposition 0.1 because $H_0 = L_v^p$. If m is not trivial then H_m is genuinely a hyperplane in L_v^p. Note that multiplying m by a non-zero constant has no affect on H_m. The case $m \equiv 1$ (or any non-zero constant) was solved by P. Gurka (see [5, Chap. 1, Sect. 8]) for $1 < p \le q < \infty$ and in [4] for all p and q. With the aid of the following lemma we will be able to answer Question 1.1 in the case that both $\{x : m(x) > 0\}$ and $\{x : m(x) < 0\}$ are of positive measure.

Lemma 1.2 If T is a non-negative, linear operator that satisfies

$$\| Th \|_{L_v^q} \le C_0 \| h \|_{L_v^p}, h \in H.$$

for some C_0 then

$$\| Tg \|_{L_u^q} \le C_1 \| g \|_{L_v^p}, g \in L_v^p,$$

where

$$C_1 = C_0 \left(1 + \| m \|_{L_w^{p'}} / \min\left(\| m\chi_{m>0} \|_{L_w^{p'}} \cdot \| m\chi_{m<0} \|_{L_w^{p'}} \right)\right).$$

(Recall that $w = v^{1-p'}$.)

Proof If $\| m\chi_{m>0} \|_{p',w} = 0$ or $| m\chi_{m<0} \|_{p',w} = 0$ then $C_1 = \infty$ and the conclusion holds trivially. Otherwise, fix $g \in L_v^p$ and define h by

$$h = | g | + \left(\int_0^1 | g | m \right) \chi_{m<0} | m |^{p'-1} w \Big/ \int_{m<0} | m |^{p'} w \text{ if } \int_0^1 | g | m \geq 0, \text{ and}$$

$$h = | g | - \left(\int_0^1 | g | m \right) \chi_{m>0} | m |^{p'-1} w \Big/ \int_{m>0} | m |^{p'} w \text{ if } \int_0^1 | g | m < 0.$$

In either case we clearly have $h \geq | g | \geq g$.

If $\int_0^1 | g | m \geq 0$ then

$$\int_0^1 hm = \int_0^1 | g | m + \left(\int_0^1 | g | m \right) \int_{m<0} | m |^{p'-1} mw \Big/ \int_{m<0} | m |^{p'} w = 0$$

and if $\int_0^1 | g | m < 0$ then

$$\int_0^1 hm = \int_0^1 | g | m - \left(\int_0^1 | g | m \right) \int_{m>0} | m |^{p'-1} mw \Big/ \int_{m>0} | m |^{p'} w = 0$$

so in either case $h \in H$.

Since T is a non-negative operator and $g \leq h$ we have

$$\| Tg \|_{L_u^q} \leq \| Th \|_{L_u^q} \leq C_0 \| h \|_{L_v^q}$$

so we may complete the proof by estimating $\| h \|_{L_v^p}$. If $\int_0^1 | g | m \geq 0$ then

$$\| h \|_{L_v^p} \leq \| g \|_{L_v^p} + \left(\int_0^1 | g | m \right) \| m^{p'-1} \chi_{m<0} \|_{L_v^p} \Big/ \int_{m<0} | m |^{p'} w$$

$$= \| g \|_{L_v^p} + \left(\int_0^1 | g | m \right) \Big/ \| m\chi_{m<0} \|_{L_w^{p'}}$$

$$\leq \| g \|_{L_v^p} + \| g \|_{L_v^p} \| m \|_{L_w^{p'}} \Big/ \| m\chi_{m<0} \|_{L_w^{p'}}$$

Similarly, if $\int_0^1 |g|m < 0$ then

$$\|h\|_{L^p_v} \le \|g\|_{L^p_v} + \|g\|_{L^p_v} \|m\|_{L^{p'}_w} / \|m\chi_{m>0}\|_{L^{p'}_w}$$

The conclusion follows.

Corollary 1.3 Suppose that both $\{x : m(x) > 0\}$ and $\{x : m(x) < 0\}$ have positive w-measure. If there exists a finite constant C such that (1.1) holds for all $f \in H_m$ then there exists a (different) finite constant C such that (1.1) holds for all $f \in L^p_v$. In particular, Question 1.1 reduces to the usual Hardy inequality (0.1).

2. Non-Constant Indices

Let p, q, u, and v be non-negative, measurable functions and consider the inequality

$$\int_0^1 \left| \int_0^x f(t)\, dt \right|^{q(x)} u(x)\, dx \le C \int_0^1 |f(t)|^{p(t)} v(t)\, dt. \qquad (2.1)$$

If p and q are constant functions and take the same value then (2.1) reduces to the familiar weighted Hardy inequality. The theorem which follows shows that (2.1) never holds otherwise.

Definition 2.1 Suppose that (X, μ) and (T, ν) are σ-finite measure spaces. A $\mu \times \nu$-measurable function $k(x, t)$ is called a proper kernel on $X \times T$ provided that if X_0 and X_1 are μ-measurable subsets of X and T_0 and T_1 are ν-measurable subsets of T such that

$$k(x,t) = k(x,t)(\chi_{X_0 \times T_0}(x,t) + \chi_{X_1 \times T_1}(x,t))$$

then either $\mu(X_0) = \nu(T_0) = 0$ or $\mu(X_1) = \nu(T_1) = 0$.

Theorem 2.2 Suppose that (X, μ) and (T, ν) are σ-finite measure spaces and $k(x, t)$ is a proper kernel on $X \times T$. Let $p(t)$ and $q(x)$ be non-negative, measurable functions on T and X respectively. If there exists a constant C such that

$$\int_X \left(\int_T k(x,t) f(t)\, d\nu(t) \right)^{q(x)} d\mu(x) \le C \int_T f(t)^{p(t)} d\nu(t) \qquad (2.2)$$

holds for all non-negative ν-measurable functions f then $p(t)$ is constant ν-almost everywhere, $q(x)$ is constant μ-almost everywhere, and the two functions take the same value.

Proof Since (T, ν) is a σ-finite measure space there exists a positive function φ such that $\int_T \varphi(t)\, d\nu(t) < \infty$. Fix such a function φ. For each $\lambda > 0$ set

$T_0(\lambda) = \{t : p(t) < \lambda\}$, $X_1(\lambda) = \{x : q(x) \geq \lambda\}$, and $f_\lambda(t) = \varphi(t)^{1/p(t)} \chi_{T_0(\lambda)}(t)$.

For any $m \geq 1$ we have

$$\int_{X_1(\lambda)} \left(\int_{T_0(\lambda)} k(x,t) f_\lambda(t)\, d\nu(t) \right)^{q(x)} d\mu(x) \tag{2.3}$$

$$\leq m^{-\lambda} \int_X \left(\int_T k(x,t) m f_\lambda(t)\, d\nu(t) \right)^{q(x)} d\mu(x)$$

$$\leq m^{-\lambda} C \int_T (m f_\lambda(t))^{p(t)}\, d\nu(t) = C \int_{T_0(\lambda)} m^{p(t)-\lambda} \varphi(t)\, d\nu(t).$$

For $t \in T_0(\lambda)$, $m^{p(t)-\lambda} \to 0$ as $m \to \infty$ so, by the Dominated Convergence Theorem, the last integral tends to zero as $m \to \infty$. It follows that the integral (2.3) is zero. Since $f_\lambda(t) > 0$ for $t \in T_0(\lambda)$ we see that $k(x,t) = 0$ $\mu \times \nu$-almost everywhere on $X_1(\lambda) \times T_0(\lambda)$.

This time we set

$$T_1(\lambda) = \{t : p(t) > \lambda\},\ X_0(\lambda) = \{x : q(x) < \lambda\}, \text{ and } g_\lambda(t) = \varphi(t)^{1/p(t)} \chi_{T_1(\lambda)}(t)$$

For any $m \leq 1$ we have

$$\int_{X_0(\lambda)} \left(\int_{T_1(\lambda)} k(x,t) g_\lambda(t)\, d\nu(t) \right)^{q(x)} d\mu(x) \tag{2.4}$$

$$\leq m^{-\lambda} \int_X \left(\int_T k(x,t) m g_\lambda(t)\, d\nu(t) \right)^{q(x)} d\mu(x)$$

$$\leq m^{-\lambda} C \int_T (m g_\lambda(t))^{p(t)}\, d\nu(t) = C \int_{T_1(\lambda)} m^{p(t)-\lambda} \varphi(t)\, d\nu(t).$$

For $t \in T_1(\lambda)$, $m^{p(t)-\lambda} \to 0$ as $m \to 0$ so, by the Dominated Convergence Theorem, the last integral tends to zero as $m \to 0$. It follows that the integral (2.4) is zero. Since $g_\lambda(t) > 0$ for $t \in T_1(\lambda)$ we see that $k(x,t) = 0$ $\mu \times \nu$-almost everywhere on $X_0(\lambda) \times T_1(\lambda)$.

If $p(t)$ is not constant as a ν-measurable function then we can find a $\bar{\lambda} > 0$ such that $\nu(T_0(\bar{\lambda})) > 0$ and $\nu(T_1(\bar{\lambda})) > 0$. Moreover, using σ-finiteness again, we can choose such a $\bar{\lambda}$ satisfying $\nu(\{t : p(t) = \bar{\lambda}\}) = 0$. Since $k(x,t)$ is zero $\mu \times \nu$-almost everywhere on $X_1(\bar{\lambda}) \times T_0(\bar{\lambda})$ and on $X_0(\bar{\lambda}) \times T_1(\bar{\lambda})$ we have

$$k(x,t) = k(x,t)\,(\chi_{X_0(\bar{\lambda}) \times T_0(\bar{\lambda})}(x,t) + \chi_{X_1(\bar{\lambda}) \times T_1(\bar{\lambda})}(x,t))$$

contradicting our hypothesis that $k(x,t)$ is a proper kernel on $X \times T$. Thus $p(t)$ is constant ν-almost everywhere. We denote its constant value by p.

If $\lambda > p$ then $T_0(\lambda)$ has full ν-measure in T and, since $k(x, t)$ is zero $\mu \times \nu$-almost everywhere on $X_1(\lambda) \times T_0(\lambda)$ we see that

$$k(x,t) = k(x,t)\,(\chi_{X_0(\lambda)\times T_0(\lambda)}(x,t) + \chi_{X_1(\lambda)\times T_1(\lambda)}(x,t)).$$

Since $k(x, t)$ is proper we have $\mu(X_1(\lambda)) = 0$ so $q(x) < \lambda$ μ-almost everywhere. As $\lambda \to p^+$ we see that $q(x) \geq p$ μ-almost everywhere.

If $\lambda < p$ then $T_1(\lambda)$ has full ν-measure in T and, since $k(x, t)$ is zero $\mu \times \nu$-almost everywhere on $X_0(\lambda) \times T_1(\lambda)$ we see that

$$k(x,t) = k(x,t)\,(\chi_{X_0(\lambda)\times T_0(\lambda)}(x,t) + \chi_{X_1(\lambda)\times T_1(\lambda)}(x,t)).$$

Since $k(x, t)$ is proper we have $\mu(X_0(\lambda)) = 0$ so $q(x) \leq \lambda$ μ-almost everywhere. As $\lambda \to p^-$ we see that $q(x) \leq p$ μ-almost everywhere. Thus $q(x) = p$ μ-almost everywhere, completing the proof.

Corollary 2.3 If there exists a constant C such that (2.1) holds for all non-negative f then there exists $b \in [0, 1]$ such that $v(t) > 0$ for almost every $t < b$ and $u(x) = 0$ for almost every $x > b$, and $p(t)$ and $q(x)$ take the same constant value almost everywhere for $t \in (0, b)$ and $x \in (0, b) \cap \{u \neq 0\}$.

Proof Define b to be the essential infimum in [0, 1] of the set $\{\bar{b} : u(x) = 0$ almost everywhere on $(\bar{b}, 1)\}$. Since (2.1) holds it is easy to see that $u = 0$ almost everywhere on the interval

$$(\text{ess inf}\ \{t : v(t) = 0\},\ 1)$$

so $v(t) > 0$ for almost every $t < b$. To complete the proof we observe that $\chi_{(0,x)}(t)$ is a proper kernel on $[(0, b) \cap \{u \neq 0\}] \times (0, b)$ and apply Theorem 2.2.

The reason for the failure of (2.1) for non-constant p and q is evident from the proof of Theorem 2.1—homogeneity fails in a disastrous way. There is, however, a standard way to restore lost homogeneity. Set

$$\| f \|_{p(t),v(t)} = \inf\ \{\eta > 0 : \int_0^\infty | f(t)/\eta |^{p(t)} v(t)\, dt \leq 1\}.$$

Question 2.4 Set $If(x) = \int_0^x f(t)\, dt$ and consider the inequality

$$\| If \|_{q(x),u(x)} \leq C \, \| f \|_{p(t),v(t)}. \tag{2.5}$$

Does there exist pair of functions $p(t)$ and $q(x)$, not both constant, and a constant C such that (2.5) holds for all non-negative f? For which $p(t)$ and $q(x)$ does such *a* C exist?

3. More General Limits of Integration

In [3] and [1] the operator

$$\int_{a(x)}^{b(x)} f(t)\,dt \tag{3.1}$$

is studied, where a and b are non-decreasing functions with $a(x) \le b(x)$. The results in [3] are for increasing, differentiable a and b and include necessary and sufficient conditions for the boundedness of the operator from $L_v^p(0, \infty)$ to $L_u^q(0, \infty)$ for $1 < p < \infty$ and $0 < q < \infty$ while the results of [1] are for all a and b described above and include necessary and sufficient conditions for the operator and related operators to be bounded between pairs of Banach function spaces satisfying Berezhnoi's l-condition. Such pairs include ($L_v^p(0, \infty)$, $L_u^q(0, \infty)$) for $1 < p \le q < \infty$ but not for $0 < q < p$, $1 < p < \infty$.

Question 3.1 Can the monotonicity restriction on a and b be removed?

This question seems to be a difficult one but we are able to characterize the boundedness from $L_v^p(0, \infty)$ to $L_u^q(0, \infty)$ for $1 < p < \infty$ and $0 < q < \infty$ of the operator (3.1) with $a = 0$ and b non-negative and measurable.

Theorem 3.2 Suppose b is *a* non-negative, measurable function, $0 < q < \infty$, $1 < p < \infty$, and u and v are weights. Suppose also that either $q > 1$ or $v^{1-p'}$ is locally integrable on $[0, \infty)$. Then

$$\left(\int_0^\infty \left(\int_0^{b(x)} f(t)\,dt\right)^q u(x)\,dx\right)^{1/q} \le C\left(\int_0^\infty f(t)^p\, v(t)\,dt\right)^{1/p} \tag{3.2}$$

holds for all non-negative functions f if and only if either $p \le q$ and

$$\sup_{y>0} \left(\int_0^y v^{1-p'}\right)^{1/p'} \left(\int_{\{x:b(x)\ge y\}} u(x)\,dx\right)^{1/q} < \infty,$$

or $q < p$, $1/r = 1/q - 1/p$, and

$$\int_0^\infty \left(\int_{\{x:b(x)\ge y\}} u(x)\,dx\right)^{r/q} \left(\int_0^y v^{1-p'})^{r/q'} v(y)^{1-p'}\,dy\right)^{1/r} < \infty.$$

Before we prove Theorem 1.1 we need the following lemma.

Lemma 3.3 Let b be *a* non-negative, measurable function, u be a non-negative, integrable function, and $q > 0$. Then there exists a regular Borel measure μ on $(0, \infty)$ such that

$$\int_y^\infty d\mu = \int_{\{x:b(x)\ge y\}} u(x)\,dx \tag{3.3}$$

and

$$\int_0^\infty \left(\int_0^{b(x)} f(t)\,dt\right)^q u(x)\,dx = \int_0^\infty \left(\int_0^z f(t)\,dt\right)^q d\mu(z)$$

for all non-negative functions f.

Proof Since u is integrable, the expression $\int_{\{x:b(x)\geq y\}} u(x)\,dx$ is a non-negative, non-increasing function of y which tends to zero as y tends to infinity. Using [6, Theorem 12, page 262] we see that there exists a finite Borel measure μ satisfying (3.3). Following the construction in the book it is an exercise to show that μ is regular. To complete the proof we calculate as follows:

$$\int_0^\infty \left(\int_0^{b(x)} f(t)\,dt \right)^q u(x)\,dx$$

$$= \int_0^\infty \left(\int_0^{b(x)} q \left(\int_0^t f(s)\,ds \right)^{q-1} f(t)\,dt \right) u(x)\,dx$$

$$= \int_0^\infty \int_0^\infty q \left(\int_0^t f(s)\,ds \right)^{q-1} f(t) u(x)\, \chi_{(0,b(x))}(t)\,dx\,dt$$

$$= \int_0^\infty q \left(\int_0^t f(s)\,ds \right)^{q-1} f(t) \int_{\{x:t\leq b(x)\}} u(x)\,dx\,dt$$

$$= \int_0^\infty q \left(\int_0^t f(s)\,ds \right)^{q-1} f(t) \int_t^\infty d\mu(z)\,dt$$

$$= \int_0^\infty \int_0^z q \left(\int_0^t f(s)\,ds \right)^{q-1} f(t)\,dt\,d\mu(z) = \int_0^\infty \left(\int_0^z f(t)\,dt \right)^q d\mu(z).$$

Proof of Theorem 3.2 It is enough to establish the claim in the case that u is integrable since the general result then follows using the Monotone Convergence Theorem. If μ is the regular Borel measure given by Lemma 3.3, inequality (3.2) becomes

$$\left(\int_0^\infty \left(\int_0^z f(t)\,dt \right)^q d\mu(z) \right)^{1/q} \leq C \left(\int_0^\infty f(t)^p v(t)\,dt \right)^{1/p}$$

Using Proposition 0.1 and the remark on page 93 in [9] we see that this inequality holds if and only if

either $p \leq q$ and $$\sup_{y>0} \left(\int_0^y v^{1-p'} \right)^{1/p'} \left(\int_y^\infty d\mu \right)^{1/q} < \infty,$$

or $q < p$, $1/r = 1/q - 1/p$, and

$$\left(\int_0^\infty \left(\int_y^\infty d\mu \right)^{r/q} \left(\int_0^y v^{1-p'} \right)^{r/q'} v(y)^{1-p'}\,dy \right)^{1/r} < \infty.$$

Replacing μ by u according to (3.3) completes the proof.

If a and b are similarly ordered in the sense of [2, page 43] the argument of Theorem 3.2 should extend to the operator (3.1). If not then Question 3.1 is quite a different sort of problem than the Hardy operator because the sections of the kernel are no longer a totally ordered set.

4. The Higher Order Hardy Inequality with One Weight Fixed

We look at the inequality

$$\left(\int_0^1 \left(\int_0^x (x-t)^k f(t)\,dt\right)^q d\mu(x)\right)^{1/q} \le C\left(\int_0^1 f(t)^p v(t)\,dt\right)^{1/p}, \quad (4.1)$$

for a fixed weight v. Here k is a non-negative integer, $1 < p < \infty$ and $0 < q < \infty$. The weights for which (4.1) holds have been characterized, see [10].

Let W_k denote the collection of those non-negative, regular measures μ for which there exists a constant C such that (4.1) holds for all non-negative f.

Definition 4.1 If μ_1 and μ_2 are non-negative, regular measures on [0, 1] we say that $\mu_1 \preceq_k \mu_2$ provided

$$\int_y^1 (y-x)^k\,d\mu_1(x) \le \int_y^1 (y-x)^k d\mu_2(x)$$

for all $y \in [0, 1]$.

There is a natural connection between the partial order $\preceq_k$ and the class W_k.

Lemma 4.2 Suppose $1 < p < \infty$, $0 < q < \infty$, and $0 \le k \le q$. If $\mu_1 \in W_k$ and $\mu_2 \preceq_k c\mu_1$ for some $c > 0$ then $\mu_2 \in W_k$.

Proof It is enough to verify (4.1), with μ replaced by μ_2, for continuous functions f since they are dense in $L_v^p(0, 1)$. For a non-negative, continuous function f it is easy to check that, since $q \ge k$,

$$F(x) = \left(\int_0^x (x-t)^k f(t)\,dt\right)^q$$

satisfies $F^{(j)}(x) \ge 0$ for $0 \le j \le k+1$ and $F^{(j)}(0) = 0$ for $0 \le j \le k$.

Therefore $$F(x) = \int_0^x (x-t)^k F^{(k+1)}(t)/(k!)\,dt.$$

We have

$$\left(\int_0^1 F(x)\,d\mu_2(x)\right)^{1/q} = \left(\int_0^1 \int_t^1 (x-t)^k\,d\mu_2(x)\,F^{(k+1)}(t)/(k!)\,dt\right)^{1/q}$$

$$\le c^{1/q}\left(\int_0^1 \int_t^1 (x-t)^k\,d\mu_1(x)\,F^{(k+1)}(t)/(k!)\,dt\right)^{1/q}$$

$$= c^{1/q}\left(\int_0^1 F(x)\,d\mu_1(x)\right)^{1/q}$$

$$\le c^{1/9} C\left(\int_0^1 f(t)^p v(t)\,dt\right)^{1/p}$$

which completes the proof.

In the case $p \le q$ the weight class W_0 has a largest element (up to constant multiples) with respect to the partial order $\preceq_0$. In view of Theorem 4.2 this shows that W_0 is completely determined by this maximum element. Moreover, the maximum measure can be expressed in terms of p, q, and the fixed weight v. Define the measure ω_0 by

$$d\omega_0(x) = (q/p')\left(\int_0^x v^{1-p'}\right)^{-1-q/p'} v(x)^{1-p'}\,dx + \left(\int_0^1 v^{1-p'}\right)^{-q/p'} d\delta_1(x).$$

Here δ_1 is the Dirac measure at 1.

Theorem 4.3 Suppose that $1 < p \le q < \infty$. Then $\mu \in W_0$ if and only if $\mu \preceq c\omega_0$ for some $c \ge 0$.

Proof The measure ω_0 was defined so that

$$\int_y^1 d\omega_0(x) = -\left(\int_0^x v(t)^{1-p'}\,dt\right)^{-q/p'}\Bigg|_y^1 + \left(\int_0^1 v(t)^{1-p'}\,dt\right)^{-q/p'}$$

$$= \left(\int_0^y v(t)^{1-p'}\,dt\right)^{-q/p'}$$

which shows that

$$\left(\int_y^1 d\omega_0(x)\right)^{1/q}\left(\int_0^y v(t)^{1-p'}\,dt\right)^{-1/p'} = 1.$$

By Proposition 0.1 the Hardy inequality (4.1) (with $k = 0$) holds with μ replaced by ω_0. That is, $\omega_0 \in W_0$. Theorem 4.2 shows that if $\mu \preceq_0 c\omega_0$ for some $c \ge 0$ then $\mu \in W_0$.

To prove the converse we suppose that $\mu \in W_0$, fix $y \in (0, 1)$, and substitute $f(t) = v(t)^{-p'}\chi_{(0,y)}(t)$ into (4.1) to see that

$$\left(\int_y^1\left(\int_0^y v(t)^{1-p'}\,dt\right)^q d\mu(x)\right)^{1/q} \le C\left(\int_0^y v(t)^{1-p'}\,dt\right)^{1/p}$$

It follows that

$$\int_y^1 d\mu(x) \le C^q\left(\int_0^y v(t)^{1-p'}\,dt\right)^{-q/p'} = C^q\int_y^1 d\omega_0(x)$$

so that $\mu \preceq_0 C^q\omega_0$ as required.

Question 4.4 Is there a measure ω_k such that $\mu \in W_k$ if and only if $\mu \preceq_k c\omega_k$ for some constant $c \geq 0$?

We have already answered the question in the case $k = 0$ and $1 < p \leq q < \infty$ and we can also answer it in the case $k = 0$ and $0 < q < p$, $1 < p < \infty$. The next theorem shows that if $q < p$ then W_0 has no maximal element.

Theorem 4.5 Suppose that $0 < q < p$, $1 < p < \infty$ and that either $q > 1$ or $v^{1-p'}$ is locally integrable. If $\mu \in W_0$ then there exists a measure $\mu^+ \in W_0$ such that $\mu^+ \not\preceq_0 c\mu$ for any constant $c \geq 0$.

Proof Proposition 0.1 and the remark on page 93 of [9] shows that (4.1) holds if and only if

$$\left(\int_0^1 \left(\int_y^1 d\mu\right)^{r/q} \left(\int_0^y v^{1-p'}\right)^{r/q'} v(y)^{1-p'}\, dy\right)^{1/r} < \infty,$$

where r is defined by $1/r = 1/q - 1/p$. If we set $V(y) = \left(\int_a^y v^{1-p'}\right)^{r/q'} v(y)^{1-p'}$

then we see that $\mu \in W_0$ if and only if $\int_0^1 \left(\int_y^1 d\mu\right)^{r/q} V(y)\, dy < \infty$.

Fix a measure $\mu \in W_0$. Our object is to construct a measure μ^+ such that

$\int_0^1 \left(\int_y^1 dy^+\right)^{r/q} V(y)\, dy < \infty$ and $\int_y^1 dy^+ \Big/ \int_y^1 d\mu$ is an unbounded function of y.

Set $F(y) = (\int_y^1 d\mu)^{r/q}$. Since $V(y)\, dy$ is non-atomic we can choose a decreasing sequence $y_0 = 1, y_1, y_2, \ldots$, converging to 0, such that $\int_0^{y_k} FV = 2^{-k} \int_0^1 FV$. Now let h be the function whose graph is the polygonal path connecting the points (y_k, k). Clearly, h is a continuous, non-increasing function on $(0, 1)$. Since $(hF)^{q/r}$ is non-decreasing there exists a Borel measure μ^+ satisfying $\int_y^1 d\mu^+ = (h(y)F(y))^{q/r}$ for almost every y. Because h is unbounded it is clear that $\mu^+ \not\preceq_0 \mu$ and for the other requirement we estimate as follows:

$$\int_0^1 \left(\int_y^1 d\mu^+\right)^{r/q} V(y)\, dy = \int_0^1 hFV.$$

$$\leq \sum_{k=1}^{\infty} k \int_{y_k}^{y_{k-1}} FV = \sum_{k=1}^{\infty} k2^{-k} \int_0^1 FV < \infty$$

This completes the proof.

References

1. A. Gogatishvili and J. Lang, *The generalized Hardy operator with kernel and variable integral limits in Banach function spaces* (to appear).
2. G. Hardy, J. Littlewood, and G. Pólya, *Inequalities, Second Edition,* Cambridge University Press, Cambridge, 1952.
3. H.P. Heinig and G. Sinnamon, *Mapping properties of integral averaging operators,* Studia Mathematica **129**(1998), 157–177.
4. A. Kufner and G. Sinnamon, *Overdetermined Hardy Inequalities,* Journal of Mathematical Analysis and Applications **213** (1997), 468–486.
5. B. Opic and A. Kufner, *Hardy-type Inequalities,* Longman, Harlow/Essex, England, 1990.
6. H.L. Royden, *Real Analysis, Second Edition,* Macmillan, New York, 1968.
7. G. Sinnamon, *Operators on Lebesgue Spaces with General Measures,* PhD Thesis, McMaster University, Hamilton, Canada, 1987.
8. G. Sinnamon, *Spaces defined by the level function and their duals,* Studia Mathematica **111** (1994), 19–52.
9. G. Sinnamon and V. Stepanov, *The weighted Hardy inequality: New proofs and the case* $p = 1$, J. London Math. Soc. (2) **54** (1996), 89–101.
10. V.D. Stepanov, *Weighted norm inequalities of Hardy type for a class of integral operators,* J. London Math. Soc. (2) **50** (1994), 105–120.

Function Spaces and Applications
D.E. Edmunds et al (Eds)

19. Optimal Inequalities on Quasinormed Function Spaces

R. Kerman*

Department of Mathematics, Brock University, St. Cathariness, Ontario, L2S 3A1, Canada

1. Introduction

Suppose Ω is a domain in $\mathbb{R}^n$. Denote by $M(\Omega)$ the class of real-valued measurable functions on Ω. Let T be a linear operator satisfying

$$\rho_R(|Tf|) \le C\rho_D(|f|), \tag{1.1}$$

where ρ_R and ρ_D are quasinorms on $M_+(\Omega)$, the nonnegative functions in $M(\Omega)$. We are interested in the question of when ρ_R and/or ρ_D are optimal in a certain class Q (dependent on the problem at hand) in the sense that there is no quasinorm in Q essentially larger than ρ_R and/or essentially smaller than ρ_D for which (1.1) holds.

The focus is on operators $T = T_K$ of Hardy type:

$$(Tf)(x) = \int_0^x K(x, y) f(y) dy,$$

where the kernel K is nonnegative, nondecreasing in x, nonincreasing in y and satisfies the triangle inequality

$$K(x, y) \le D[K(x, z) + K(z, y)], \qquad y < z < x;$$

see [2] or [10].

We look at two problems. In Section 3 we consider weighted versions of the classical Hardy operator ($K(x, y) = 1$) and its dual, with Q the set of rearrangement invariant quasinorms. This problem arose in joint work with D.E. Edmunds and L. Pick on optimal Sobolev imbeddings. Section 4 deals with general Hardy type operators on weighted Orlicz and Orlicz-Lorentz spaces, work done with S. Bloom and M. Goldman, respectively. Here an element of Q corresponds to a weight.

---

*Research supported in part by NSERC grant A4021.

2. Quasinormed Function Spaces

The distribution function, μ_f, of $f \in M(\Omega)$ at $\lambda > 0$ is

$$\mu_f(\lambda) = |\{ x \in \Omega : |f(x)| > \lambda \}|.$$

Definition 2.1 A quasinorm ρ on $M_+(\Omega)$ is defined by the following five axioms:

(A_1) $\rho(f) \geq 0$ with $\rho(f) = 0$ if and only if $f = 0$ a.e.;
(A_2) $\rho(cf) = c\rho(f)$ $c \geq 0$;
(A_3) $\rho(f + g) \leq C[\rho(f) + \rho(g)]$, $C \geq 1$;
(A_4) $f_n \uparrow f$ implies $\rho(f_n) \uparrow \rho(f)$;
(A_5) $\rho(\chi_E) < \infty$ whenever $E \subset \Omega$, $|E| < \infty$.

In Section 3, where ρ is defined on $M_+(0, R)$, $0 < R \leq \infty$, we further suppose

(A_6) To each s, $0 < s < 1$, there corresponds $C = C(s) > 0$, independent of $f \in M_+(0, R)$, such that

$$\rho(E_s f) \leq C\rho(f), \qquad (E_s f)(t) = f(st), \qquad 0 < t < R.$$

If, in addition to (A_1)–(A_5), ρ satisfies

(A_7) $$\rho(f) = \rho(g) \text{ whenever } \mu_f(\lambda) = \mu_g(\lambda), \ \forall\, \lambda > 0$$

we say ρ is a rearrangement invariant (r.i.) quasinorm. The nonincreasing rearrangement, f^*, of $f \in M(\Omega)$ on $(0, |\Omega|)$ is

$$f^*(t) = \inf \{\lambda > 0 : \mu_f(\lambda) \leq t\}, \qquad 0 < t < |\Omega|.$$

It satisfies $\mu_{f^*}(\lambda) = \mu_f(\lambda)$, $\lambda > 0$. One can always define an r.i. quasinorm on $M_+(\Omega)$ in terms of an r.i. quasinorm σ on $M_+(0, |\Omega|)$; namely, by

$$\rho(f) = \sigma(f^*), \qquad f \in M_+(\Omega).$$

Definition 2.2 Let ρ be an r.i. quasinorm on $M_+(\Omega)$. Then, ρ is said to be a rearrangement invariant (r.i.) norm if we can take $C = 1$ in (A_3) and if there exists $C > 0$ such that

$$\int_E f(x)dx \leq C\rho(f), \qquad f \in M_+(\Omega). \tag{2.1}$$

A functional ρ satisfying (A_1)–(A_5), with $C = 1$ in (A_3), and (2.1) is called a Banach function norm.

The dual of a quasinorm ρ is the functional

$$\rho'(g) = \sup_{\rho(h)=1} \int_\Omega g(x)h(x)dx, \qquad g, h \in M_+(\Omega).$$

When ρ is an r.i. quasinorm on $M_+(0, R)$ there is also the "down" dual

$$\rho'_d(g) = \sup_{\rho(h)=1} \int_0^R g(t)h^*(t)dt, \qquad g, h \in M_+(0, R);$$

in this case, $\rho'_d(g) = \rho'(g^*)$.

Both ρ' and ρ'_d obey axioms (A_1)–(A_4) and, moreover, we can take $C = 1$ in (A_3). Again, (A_5) is verified by either ρ' or ρ'_d if and only if (2.1) holds for ρ. Finally, if ρ is an r.i. quasinorm, then ρ' is an r.i. norm if and only if one has the L^1-imbedding (2.1) for ρ.

Examples Let v be a nonnegative measurable (weight) function on $(0, R)$, $0 < R \leq \infty$. The weighted Lebesgue functionals on $M_+(0, R)$

$$\rho_{p,v}(f) = \left[\int_0^R (f(t)v(t))^p\,dt\right]^{1/p}, \qquad 0 < p < \infty, \tag{2.2}$$

and

$$\rho_{\infty,v}(f) = \sup_{0<t<R} f(t)v(t)$$

are quasinorms if

$$\int_0^t v(s)^p\,ds \leq C \int_0^{t/2} v(s)^p\,ds, \qquad 0 < t < R, \tag{2.3}$$

when $0 < p < \infty$, and if

$$v(t) \leq Cv(t/2), \qquad 0 < t < R, \tag{2.4}$$

when $p = \infty$ (see [5], for example). Moreover, they are norms when $1 \leq p \leq \infty$. When $v = 1$, we use the abbreviated notation ρ_p.

The classical Lorentz functionals, $\rho_{\Lambda_p}(v)$, are defined in terms of $\rho_{p,v}$ by

$$\rho_{\Lambda_p(v)}(f) = \rho_{p,v}(f^*), \qquad f \in M_+(0, R). \tag{2.5}$$

They are r.i. quasinorms, given (2.3) when $0 < p < \infty$ and (2.4) when $p = \infty$. The functional $\rho_{\Lambda_p(v)}$ is an r.i. norm if and only if $1 \leq p < \infty$ and v is nonincreasing (see [9]). The Lorentz-Karamata (L-K) quasinorms are of particular interest to us. We suppose $R = 1$, $1 \leq p,\ q \leq \infty$ and take $v(t) = t^{1/p-1/q}b(t^{-1})$ in (2.5), where $b(t)$ is slowly varying in $(1, \infty)$, in the sense that for each $\varepsilon > 0$, $t^{\varepsilon}b(t)$ is eventually increasing and $t^{-\varepsilon}b(t)$ is eventually decreasing.

One generalization of (2.2) is the Orlicz functional, in which t^p is replaced by an unbounded increasing function $\Phi(t)$, $\Phi(0+) = 0$, with

$$\rho_{\Phi,v}(f) = \inf\left\{\lambda > 0 : \int_0^R \Phi\left(\frac{f(t)}{\lambda}\right) v(t)dt \leq 1\right\}, \quad f \in M_+(0, R). \tag{2.6}$$

Given such a Φ, we define the Orlicz-Lorentz functional

$$\rho_{\Lambda_\Phi(v)}(f) = \rho_{\Phi,v}(f^*), \qquad f \in M_+(0, R). \tag{2.7}$$

The functionals (2.6) and (2.7) are quasinorms if (2.3) holds with $p = 1$; indeed, (2.7) is then an r.i. quasinorm. In case Φ is convex, $\rho_{\Phi,v}$ is the

Luxembourg norm, which is equivalent to the classical Orlicz norm, written simply as ρ_Φ, if $v = 1$. Finally, $\rho_{\Lambda_\Phi(v)}$ is an r.i. norm if, in addition to the convexity of Φ, v is nonincreasing.

3. Optimal Sobolev Imbeddings

Let m and n be positive integers satisfying $n \geq 2$ and $1 \leq m \leq n-1$. As usual, denote by $C_0^m(\Omega)$ the set of m-times continuously differentiable functions with compact support in the domain $\Omega \subset \mathbb{R}^n$. The mth order gradient, $\nabla^m u$, of a function $u \in C_0^m(\Omega)$ is defined in terms of the standard first order gradient $\nabla = \left(\frac{\partial}{\partial x_1}, \ldots \frac{\partial}{\partial x_n}\right)$ and the Laplacian $\Delta = \frac{\partial^2}{\partial x_1^2}, + \ldots + \frac{\partial^2}{\partial x_n^2}$ as follows:

$$\nabla^m u = \begin{cases} \Delta^k u & \text{when } m = 2k \\ \nabla(\Delta^k u) & \text{when } m = 2k+1, \end{cases}$$

where $\Delta^j u = \Delta(\Delta^{j-1} u)$, $j = 2, \ldots, \left[\frac{m}{2}\right]$.

One form of the classical Sobolev inequality asserts that, given $1 < p < \frac{m}{n}$ and setting $q = \frac{np}{n - mp}$, there exists $C > 0$ such that

$$\left[\int_\Omega |u(x)|^q \, dx\right]^{1/q} \leq C\left[\int_\Omega |(\nabla^m u)(x)|^p \, dx\right]^{1/p}, \tag{3.1}$$

$u \in C_0^m(\Omega)$.

Standard examples show that in the limiting case $p = n/m$ one cannot take the L^∞ norm $\operatorname{ess\,sup}_{x\in\Omega} |u(x)|$ on the left side of (3.1). Trudinger [13], among others, in the case $m = 1$, and Strichartz [11] and, later, Adams [1], in the case $m > 1$, showed that if Φ_n is convex function on $(0, \infty)$ equal to $\exp(t^{\frac{n}{n-m}})$ for large t, there corresponds to each bounded domain $\Omega \subset \mathbb{R}^n$ a constant $C = C(|\Omega|) > 0$, independent of $u \in C_0^m(\Omega)$, with

$$\inf\left\{\lambda > 0: \int_\Omega \Phi_n\left(\frac{|u(x)|}{\lambda}\right) dx \leq 1\right\} \leq C\left[\int_\Omega |(\nabla^m u)(x)|^{n/m} \, dx\right]^{m/n}, \tag{3.2}$$

$u \in C_0^m(\Omega)$.

Brezis and Wainger [4], again among others, improved on (3.2), replacing the Orlicz exp $L^{\frac{n}{n-m}}$ norm on the left side of (3.2) by a larger, classical Lorentz norm to get

$$\left[\int_0^{|\Omega|} u^*(t)^{n/m}\left(\log\frac{|\Omega|}{t}\right)^{-n/m}\frac{dt}{t}\right]^{m/n} \leq C\left[\int_\Omega |(\nabla^m u)(x)|^{n/m}\,dx\right]^{m/n},$$

$u \in C_0^m(\Omega)$.

The focus in [6] is on when such Sobolev inequalities are optimal. Thus, given r.i. quasinorms ρ_R and ρ_D on $M_+(0, 1)$ with the property that to each bounded domain $\Omega \subset \mathbb{R}^n$ there corresponds $C = C(|\Omega|) > 0$ such that

$$\rho_R(u^*(|\Omega|t) \leq C\rho_D(|\nabla^m u|^*(|\Omega|t)), \quad u \in C_0^m(\Omega), \tag{3.3}$$

we want to know when ρ_R and ρ_D are optimal in the class Q of r.i. quasinorms. When $m = 1$ we deal, in fact, with a closely related imbedding inequality of Talenti [12], in which ρ_D need not be rearrangement invariant:

$$\rho_R(u^*(|\Omega|t)) \leq C\rho_D\left(\frac{d}{dt}\int_{\{x\in\Omega:|u(x)|>u^*(|\Omega|t)\}}|(\nabla u)(x)|dx\right), \tag{3.4}$$

$u \in C_0^1(\Omega)$.

A key step in the analysis is the reduction of inequalities (3.3) and (3.4) to the weighted boundedness of certain Hardy operators. When $m = 1$, the reduction depends on an appropriate version of the Polya-Szegö inequality[1] due to Talenti.

Theorem 3.1 ([12], p. 203) Let $u \in C_0^1(\mathbb{R}^n)$. Then,

$$-t^{1/n'}\frac{du^*}{dt} \leq C_n\frac{d}{dt}\int_{\{x\in\mathbf{R}^n:|u(x)|>u^*(t)\}}|(\nabla u)(x)|dx \tag{3.5}$$

for a.a. $t \in (0, \infty)$, where $n' = \dfrac{n}{n-1}$ and $C_n = n^{-1}K_n^{-1/n}$, $K_n = \pi^{n/2}\Gamma\left(\frac{n}{2}+1\right)^{-1}$ being the measure of the n-dimensional unit ball.

When $n \geq 3$ and $2 \leq m \leq n-1$, we require the following consequence of a convolution inequality of O'Neil [10]:

$$u^*(t) \leq C\left[\frac{n}{m}t^{m/n-1}\int_0^t|\nabla^m u|^*(s)\,ds + \int_t^{|\mathrm{supp}\,u|}|\nabla^m u|^*(s)s^{\frac{m}{n}-1}ds\right],$$

$0 < t < |\mathrm{supp}\,u|$. Adams [1] has shown the least possible value C can have

[1]I would like to thank A. Cianchi for very useful discussions concerning the Polya-Szegö inequality.

in (3.6) is $\beta_0^{\frac{m}{n}-1}$, where $\beta_0 = \dfrac{n}{\omega_{n-1}}\left[\dfrac{\pi^{n/2}\,4^k\,\Gamma(k)}{\Gamma\left(\frac{n}{2}-k\right)}\right]^{n/n-2k}$ when $m = 2k$ and

$\beta_0 = \dfrac{n}{\omega_{n-1}}\left[\dfrac{\pi^{n/2}\,2\cdot 4^k\,\Gamma(k+1)}{\Gamma\left(\frac{n}{2}-k\right)}\right]^{n/n-2k-1}$ when $m = 2k + 1$; here, ω_{n-1} is the area of the surface of the n-dimensional unit ball.

The inequalities (3.5) and (3.6) allow us to prove

Theorem 3.2 Fix positive integers m and n satisfying ≥ 2 and $1 \leq m \leq n-1$. Let ρ_R be an r.i. quasinorm on $M_+(0, 1)$. Then, when $m = 1$, a necessary and sufficient condition that (3.4) hold with ρ_R and a quasinorm ρ_D on $M_+(0, 1)$ is the existence of $K > 0$ for which

$$\rho_R\left(\int_t^1 f(s)s^{\frac{1}{n}-1}ds\right) \leq K_{\rho_D}(f), \quad f \in M_+(0, 1).$$

When $n \geq 3$ and $2 \leq m \leq n-1$, a necessary and sufficient condition that (3.3) hold for ρ_R and another r.i. quasinorm ρ_D on $M_+(0, 1)$ is the existence of $K > 0$ for which

$$\rho_R\left(\int_t^1 (Pf)s^{\frac{m}{n}-1}ds\right) \leq K_{\rho_D}(f);$$

here $f \downarrow 0$ on $(0, 1)$ and $(Pf)(s) = s^{-1}\displaystyle\int_0^s f(y)dy$;

Theorem 3.2 can be used, as in [8], to associate to a given r.i. quasinorm ρ_R the essentially smallest quasinorm ρ_D for which (3.4) holds when $m = 1$ and (3.3) holds when $n \geq 3$ and $2 \leq m \leq n-1$; in the latter case, ρ_D is rearrangement invariant.

Theorem 3.3 Fix positive integers m and n with $n \geq 2$ and $1 \leq m \leq n-1$. Let ρ_R be an r.i. quasinorm on $M_+(0, 1)$. For $f \in M_+(0, 1)$ define

$$\rho_D(f) = \rho_R\left(\int_t^1 f(s)s^{\frac{1}{n}-1}ds\right), \quad \text{when } m = 1, \tag{3.7}$$

and

$$\rho_D(f) = \rho_R\left(\int_t^1 (Pf^*)(s)s^{\frac{m}{n}-1}ds\right), \text{ when } n \geq 3 \tag{3.8}$$

and $2 \leq m \leq n-1$.

Then, ρ_D is a quasinorm on $M_+(0, 1)$ with the property that to each bounded domain Ω in $\mathbb{R}^n$ there corresponds $C = C(|\Omega|) > 0$ such that (3.4) holds when $m = 1$ and (3.3) holds when $n \geq 3$ and $2 \leq m \leq n - 1$. Moreover, it is the smallest such quasinorm when $m = 1$ and the smallest such r.i. quasinorm when $n \geq 3$ and $2 \leq m \leq n - 1$, in the sense that if ρ is another, then there exists $K > 0$ for which

$$\rho_D(f) \leq K\rho(f), \qquad f \in M_+(0, 1).$$

Arguing by duality from Theorem 3.3 we obtain the following two results.

Theorem 3.4 Let n be a positive integer, $n \geq 2$, and suppose ρ is a quasinorm on $M_+(0, 1)$ such that

$$\int_0^1 f(t)t^{1/n}\,dt \leq C\rho(f), \qquad f \in M_+(0, 1).$$

Then, the functional σ defined

$$\sigma(g) = \rho'(t^{1/n}(Pg^*)(t))$$

is an r.i. norm on $M_+(0, 1)$. Moreover, (3.4) holds for $\rho_R = \sigma'$ and $\rho_D = \rho$, σ' being the largest such r.i. range norm.

Theorem 3.5 Let m and n be positive integers with $n \geq 3$ and $2 \leq m \leq n - 1$. Suppose ρ is a quasinorm on $M_+(0, 1)$ for which the L^1-imbedding (2.1) holds. Then, the functional σ defined at $g \in M_+(0, 1)$ by

$$\sigma(g) = \rho'\left(\int_t^1 (Pg^*)(s)s^{\frac{m}{n}-1}\,ds\right)$$

is an r.i. norm on $M_+(0, 1)$. Moreover, if ρ is rearrangement invariant, then (3.3) holds for $\rho_R = \sigma'$ and $\rho_D = \rho$, σ' being the largest such r.i. range norm.

We now consider, for the class of Lorentz-Karamata quasinorms, the question of when the optimal domain quasinorm ρ_D associated to ρ_R in Theorem 3.3 has ρ_R as its optimal range quasinorm.

Theorem 3.6 Fix p and q with $1 \leq p, q \leq \infty$. Suppose b is a slowly varying function on $(1, \infty)$, which is such that $\phi(t) = t^{\frac{1}{p}-\frac{1}{q}}b(t^{-1})$ satisfies $\rho_q(\phi) < \infty$. Let

$$\rho_R(f) = \begin{cases} \rho_q(\phi f^*) & \text{when } p > q, \\ \rho_q(\phi P f^*) & \text{when } p \leq q, \end{cases}$$

and

$$\rho_D(f) = \rho_R\left(\int_t^1 f(s)s^{\frac{1}{n}-1}\,ds\right), \qquad n \text{ an integer, } n \geq 2.$$

Then, ρ_R and ρ_D are optimal in (3.4) as an r.i. norm and a Banach function norm, respectively.

Theorem 3.7 Let p, q, ϕ and ρ_R be as in Theorem 3.6. Suppose m and n are positive integers such that $n \geq 3$ and $2 \leq m \leq n - 1$. Given $f \in M_+(0, 1)$, define

$$\rho_D(f) = \rho_R\left(\int_t^1 (Pf^*)(s)s^{\frac{m}{n}-1} ds\right).$$

Then, ρ_R and ρ_D are r.i. norms which are optimal in (3.3) if $1 \leq p' < n/m$; otherwise, ρ_D is optimal, but ρ_R is not, except, possibly, when $p' = n/m$, $q = \infty$ and b is bounded away from zero.

4. Weighted Orlicz and Orlicz-Lorentz Modular Inequalities for Operators of Hardy Type

Consider an operator $T = T_K$ of Hardy type. Let Φ be an N-function in the sense that it is unbounded and convex on $\mathbb{R}_+ = (0, \infty)$ with right-hand limit zero at 0. Our first result gives necessary and sufficient conditions in order that the weight functions t, u, v, w on $\mathbb{R}_+$ satisfy

$$\int_{\mathbb{R}+} \Phi(t(x)(Tf)(x))\omega(x)\, dx \leq \int_{\mathbb{R}} \Phi(Cu(y)f(y))v(y)\, dy, \tag{4.1}$$

$f \in M_+(\mathbb{R}_+)$.

Recall that if $\Phi(t) = \int_0^t \phi(s)\, ds$ is an N-function, then its complementary N-function, Ψ, is given by $\Psi(t) = \int_0^t \phi^{-1}(s)ds$. In collaboration with S. Bloom we proved, in [3],

Theorem 4.1 Let Φ be an N-function with complementary N-function Ψ. Then, the operator $T = T_K$ of Hardy type satisfies (4.1) for the weights t, u, v and w if and only if there exists $C > 0$, independent of ε, $x > 0$ such that

$$\int_0^x \Psi\left[\frac{\alpha(\varepsilon, x)k(x, y)}{C\varepsilon u(y)v(y)}\right]v(y)dy \leq \alpha(\varepsilon, x)$$

$$\text{and} \qquad \int_0^x \Psi\left[\frac{\beta(\varepsilon, x)}{C\varepsilon u(y)v(y)}\right]v(y)dy \leq \beta(\varepsilon, x), \tag{4.2}$$

where $\alpha(\varepsilon, x) = \int^\infty \Phi(\varepsilon t(y))w(y)dy$ and $\beta(\varepsilon, x) = \int^\infty \Phi(\varepsilon t(y)k(y, x))w(y)dy$.

As an illustration of how (4.2) can be used to construct weights, which we conjecture are optimal, consider the classical Hardy operator, I, for which $K = 1$.

Theorem 4.2 Let Φ be an N-function satisfying the Δ_2 condition $\Phi(2x) \leq C\Phi(x)$, $x \in \mathbb{R}_+$ and let Ψ be its complementary N-function. Given weights t, v and w on $\mathbb{R}_+$, define

$$u(x) = \sup_{\varepsilon>0} \frac{\alpha(\varepsilon, x)}{\varepsilon v(x)\Psi^{-1}\left(\dfrac{-\alpha'(\varepsilon, x)}{v(x)}\right)}$$

Then, (4.1) holds for t, u, v and w. Again, given the weights t, u and w on $\mathbb{R}_+$ and setting

$$v(x) = \sup_{\varepsilon>0} \frac{\alpha(\varepsilon, x)}{u(x)\phi\left(\dfrac{-\varepsilon\alpha'(\varepsilon, x)u(x)}{\alpha(\varepsilon, x)}\right)},$$

(4.1) holds for t, u, v and w.

There are analogues of Theorems 4.1 and 4.2 in the context of Orlicz-Lorentz spaces which we do not go into here. They are consequences of the following characterization of the dual of an Orlicz-Lorentz space proved in [7] by M. Goldman and the author.

Theorem 4.3 Let Φ be an N-function with complementary N-function Ψ and suppose v is a weight on $\mathbb{R}_+$. Set $V(t) = \int_0^t v(s)ds$ and $w(t) = V^{-1}(t)$. Then, for fixed $C > 1$ and all $f \in M_+(\mathbb{R}_+)$,

$$\rho'_{\Lambda_\Phi(v)} \asymp \rho_{\Psi,v}\left(\frac{1}{V(t)}\int_t^{w(CV(t))} f^*(s)ds\right),$$

if $V(\infty) = \infty$, while if $V(\infty) = 1$, say,

$$\rho'_{\Lambda_\Phi(v)} \asymp \rho_{\Psi,v}(\xi(t)),$$

where

$$\xi(t) = \begin{cases} \dfrac{1}{V(t)}\displaystyle\int_t^{w(CV(t))} f^*(s)ds & 0 < t \leq w(1/C) \\ \xi(w(1/C)) = C\displaystyle\int_{w(1/C)}^{\infty} f^*(s)ds & t > w(1/C) \end{cases}$$

References

1. David R. A. Adams, A sharp inequality of J. Moser for higher order derivatives, *Annals of Math.* **38** (1988), 385–398.

2. S. Bloom and R. Kerman, Weighted norm inequalities for operators of Hardy type, *Proc. Amer. Math. Soc.* **113** (1991), 135–141.
3. S. Bloom and R. Kerman. Weighted L_Φ integral inequalities for operators of Hardy type, *Studia Math.* **110** (1994), 35–52.
4. H. Brezis and S. Wainger. A note on limiting cases of Sobolev embeddings and convolution inequalities, *Comm. Partial Diff. Eq.* **5** (1980), 773–789.
5. M.J. Carro and J. Soria. Weighted Lorentz spaces and the Hardy operator, *J. Funct. Analysis* **112** (1993), 480–494.
6. D.E. Edmunds, R. Kerman and L. Pick. Optimal Sobolev imbeddings involving rearrangement invariant quasinorms, *J. Funct. Analysis* (to appear).
7. M. Goldman and R. Kerman. On the principle of duality in Orlicz-Lorentz spaces and its application to modular inequalities, in preparation.
8. R.A. Kerman. Function spaces continuously paired by operators of convolution-type, *Canad. Math. Bull.* **22** (1979), 499–507.
9. G.G. Lorentz, On the theory of spaces Λ, *Pacific J. Math.* **1** (1951), 411–429.
10. R. Oinarov, Weighted inequalities for a class of integral operators (Russian), *Dokl. Akad. Nauk. SSSR* **319** (1991), no. 5, 1076–1078; translation in *Soviet Math. Dokl.* **44** (1992), no. 1, 291–293.
11. R.S. Strichartz, A note on Trudinger's extension of Sobolev's inequalities, *Ind. U. Math. J.* **21** (1972), 841–842.
12. G. Talenti, Inequalities in rearrangement-invariant function spaces, *Nonlinear Analysis, Function Spaces and Applications,* Vol. 5, Prometheus, Prague (1995), 177–230.
13. N.S. Trudinger. On imbeddings into Orlicz spaces and some applications, *J. Math. Mech.* **17** (1967), 473–483.

Index